C for you

C for you

B Thiagarajan

Professor and Head
Department of Bioinformatics
Dr G R Damodaran College of Science
Coimbatore

MJP Publishers

Cataloguing-in-Publication Data

Thiagarajan, B. (1975-).

C for You / by B. Thiagarajan. –
Chennai : MJP Publishers, 2010
xviii, 442 p.; 24 cm.
Includes References and Index.
ISBN 978-81-8094-082-8 (pb.)
1. Programming languages 2. C programming languages
I. Title.
005.13 dc22 THI MJP 077

ISBN 978-81-8094-082-8

Printed and bound in India

MJP PUBLISHERS
47, Nallathambi Street
Triplicane
Chennai 600 005

Publisher : J.C. Pillai
Managing Editor : C. Sajeesh Kumar
Marketing Manager : S.Y. Sekar
Project Editor : P. Parvath Radha
Acquisitions Editor : C. Janarthanan
Editorial Team : Lissy John, N. Yamuna Devi, M. Gnanasoundari, N. Thilagavathi
CIP Data : Prof. K. Hariharan, Librarian
RKM Vivekananda College, Chennai.

To

My teachers,

my family members,

my loving wife,

my dearest pappu,

my dear friends and students.

FOREWORD

I am immensely pleased to write a foreword for the project *C for You*. I appreciate the efforts of Mr. B. Thiagarajan in bringing out a book on C Programming, which is very lucid in all the ways. Upon reading the Chapters of the book, I am very much convinced that there is no royal road to language mastery. The author has been very keen on making the readers to be in his/her comfort zone, when they read through the book.

The more different your background from the language you need to learn, the more time you need and the more difficulties you will encounter. Generally, any user of a new programming language suffers a lot even if he/she is an accomplished programmer in other languages. I would advise you not to expect becoming expert in a programming language in a semester or two. In order to master the language, you need to be in constant touch with the language and continuously learn and program in that. I am sure this book would be of profound use for those who begin their journey with C programming language. This book is different from other, often enthusiastic, books on C, since the flow of the concepts that are explained and the examples that are provided are profuse and more than understandable.

The book is an indicator that the author is a good educator and he wants us to think and to work. He is aware of the fact that if you and I are ready to learn, we have to practice a lot and hence he had provided a separate section for exercises, apart from providing the mind blowing Crack the Nut at the end of every Chapter, especially at the end of Arrays, Structures and Unions and Pointers. This, I would say, is an excellent addition to your bag of tricks and techniques. Read and study the examples given by him throughout the book, type them in to drill them into your mind and then follow his lead as you work the interesting examples. You will be at cloud nine once you are done.

As readers, you will learn more interesting things about C that you probably didn't know, even if you are not an absolute beginner. I thoroughly enjoyed reading this book and I am sure all the readers would also enjoy reading and appreciate the value of it.

I congratulate Mr. B. Thiagarajan for bringing out such an excellent book on C programming and unarguably a great gift to the student community. It would add motivation to the members of staff of our College and other Colleges too. The kind of dedication shown by him has added more reputation to our College. I wish him to script many more in the years to come and serve the students forever.

Dr. A. Ponnusamy

Principal
Dr. G.R. Damodaran College of Science
Coimbatore
Tamil Nadu

PREFACE

Programming has not always been the *cup of tea* for the students of computer science and information technology. Often people have ridiculed with the programming languages, though they had mastered them with ease. This book is basically for those who are interested in learning computer programming. This book takes the reader quite far into the world of computer programming with the C language, but before that, it presents some elementary aspects related to computers. This book does not assume any previous knowledge about computer programming. So this book is intended for beginners in the field of computing. I have, though, assumed that a person who reads this book has a basic knowledge of mathematics.

A person studying with this book should have a computer at his/her disposal. It is an advantage if the reader is already able to use the computer for different purposes. The reader does not necessarily have to purchase any programming tools for his or her computer. In this book, we use programming tools that are freely available on the Internet. This book should be easy to read and understand because all computer programs in this book are written with so-called natural names. This means that the computer programs of this book contain a lot of familiar English words. Above all, the style of the script is very interactive and hence the readers would surely feel like being in a classroom, listening to a lecture.

The text, figures, tables and example programs are presented in a unique and most readable manner in this book. The example computer programs are explained graphically with memory pictures and maps. This book should be somewhat entertaining because the example programs in the book have been carefully written and selected.

This book teaches computer programming by presenting the C programming language to the reader. These days, the Java and C# programming languages might be alternative languages to study. I believe, though, that C is a fine programming language for a beginner in computer programming, and will be an important programming language for a long time. With C, you can get *closer to the machine* and learn how a computer operates. It is easy to write simple computer programs that read something from the keyboard and print something to the screen. These kinds of programs are needed to learn the basics of computer programming. For these reasons, the programming language of this book is C, but above all, this book teaches computer programming in general terms. Once you have learned C and the principles of computer operation by reading this book, you have acquired a robust basis for continuing your studies and learning other programming languages.

I have written this book because I believe that the way things are presented in this book makes learning easier. I believe also that the way computer programs are written in this book, is a very good way to write computer programs. I have written computer programs this way for more close to a decade. Therefore, I think that this book is an excellent way to study computer programming.

How Best You Can Use This Book?

Although I have tried my best while writing this book and I believe this is a very good textbook for a person who is starting to learn computer programming from the scratch, it remains a fact that computer programming can be a difficult subject to start with. When you read about something new in this book, you may not understand it immediately. So do not worry if a particular concept is not clear to you. Remember that always you can learn any concept with ease. It often happens that when you study something that appears to be difficult in the beginning, the difficult thing is only understood later, perhaps the following day when you reread the text which turned out to be difficult. Therefore, it is of prime importance to learn to accept that there may be difficult things which cannot be understood quickly.

You just can't learn computer programming without writing programs by yourself. There are plenty of programming exercises towards the end of this book. I would advice you to attempt those exercises initially with a pen and paper and later with a computer. Also, to write programs swiftly and effectively, be friendly with your keyboard and use all the ten fingers when you edit a program. Though there is no hard and fast rule that you must be a master in handling the keyboard while you learn programming, make sure that you make a good effort in handling the keyboard with a good fingering.

B. Thiagarajan

ACKNOWLEDGEMENTS

A year has gone by since my first project was published and I have been able to script my second project. This is really eventful and I place my sincere prayers to the Almighty, who had given me enough strength, skill and patience to execute the assignment.

I would like to express my gratitude to my Parents, Aunt and Paternal Uncle, for their support and good wishes, due to which I could do everything without any hiccups.

I would like to place my sincere thanks on the lotus feet of all my teachers of school days, with a special reference to K. Swaminathan, V. Subramanian, S. Ganapathy Subramanian, K. Chandrasekar and S. Nagarajan.

Special thanks to Prof. S.T. Rajan, Department of Computer Science and Applications, St. Joseph's College, Trichy, from whom I had acquired most of the skills in C programming.

I would like to express my gratefulness to my College Management, especially Dr. D. Padmanaban, Correspondent, and Mrs. Geetha Padmanaban, Secretary, for giving me a platform to prove myself and constantly motivating me in all my endeavours.

I am more than pleased to express my heartfelt gratitude to our Adviser Prof. A.N. Lakshmanan who has been a guiding force behind all my efforts and has been consistently encouraging me in my academic accomplishments.

I would be happy to give away a special word of gratitude to my beloved Principal Dr. A. Ponnusamy, who constantly supported all my initiatives and guided me from the front with his rich experience. I would specially thank him for writing foreword for this book.

Dr. K.K. Ramachandran, Vice-Principal and Director, GRDSCIB, deserves a big word of thanks, since he was a silent motivator and a catalyst for my academic achievements and has always been my well-wisher.

I would like to record my sincere gratitude to my mentors in my academic life Dr. K. Mani, Reader, Department of Plant Biotechnology, PSG College of Arts and Science, Coimbatore and Dr. N. Vijayaraj, National Institute of Health, USA.

I would also like to express my heartfelt gratitude to Mr. V. Sennaraj, Deputy Registrar, Indian Institute of Space Science and Technology, Trivandrurm, a close friend and associate of mine for constantly providing me with his opinions and suggestions during the days when I scripted this book.

I am grateful to Dr. M.N. Sundarraman, Former Head, Department of English, St. Joseph's College, Trichy, Dr. K. Palaniappan, Secretary, Sri Krishna Arts and Science College, Coimbatore, Dr. R. Saravana Selvan, Reader, School of English and Foreign Languages, Bharathiar University, Coimbatore, Dr. S. Manian, Head, Department of Botany, Bharathiar University, Coimbatore and Dr. K. Saminathan, Head, Department of Biotechnology, Bharathiar University, Coimbatore, for their constant support throughout my academic life.

I would like to place my heartfelt thanks to my M Phil research supervisor Dr. R. Amalraj, Assistant Professor, Sri Vasavi College, Erode, who encouraged me a lot during the initial stages of my research.

I am grateful to my PhD research supervisor Dr R Bremananth, Professor and Head, Department of Computer Applications, Sri Ramakrishna Engineering College, Coimbatore, for his consistent motivation and guidance.

My colleagues R. Subhashini and V.S. Shafna Asmy deserve a special word of appreciation for getting in touch with me every now and then regarding the development of this book.

Students are always the real inspiration to me in my career spanning more than nine years and I would like to attribute my growth to their constant and constructive criticisms.

At this moment, I would like to reciprocate the love and gratitude to all my friends of both school and college days for offering me a great support.

I would like to thank J.C. Pillai, C. Sajeesh Kumar and their team of technical reviewers for taking up a Herculean effort in editing this project, constantly giving their opinions during the course of time and who have given a final shape to the project. I would also like to thank all the anonymous reviewers of this book, who have been patient enough to read the manuscript and give their queries and suggestions.

My wife R. Mahalakshmi deserves a greater word of appreciation for constantly helping me out, right from the day one of this project. I would love to attribute the achievements in my life to my kid Akshaye, who has always showed me the lighter part of this world with his smile and mischief.

I would like to thank all those whose names might have missed from my mind when this acknowledgement was written.

To put in a nutshell, a big THANKS to all.

B. Thiagarajan

CONTENTS

1 INTRODUCTORY IDEAS

1.1 WHAT IS A COMPUTER PROGRAM?

A computer program is just like a schedule everyone makes use of in their routine life. When we are scheduling our day, we do have a paper and pencil and chart it out clearly and execute it. We refer to this routine affair as follows: *Plan your work and then work your plan.* Once the schedule goes smoothly, we get the required output in the form of a physical product or get a logical conclusion out of the execution. A similar thing happens with the computer program. We first write a series of instructions and check for the validity of all those and then we execute the instructions in the same order and get the desired results.

One common attribute between the real-life programming and computer programming is that, in both, we arrive at logic before actually implementing it. If the logic goes right, we get the desired results.

For writing programs that a computing system could execute and give you the results, we have to learn the programming languages. There are umpteen programming languages available, which we will be discussing very shortly. In fact, programming languages are the interfaces between human beings and the computing systems for a better interaction.

1.2 IMPORTANCE OF LEARNING A PROGRAMMING LANGUAGE

You must be surprised and would in fact be ready to ask the question, "why should I learn a programming language?" The answer is very simple. Just think of a similar question, "why should one learn languages such as English, French, etc?" Yes. You are right. Just to interact with fellow human beings. You will now be able to get the answer for the former question—We have to interact with the computing system.

Immediately you will be tempted to know whether you will be able to work with the computing system without knowing any programming languages. Yes, but then you have to be in one-way traffic, in the sense that you will be just remain as a user and not as a creator. So, if you want to be a creator, learning programming languages is a must.

1.3 EVOLUTION OF PROGRAMMING LANGUAGES

The introduction of the Von Neumann architecture was a crucial step in the development of electronic computers. The basic idea is that instructions can be encoded as data and stored in the memory of the computer. The first consequence of this idea is that changing and modifying the stored program is simple and efficient. In fact, changes can take place at electronic speeds, a very different situation from earlier computers that were programmed by plugging wires into panels. The second, and ultimately more far-reaching consequence is that computers can process programs themselves, under program control. In particular, a computer can translate a program from one notation to another. Thus the stored-program concept led to the development of programming languages (PLs).

The history of Programming Languages, like the history of any technology, is complex. There are advances and setbacks; ideas that enter the mainstream and ideas that end up in a backwater; even ideas that are submerged for a while and later surface in an unexpected place. With the benefit of hindsight, we can identify several strands in the evolution of PLs. These strands are commonly called paradigms.

1.3.1 FORTRAN

FORTRAN was introduced in 1957 at IBM by a team led by John Backus. It is interesting to note that, 20 years later, Backus (1978) criticized FORTRAN and similar languages as "lacking useful mathematical properties". He saw the assignment statement as a source of inefficiency: "The Von Neumann bottleneck". The solution, however, was very similar to the solution he advocated in 1957: programming must become more like mathematics: "we should be focusing on the form and content of the overall result".

Although FORTRAN did not eliminate programming, it was a major step towards the elimination of assembly language coding. The designers focused on efficient implementation rather than elegant language design, knowing that acceptance depended on the high performance of compiled programs. FORTRAN has value semantics. Variable names stand for memory addresses that are determined when the program is loaded.

The major achievement of FORTRAN is its efficient compilation. It has many limitations and a few are listed below:

Flat and uniform structure There is no concept of nesting in FORTRAN. A program consists of a sequence of subroutines and a main program. Variables are either global or local to subroutines. In other words, FORTRAN programs are rather similar to assembly language programs: the main difference is that a typical line of FORTRAN describes evaluating an expression and storing its value in memory whereas a typical line of assembly language specifies a machine instruction.

Limited control structures The control structures of FORTRAN are IF, DO, and GOTO. Since there are no compound statements, labels provide the only indication that a sequence of statements forms a group.

Unsafe memory allocation FORTRAN borrows the concept of COMMON storage from assembly language program. This enables different parts of a program to share regions of memory, but the compiler does not check for consistent usage of these regions. One program component might use a region of memory to store an array of integers, and another might assume that the same region contains real values. To conserve precious memory, FORTRAN also provides the EQUIVALENCE statement, which allows variables with different names and types to share a region of memory.

No recursion FORTRAN allocates all data, including the parameters and local variables of subroutines, statically. Recursion is forbidden because only one instance of a subroutine can be active at one time.

1.3.2 ALGOL 60

ALGOL 60 was designed by an international committee, partly to provide a programming language that was independent of any particular company and its computers. The committee included both John Backus (chief designer of FORTRAN) and John McCarthy (designer of LISP).

The goal was a "universal programming language". In one sense, ALGOL was a failure: a few complete, high-quality compilers were written and the language was not widely used (although it was used more in Europe than in North America). In another sense, ALGOL was a huge success: it became the standard language for describing algorithms. For the better part of 30 years, the Association of Computing Machineries (ACM) required submissions to the algorithm collection to be written in ALGOL.

The major innovations of ALGOL are listed below.

It was Block Structured in the sense that the programs written were recursively structured. The recursive structure of programs means that large programs can be constructed from small programs.

It supported Dynamic Array usage. The designers of ALGOL realized that it was relatively simple to allow the size of an array to be determined at run-time. The compiler statically allocates space for a pointer and an integer (collectively called a "dope vector") on the stack. At runtime, when the size of the array is known, the appropriate amount of space is allocated on the stack and the components of the "dope vector" are initialized.

The default method of passing parameters in ALGOL was call by name and it was described by a rather complicated "copy rule". The essence of the copy rule is that the program behaves as if the text of the formal parameter in the function is replaced by the text of the actual parameter. The complications arise because it may be necessary to rename some of the variables during the textual substitution. The usual implementation strategy was to translate the actual parameter into a procedure with no arguments (called a "thunk"); each occurrence of the formal parameter inside the function was translated into a call to this function. The mechanism seems strange today because few modern languages use it. However, the ALGOL committee had several valid reasons for introducing it, the most important being the fact that it enables procedures to alter their actual parameters.

A variable in an ALGOL procedure can be declared own. The effect is that the variable has local scope (it can be accessed only by the statements within the procedure) but global extent (its lifetime is the execution of the entire program).

Some of the important aspects like records were missed out in ALGOL, even though it had above contributions to programming. In many applications, the concept of own variables itself was problematic in ALGOL, for, it was difficult to reliably initialize them. The language was quite simple, but not powerful enough.

1.3.3 ALGOL 68

ALGOL 68, which was a successor of ALGOL 60, was much more complex in nature. The main design principle of ALGOL 68 was orthogonality: the language was to be defined using a number of basic concepts that could be combined in arbitrary ways. Although it is true that lack of orthogonality can be a nuisance in programming languages, it does not necessarily follow that orthogonality is always a good thing. The highlighting features of ALGOL 68 are listed below.

The language was described in a formal notation that specified the complete syntax and semantics of the language. The fact that the Report was very hard to understand may have contributed to the slow acceptance of the language.

Operator overloading Programmers can provide new definitions for standard operators such as "+". Even the priority of these operators can be altered. ALGOL 68 has a very uniform notation for declarations and other entities.

Like ALGOL 60, ALGOL 68 was not widely used, although it was popular for a while in various parts of Europe. The ideas that ALGOL 68 introduced, however, have been widely imitated.

1.3.4 COBOL

COBOL introduced structured data and implicit type conversion. When COBOL was introduced, "programming" was more or less synonymous with "numerical computation". COBOL introduced "data processing", where data meant large numbers of characters. The data division of a COBOL program contained descriptions of the data to be processed.

Another important innovation of COBOL was a new approach to data types. The problem of type conversion had not arisen previously because only a small number of types were provided by it. COBOL introduced many new types, in the sense that data could have various degrees of precision, and different representations as text. The choice made by the designers of COBOL was radical: type conversion should be automatic.

During the early 60's, the dominant languages were ALGOL, COBOL and FORTRAN. The continuing desire for a "universal language" that would be applicable to a wide variety of problem domains led IBM (International Business Machines) to propose a new programming language (originally called NPL but changed later due to technical reasons, to PL/I) that would combine the best features of these three languages. Insiders at the time referred to the new language as "CobAlgoltran".

The design principles of PL/I included the following.

- The language should contain the features necessary for all kinds of programming;
- A programmer could learn a subset of the language, suitable for a particular application, without having to learn the entire language.

An important lesson of PL/I is that these design goals are doomed to failure. A programmer who has learned a "subset" of PL/I is likely, like all programmers, to make a mistake. With luck, the compiler will detect the error and provide a diagnostic message that is incomprehensible to the programmer because it refers to a part of the language outside the learned subset. More probably, the compiler will not detect the error and the program will behave in a way that is inexplicable to the programmer, again because it is outside the learned subset. PL/I extends the automatic type conversion facilities of COBOL to an extreme degree.

PL/I did introduce some important new features into programming languages, but they were not well-designed and their existence encouraged others to produce better designs. For example, every variable in PL/I has a storage class, that would be static, automatic, controlled or based and some of these were later incorporated into C. Other issues with PL/I are that it allows programmer-defined types, but they could not be named and it provides a simple and not-very-safe form of exception handling.

1.3.5 Pascal

Pascal was designed by Niclaus Wirth in 1996 as a reaction to the complexity of ALGOL 68, PL/I, and other languages that were becoming popular in the late 60's. The introduction of data structuring was a highlight in Pascal. The following are some of the important features of Pascal.

The language proved the fact that a programming language could be simple, yet powerful.

The type system of Pascal was based on primitives (integer, real, bool, etc.) and mechanisms for building structured types (array, record, file, set, etc.). Thus data types in Pascal form a recursive hierarchy just as blocks do in ALGOL 60.

Pascal provides no implicit type conversions other than sub-range to integer and integer to real. All other type conversions are explicit (even when no action is required) and the compiler checks type correctness.

Pascal was designed to match Wirth's ideas of program development by stepwise refinement. Pascal is a kind of "fill in the blanks" language in which all programs have a similar structure, determined by the relatively strict syntax. Programmers are expected to start with a complete but skeletal "program" and flesh it out in a series of refinement steps, each of which makes certain decisions and adds new details. The monolithic structure that this idea imposes on programs is a drawback of Pascal because it prevents independent compilation of components.

Pascal was a failure because it was too simple. Because of the perceived missing features, supersets were developed and, inevitably, these became incompatible. The first version of "Standard Pascal" was almost useless as a practical programming language and the Revised Standard described a usable language but appeared only after most people had lost interest in Pascal.

Like ALGOL 60, Pascal missed important opportunities. The record type was a useful innovation (although very similar to the ALGOL 68 struct) but allowed data only. Allowing functions in a record declaration would have paved the way to modular and even object-oriented programming.

Nevertheless, Pascal had a strong influence on many later languages. Its most important innovations were probably the combination of simplicity, data type declarations, and static type checking.

1.3.6 Modula-2

Wirth followed Pascal with **Modula-2**, which inherits Pascal's strengths and, to some extent, removes Pascal's weaknesses. The important contribution of Modula-2 was, of course, the introduction of modules. A module in Modula-2 has an interface and an implementation. The interface provides information about the use of the module to both the programmer and the compiler. The implementation contains the "secret" information about the module. This design has the unfortunate consequence that some information that should be secret must be put into the interface. For example, the compiler must know the size of the object in order to declare an instance of it. This implies that the size must be deducible from the interface which implies, in turn, that the interface must contain the representation of the object (a similar problem appears in C++ too).

Modula-2 provides a limited escape from this dilemma: a programmer can define an "opaque" type with a hidden representation. In this case, the interface contains only a pointer to the instance and the representation can be placed in the implementation module. The most striking feature of Modula-2 was the modules with separated interface and implementation descriptions.

1.3.7 C

C is a very pragmatic programming language. Dennis Ritchie designed it for a particular task "systems programming" for which it has been widely used. The enormous success of C is partly accidental. UNIX, after Bell released it to universities, became popular, with good reason. Since UNIX depended heavily on C, the spread of UNIX inevitably led to the spread of C.

C is based on a small number of primitive concepts. For example, arrays are defined in terms of pointers and pointer arithmetic. This is both the strength and weakness of C. The number of concepts is small, but C does not provide real support for arrays, strings, or Boolean operations.

C is a low-level language by comparison with the other programming languages discussed in this section. It is designed to be easy to compile and to produce efficient object code. The compiler is assumed to be rather unsophisticated and in need of hints such as register. C is notable for its concise syntax. Some syntactic features are inherited from ALGOL 68 (for example, += and other such assignment operators) and others are unique to C and C++ (for example, postfix and prefix ++ and —, which are discussed later in the Chapter 4 of this book).

Thus, mushroom of programming languages were in practice and C has emerged a successful language and stands high among them and we will discuss the programming techniques in C from the Chapter 2 of this book.

1.4 ALGORITHMS

Algorithms are nothing but a type of effective method in which a definite list of well-defined instructions for completing a task, when given a proper initial state, will proceed through a well-defined series of successive states, eventually terminating in an end-state.

Well, what is the etymology of this term? Abu Abdullah Muhammad bin Musa al-Khwarizmi, a Persian astronomer and mathematician, wrote a paper in Arabic, which was later translated to Latin, entitled as *Algoritmi de numero Indorum*, which meant "Book by Algoritmus on the numbers of the Indians". But the word "Algoritmi" in the title was treated as a Latin plural and this led to the word "Algorithm". The real meaning of the word is "Calculation Method".

Every one of us does calculation in our routine life and that may be the one related to the academy or business or personal. For the calculation, we normally adopt a formula and a method and the methods may be different for each of us, even though the result is going to be one and the same.

The concept of algorithm was introduced to define the notion of logic. Algorithms are essential to the way computers process information, because a computer program is essentially an algorithm that tells the computer what specific steps to perform and in what specific order, in order to carry out a specified task. Typically, when an algorithm is associated with processing information, data are read from an input device, written to an output device, and/or stored for further processing. Stored data are regarded as part of the internal state of the entity performing the algorithm. In practice, the state is stored in a data structure, but an algorithm requires the internal data only for specific operation sets called abstract data types.

Since an algorithm is a precise list of precise steps and hence the order of computation will almost always be critical to the functioning of the algorithm. Instructions are usually assumed to be listed explicitly, and are described as starting "from the top" and going "down to the bottom", an idea that is described more formally by flow of control.

1.4.1 How to Express an Algorithm?

Algorithms can be expressed in many kinds of notations, including natural languages, pseudocode, flowcharts and programming languages. Natural language expressions of algorithms tend to be verbose and ambiguous, and are rarely used for complex or technical algorithms. Pseudocode and flowcharts are structured ways to express algorithms that avoid many of the ambiguities common in natural language statements, while remaining independent of a particular implementation language. Programming languages are primarily intended for expressing algorithms in a form that can be executed by a computer, but are often used as a way to define or document algorithms.

1.5 FLOWCHARTS

There is a famous saying that goes like "A picture is worth thousand words" and this is exactly implied by the flowcharts, which are structured ways to express algorithms. As already mentioned, the flowcharts resolve the ambiguities that might be a common problem in using a natural language. Moreover, pictures are the easiest way to convey the messages to human beings belonging to any community or group. A flowchart is one of the seven basic tools of quality control (The other tools being histogram, Pareto chart, check sheet, control chart, cause-and-effect diagram and scatter diagram). The first structured method for documenting process flow, the flow process chart, was introduced by Frank Gilbreth (An advocate of scientific management and a pioneer of motion study).

Any vector-based drawing program can be used to create flowcharts. Some tools offer special support for flowcharts, e.g. OpenOffice.org Draw, ConceptDraw, SmartDraw, Visio, OmniGraffle and Dia.

1.5.1 Symbols Used in a Flowchart

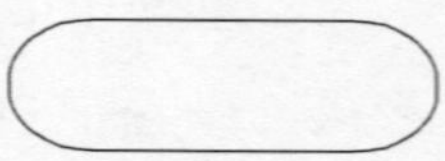

Start and End symbols, represented as lozenges, ovals or rounded rectangles, usually containing the word "Start" or "End", or another phrase signalling the start or end of a process, such as "submit enquiry" or "receive product".

Arrows show the "flow of control". An arrow coming from one symbol and ending at another symbol represents that control passes to the symbol the arrow points to.

Processing or Calculating steps are represented as rectangles.

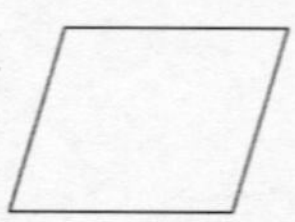

Input/Output is represented as a parallelogram.

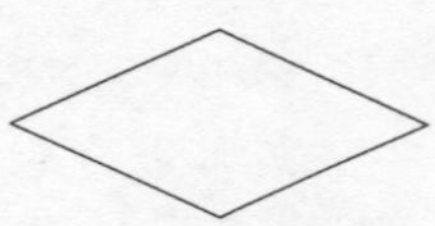

Conditional (or decision) is represented as a diamond (rhombus). These typically contain a Yes/No question or True/False test. This symbol is unique in that it has two arrows coming out of it, usually from the bottom point and right point, one corresponding to Yes or True, and one corresponding to No or False. The arrows should always be labelled. More than two arrows can be used, but this is normally a clear indicator that a complex decision is being taken, in which case it may need to be broken-down further, or replaced with the "pre-defined process" symbol.

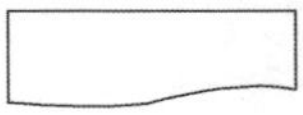

A Document is represented as a rectangle with a wavy base.

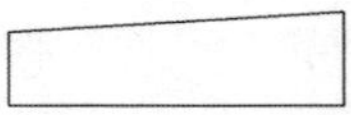

A Manual input is represented by rectangle, with the top irregularly sloping up from left to right.

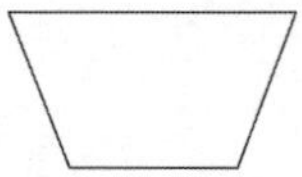

A Manual operation is represented by a trapezoid with the longest parallel side at the top, to represent an operation or adjustment to process that can only be made manually.

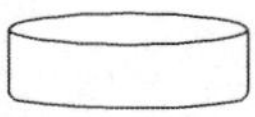

A Data File is represented by a cylinder.

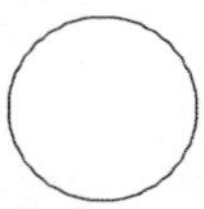

Connectors are usually represented as circles, to represent converging paths in the flowchart. Circles will have more than one arrow coming into them but only one going out. Some flowcharts may just have an arrow point to another arrow instead. These are useful to represent an iterative process. A loop may, for example, consist of a connector where control first enters, processing steps, and conditional with one arrow exiting the loop, and one going back to the connector.

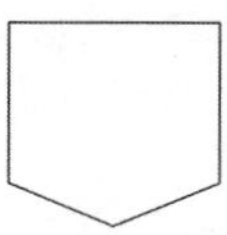

Off-page Connectors are often used to signify a connection to a (part of another) process held on another sheet or screen. It is important to remember to keep these connections logical in order. All processes should flow from top to bottom and left to right.

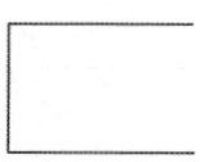

Comment Symbols are used when additional explanation or comment is required. This symbol is usually connected to the symbol it is explaining by a dashed line.

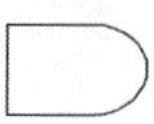

Delay—Wait for elevator, papers waiting, material waiting.

1.6 FUNDAMENTALS OF NUMBER SYSTEMS

Numbers are fundamental to everything in computers. A computer scientist shall have a good knowledge in subjects like mathematics, statistics and digital electronics. These three fields have a great influence on Computing and one cannot become a good programmer without having a brief knowledge on these. In this section, we will be discussing the different number systems and their significance and inter-conversions, which will facilitate the programmer to have a clear idea when he/she writes programs.

1.6.1 Categories of Number Systems

Number systems can be broadly categorized into two, namely weighted/positional and non-weighted/non-positional. This is based on the weight/position assigned for each of the digits in the number. Each digit will have weight/position, which is unique to itself.

Weighted/positional number systems are further classified into decimal, binary, octal and hexadecimal numbers. A good in-use example for non-positional number systems is the Roman numerals (All the number systems use what we informally call in mathematics as Arabic numerals). Our discussion will be only on the weighted/positional number systems.

1.6.2 Representation of Numbers in Computers

Computers store information in the form of numbers and that too it prefers storing in binary or hexadecimal format. No computer would possibly understand any language that is being spoken by us and it understands only the electronic signals, which are 0 and 1. Hence this section of the chapter would deal with the number systems that are important from the point of view of a novice programmer. We would consider the number systems decimal, binary, octal and hexadecimal for our discussion in the forthcoming sections and we would also learn how to convert from one system to another.

1.7 COUNTING NUMBERS IN DIFFERENT NUMBER SYSTEMS

1.7.1 Decimal Numbers (Denary Number System)

Let us begin the discussion with the decimal numbers. In Greek, 'deca' means 'ten' and hence there are ten digits in this system of numbers and the counting would begin with '0' and proceeds as 1, 2, 3, 4, 5, 6, 7, 8 and 9 (this is the first level of the decimal numbers which is very often called as single digit numbers). Once we reach '9', the next number is formed with the help of the second digit of the sequence '1', by combining it with all the other digits and hence we get 11, 12, 13, 14, 15, 16, 17, 18 and 19 (this is a part of the second level of the decimal numbers or two-digit numbers).

This process continues till all the digits in the first level are used to get the complete set of second level of decimal numbers. Once we reach 99, we are said to have completed the second level of the decimal numbers. With the help of all the numbers in the second level and combining with each of those in the first level, we get the third level of the system (three-digit numbers). Like this, the process continues without any end. Figure 1.1 shows a few of the levels of counting in the decimal number system.

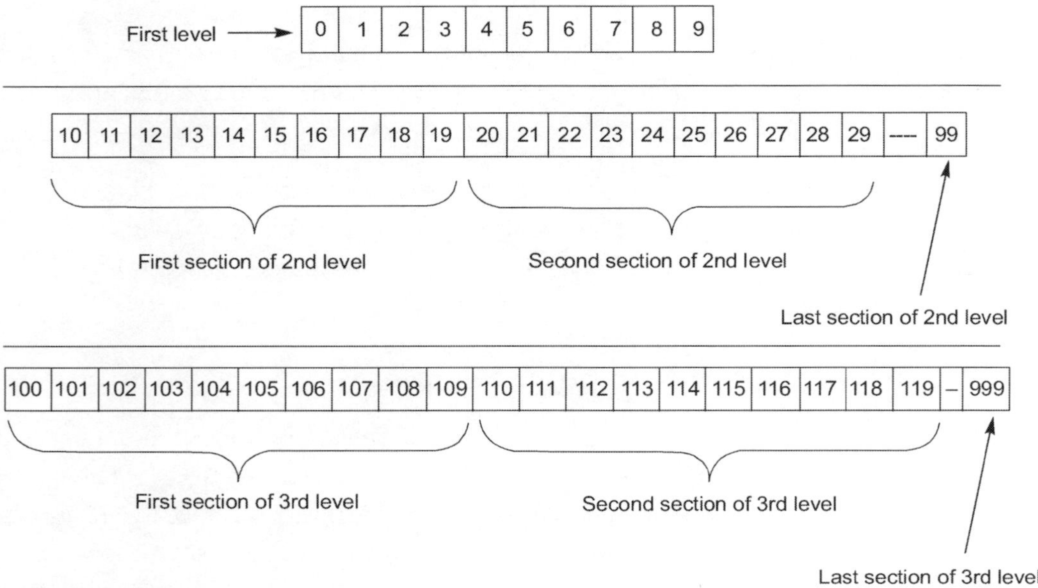

Figure 1.1 Levels of counting in decimal number system

1.7.2 Binary Numbers

In Greek, 'bi' means 'two' and hence we would have only two digits in this number system and they are 0 and 1. We would follow the same procedure for forming the other numbers in this system, given the first level of the system (0 and 1). We start with 0 and 1 in the first level (single digit numbers) and get the second level by suitably combining the digits of the first level and hence we would get the two new numbers 10 and 11 (two-digit numbers). We proceed in the same way and the process would be endless as the case with the decimal number system. Figure 1.2 shows a few of the levels of counting in the binary number system.

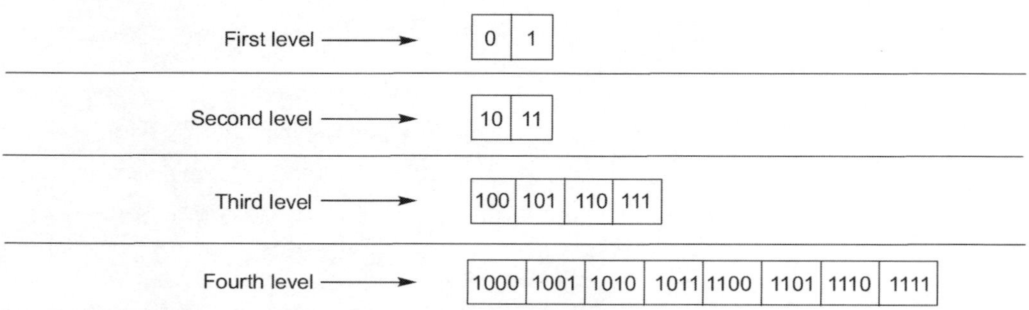

Figure 1.2 Levels of counting in binary number system

1.7.3 Octal Numbers

In Greek, 'octa' means 'eight' and hence there are eight digits in the first level of this number system. Using the combination of these eight digits, we get the second level of the octal numbers and we use the same kind of procedure as we did for the earlier two systems of numbers. The first level would contain 0, 1, 2, 3, 4, 5, 6 and 7. The Figure 1.3 shows a few of the levels of counting in octal number system.

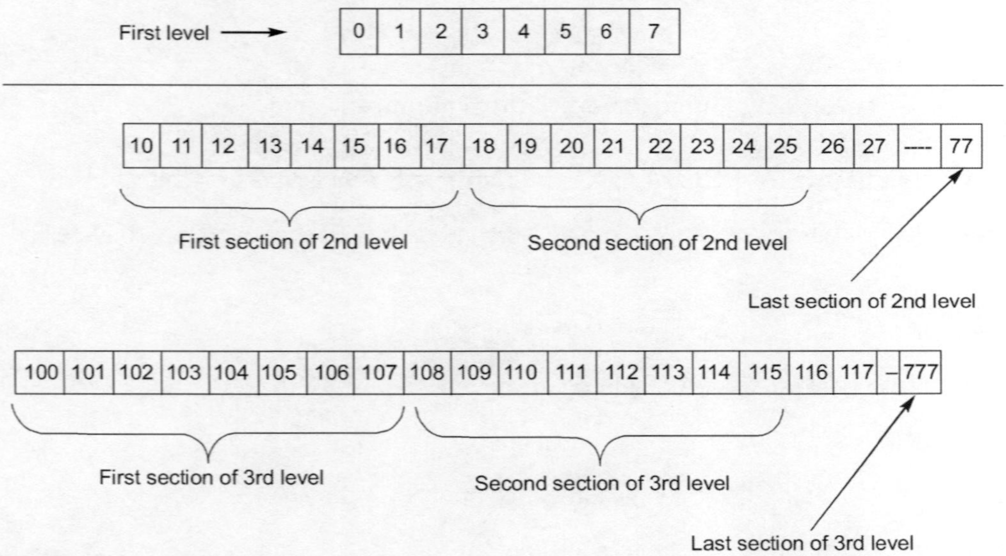

Figure 1.3 Levels of counting in octal number system

1.7.4 Hexadecimal Numbers

In Greek, 'hexa' means 'six' and we already knew that 'deca' means 'ten' and hence this system of numbers would have sixteen numbers in the first level and these would be used to generate the second, third and nth levels of the system. The speciality of this number system is that the counting begins from 0 and goes till '9' as usual. After '9', the number would be counted not as '10', but as 'A', 11 as 'B', 12 as 'C', 13 as 'D', 14 as 'E' and 15 as 'F'. The first level of counting ends with 'F' and the next level begins with the usual procedure of combining the digits of the previous level. This is again an endless process. Figure 1.4 shows a few levels in hexadecimal counting.

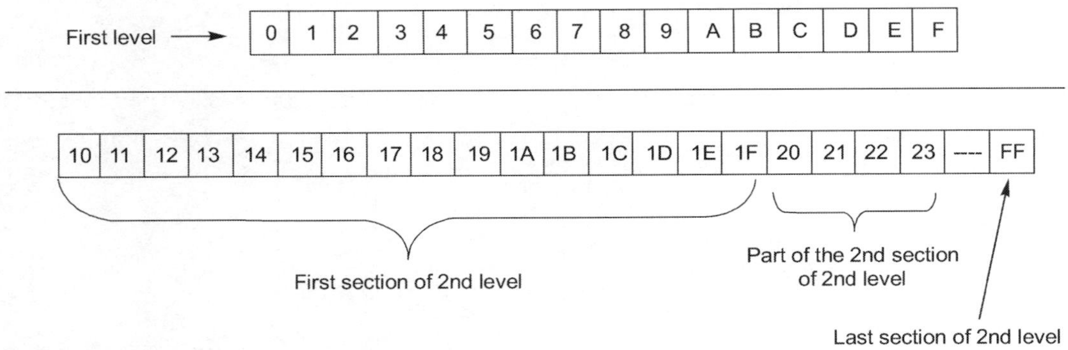

Figure 1.4 Levels of counting in hexadecimal number system

1.8 CONVERSION FROM ONE NUMBER SYSTEM TO ANOTHER

The conversion process is quite simple and we shall deal with the following conversions in this section.

Decimal integer/fraction to binary integer/fraction

Decimal integer/fraction to octal integer/fraction

Decimal integer/fraction to hexadecimal integer/fraction

Binary integer/fraction to decimal integer/fraction

Binary integer/fraction to octal integer/fraction

Binary integer/fraction to hexadecimal integer/fraction

Octal integer/fraction to decimal integer/fraction

Octal integer/fraction to binary integer/fraction

Octal integer/fraction to hexadecimal integer/fraction

Hexadecimal integer/fraction to decimal integer/fraction

Hexadecimal integer/fraction to binary integer/fraction

Hexadecimal integer/fraction to octal integer/fraction

As we do the process, you would understand clearly the conversions and to have a better practice, go through the exercises given at the end of this chapter.

1.8.1 Procedure to Convert an Integer in One Base to that in Another Base

The following are the steps to convert an integer (of any number system) to the other number system.

Step 1 Divide the (given) number by the desired base (target number system). As you divide, record the quotients down and the remainders beside.

Step 2 Stop the division process when you get the quotient zero.

Step 3 Read the remainders that you have entered from bottom to top. The resulting number is the equivalent number in the target number system for the given number.

1.8.2 Procedure to Convert a Fraction in One Base to that in Another Base

The following are the steps to convert a fraction (of any number system) to the other number system.

Step 1 Multiply the (given) number with the desired base.

Step 2 As you multiply, record the fractional part and the integer part of the product separately. Take the integer part of the result and repeat the above two steps (a kind of recursive process)

Step 3 Once the required precision is reached, stop the process of multiplication. Read the fractional parts of the products obtained in each step from top to bottom and that will be the equivalent number in the target number system.

1.8.3 Decimal Integer/Fraction to Binary Integer/Fraction

Let us consider a number such as 17 which is a decimal integer number and let us convert it into its equivalent binary by following the procedure given in the section 1.8.1. Figure 1.5 shows how the process of division is done. This method is sometimes called as ladder method of division. The remainders recorded beside each division process (shown in small squares) are read from bottom to top to get the equivalent binary number. In this case, the binary for the number 17 is 10001.

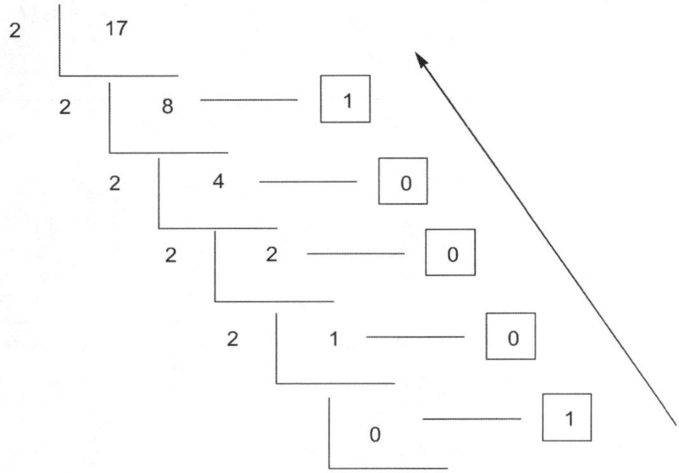

Figure 1.5 Converting a decimal integer to binary integer

Let us see what would be the result when we have a decimal fraction. Consider for example the number 0.74. Let us apply the steps that are given in the section 1.8.2 to convert this fraction into its equivalent binary fraction. Table 1.1 shows the conversion process.

Table 1.1 Conversion of decimal fraction to its equivalent binary

Multiplication	Result	Integer part	Fractional part
0.74 × 2	1.48	1	0.48
0.48 × 2	0.96	0	0.96
0.96 × 2	1.92	1	0.92
0.92 × 2	1.84	1	0.84
0.84 × 2	1.68	1	0.68

In this conversion process, we have taken the result to correspond to five places of binary point. That is the reason why we have stopped the conversion process once we get the integer part of the result in the fifth column. Now reading the column integer part from top to bottom, we get the equivalent binary fraction of the given decimal. Hence the result in binary is 0.10111 for the given decimal fraction 0.74. This is only an approximate value, since the multiplication of values is stopped abruptly. If the number is a mixed number, that is, say it is 17.74, you may follow the conversions separately for integers and fractions and combine the result. For the number 17.74, the result would be 10001.10111.

1.8.4 Decimal Integer/Fraction to Octal Integer/Fraction

The ladder method could be followed in the conversion of decimal integer to an octal fraction. Let us consider the number 17 in decimal and convert the same to octal. Figure 1.6 shows how this is done.

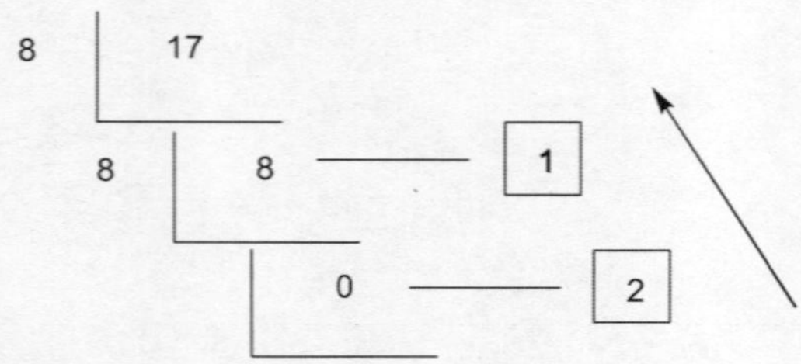

Figure 1.6 Converting a decimal integer to octal integer

From the figure, we can get the octal equivalent of the decimal number 17 as 21. Let us consider a decimal fraction 0.74 and find out the equivalent octal fraction. Table 1.2 shows the conversion process and the result could be read from the column integer part from top to bottom.

Table 1.2 Conversion of decimal fraction to its equivalent octal

Multiplication	Result	Integer part	Fractional part
0.74 × 8	5.92	5	0.92
0.92 × 8	7.36	7	0.36
0.36 × 8	2.88	2	0.88
0.88 × 8	7.04	7	0.04

From the table, we could see the result corresponding to four octal points for the decimal 0.74 is 0.5727. If the given number is 17.74, the result would be 21.5727.

1.8.5 Decimal Integer/Fraction to Hexadecimal Integer/Fraction

Let us consider the number 17 and try to find out its equivalent hexadecimal integer by applying the ladder method. Figure 1.7 shows the procedure and the result is 11 in hexadecimal for the number 17.

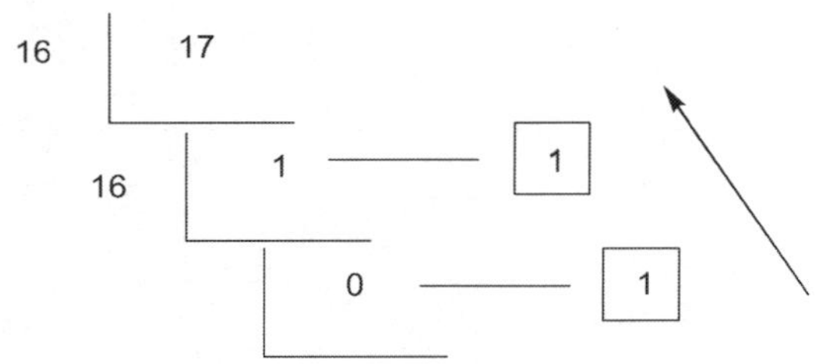

Figure 1.7 Converting a decimal integer to hexadecimal integer

Consider the number 0.74 and let us convert this decimal fraction into its equivalent hexadecimal. Table 1.3 shows the conversion process and the result.

Table 1.3 Conversion of decimal fraction to its equivalent hexadecimal

Multiplication	Result	Integer part	Fractional part
0.74 × 16	11.84	11	0.84
0.84 × 16	13.44	13	0.44
0.44 × 16	7.04	7	0.04
0.04 × 16	0.64	0	0.64
0.64 × 16	10.24	10	0.24

From the table we could see that the equivalent value is 0.(11)(13)70(10) in hexadecimal for the given decimal fraction 0.74. The value could be written also as 0.BD70A, since the numbers 10 to 15 in hexadecimal are coded with the alphabets A to F, as we had discussed earlier. If the given decimal number is 17.74, the corresponding value in hexadecimal would be 11.BD70A.

1.8.6 Binary Integer/Fraction to Decimal Integer/Fraction

This is the reverse conversion process. For example, consider the number 10001, which is a binary number. We knew that the number systems are categorized according to the weights assigned to the digits of the number. For example, consider the decimal number 189. We read this number as one hundred and eighty nine, assigning weight to each of the digits of the number. The digit '9' is assigned with 'units', '8' with 'tens' and '1' with 'hundreds'. Thus we are able to read and understand the number properly. In other words, the digit '9' is said to have 9 units and it is obtained by multiplying 9 with 10^0 and the digit '8' is said to have 8 tens and it is obtained by multiplying it with 10^1 and as we proceed towards left we have to increase the power of 10. Thus each digit is multiplied with 10^x with x taking the values from 0 to n, where n value is one less than the number of digits in the given number. If the given number has four digits, the value of x would be from 0 to 3.

The above discussion is true for any number system. The base of the number system alone changes and not the procedure. For example, consider the binary number system, whose base is 2 and if we take an example number, say, 10001, then each of the binary digit has its own unique weight and it is obtained by multiplying the binary digit with 2^x where x would take values from 0 to n, where n value would be one less than the actual number of digits in the given number. In the number that we have taken, there are five digits and hence x value would be from 0 to 4. Each digit of the binary number has to be multiplied with the corresponding weight to obtain its position in the number and once we add them, we will get the equivalent decimal number. Figure 1.8 shows the conversion process. We see that we get the result as 17.

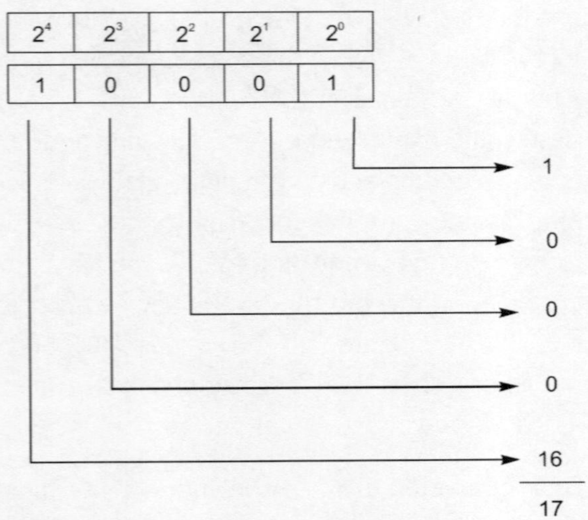

Figure 1.8 Converting a binary integer to a decimal integer

Let us consider a binary fraction 0.10001 and find out the decimal equivalent of this. We have to proceed in a similar way as shown above, except that the base has to be raised to the negative powers of 2 as we proceed towards right of the binary point. Figure 1.9 shows how the number is getting converted. We get the result as 0.53125.

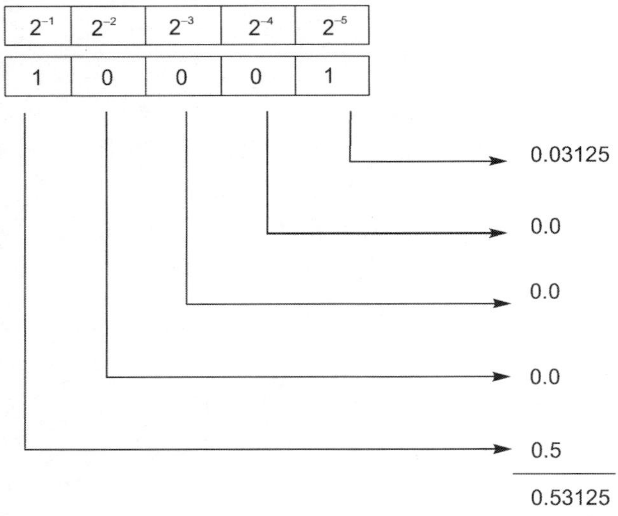

Figure 1.9 Converting a binary fraction to a decimal fraction

1.8.7 Binary Integer/Fraction to Octal Integer/Fraction

Consider the binary number 10001. If we want to convert this to its equivalent octal, we have two options. Convert the number to decimal (follow the procedure given in the section 1.8.6) and then convert the decimal to octal (as explained in the section 1.8.4). But this process will be quite time-consuming and hence I would like to suggest an easy way to do this. Consider the number and start from the right extreme position and as you proceed towards left, group the digits of the given number such that three bits are in each group. If any digit of the number is left alone, include as many leading zeroes to form a group of three digits. After you have grouped them, convert each group to their decimal equivalent and record the results down each group. Combine the results from left to right to get the final result. Figure 1.10 shows the result of converting 10001 to its octal equivalent and we see that the result is 21. (You may check this result by following the other procedure that I had mentioned in the beginning of this section).

The binary fraction is also converted to its equivalent octal in a similar way. In this case, one has to start the grouping from the left side, proceed towards right, align three digits in each group and do the conversion for each group to get the final result. Figure 1.11 shows the way the number 0.1011011 is converted to its octal equivalent. In case of non-availability of digits, you may include as many trailing zeroes. We see that the result of the conversion is 0.554.

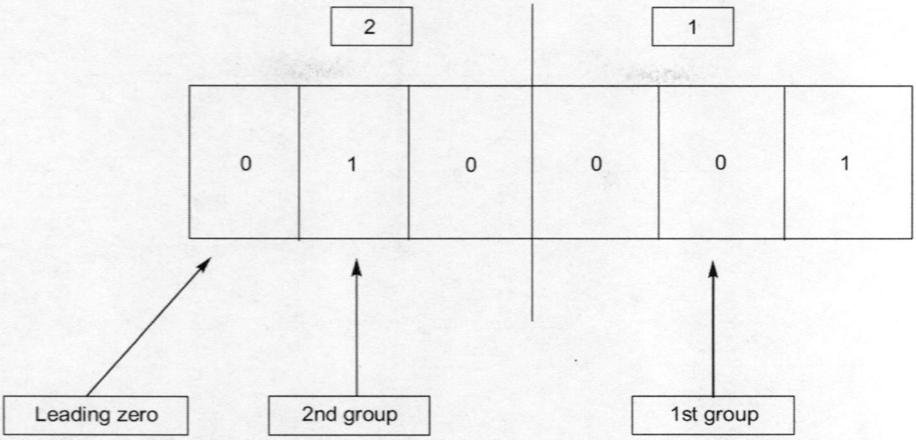

Figure 1.10 Converting a binary integer to an octal integer

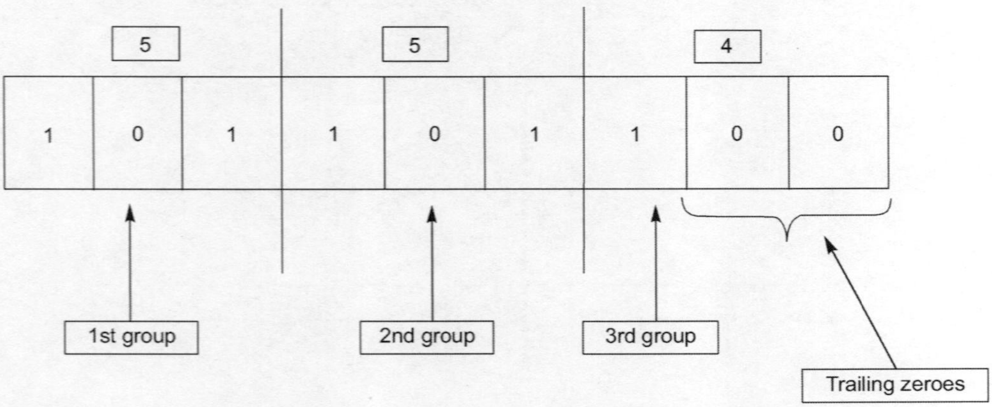

Figure 1.11 Converting a binary fraction to an octal fraction

1.8.8 Binary Integer/Fraction to Hexadecimal Integer/Fraction

We could follow a similar method discussed in the section 1.8.7 for converting a binary integer to its equivalent hexadecimal. Instead of forming a group of three, we would form a group of four and we would convert each group to decimal and combine the results to get the final result. Figure 1.12 shows how we convert the binary 10001 to its hexadecimal equivalent. We see that the result in hexadecimal is 11.

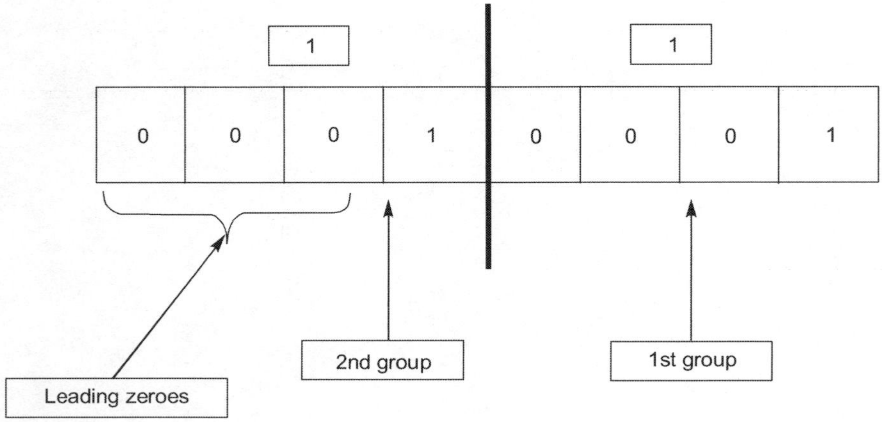

Figure 1.12 Converting a binary integer to hexadecimal integer

Consider the binary fraction 0.10111 and we follow the grouping from the left to right and convert each group to their decimal equivalent and combine the individual results to get the final answer. Figure 1.13 explains how it is done. The final answer would be 0.B8.

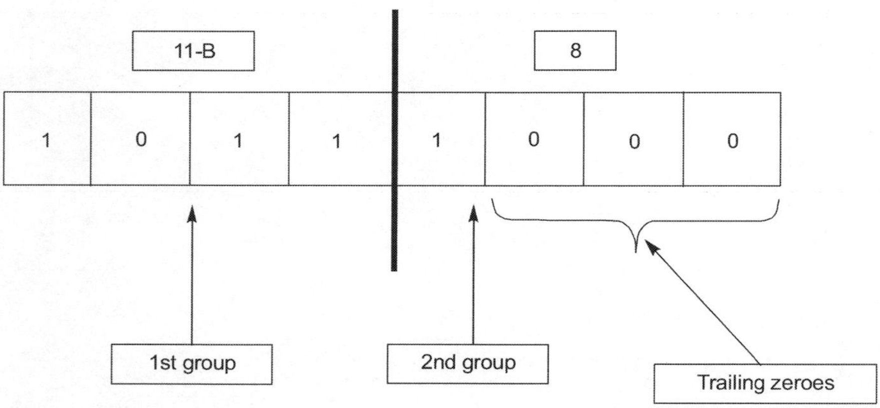

Figure 1.13 Converting a binary fraction to hexadecimal fraction

1.8.9 Octal Integer/Fraction to Decimal Integer/Fraction

Consider the octal integer 73. If we want to convert this to its decimal equivalent, multiply each of the digits with their weights (8^0, 8^1, etc.). Figure 1.14 shows how to do this and the result of the conversion is 59 in decimal.

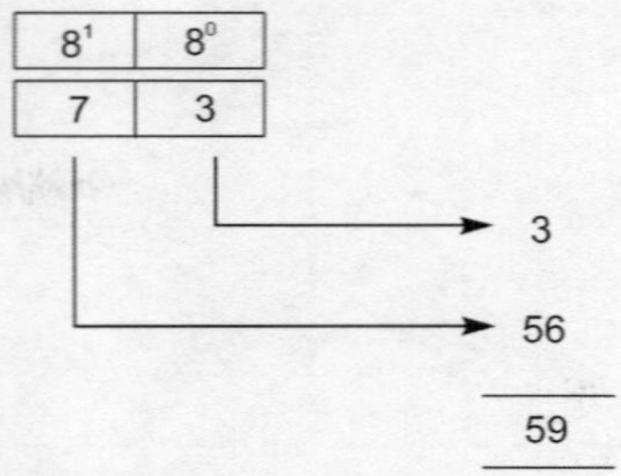

Figure 1.14 Converting an octal integer to decimal integer

1.8.10 Octal Integer/Fraction to Binary Integer/Fraction

Consider an octal number 73. Converting this to its binary is just the reverse of that discussed in the section 1.8.7. There we grouped the given binary digits in threes and then wrote the decimal equivalent of each group and got the final result by combining the individual results. In this case, we would write the binary equivalent of the given octal number such that each digit of the octal is represented by three binary digits. Then we would combine all the binary digits to get the binary equivalent of the given octal. Figure 1.15 shows how it is done for the octal 73. The result is 111011.

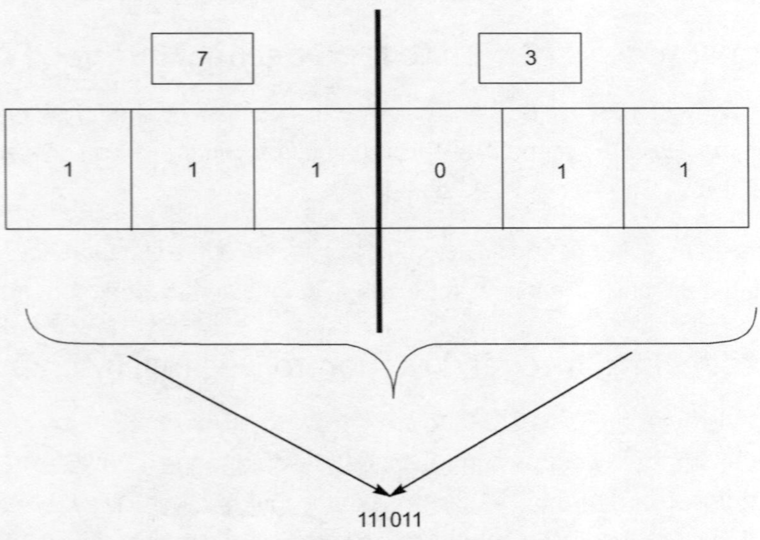

Figure 1.15 Converting an octal integer to binary integer

If we are given an octal fraction, say, 0.72, we would proceed in a similar way, by converting each of the digits towards the right of the octal point to binary digits in threes and then combine the results to get the final result in binary form. Figure 1.16 shows how it is done. The result is 0.11101.

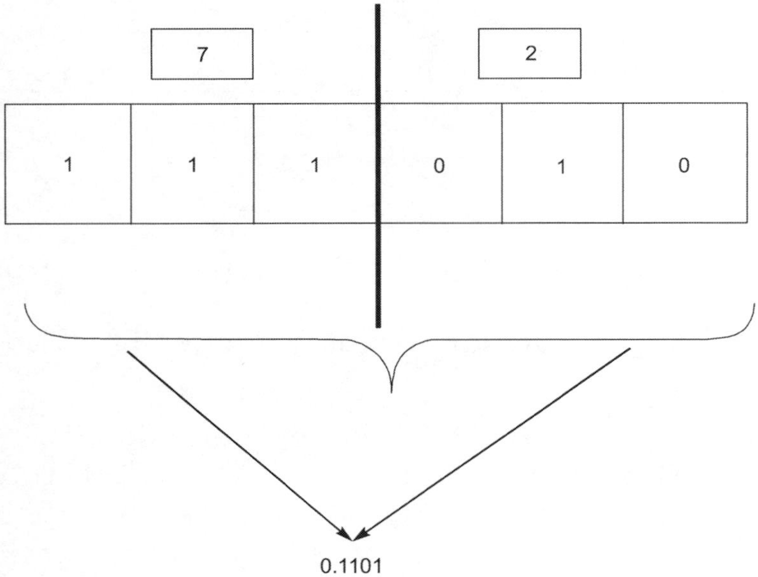

Figure 1.16 Converting an octal fraction to binary fraction

1.8.11 Octal Integer/Fraction to Hexadecimal Integer/Fraction

Consider the octal number 73. For converting this to its hexadecimal equivalent, first we have to convert it to its binary (Refer the section 1.8.10) and then from binary we can convert to hexadecimal (Refer the section 1.8.8). The result would be 3B.

For converting an octal fraction to its hexadecimal equivalent, a similar procedure as said in the sections 1.8.10 and 1.8.8 could be followed (Readers are advised to work out and cross check).

1.8.12 Hexadecimal Integer/Fraction to Decimal Integer/Fraction

Consider the hexadecimal number 1B. For converting this to its decimal equivalent, we have to multiply the digits with their weights and the weights would range from 16^0 to 16^n. The value of n depends on the number of digits in the given hexadecimal number (it will be from 0 to $n - 1$ in practice). Figure 1.17 shows how the number is converted to decimal. Before converting the above number, we must change the alphabet notation with actual number. Thus, in the above number, the alphabet B would be replaced with 11 and hence the number that is rewritten would be 1(11). The result is 27.

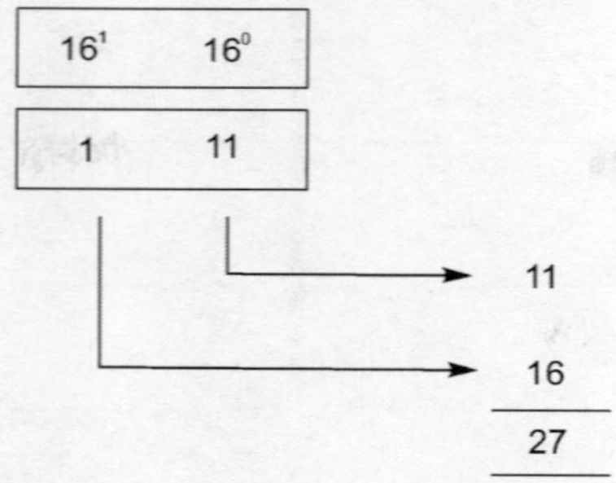

Figure 1.17 Converting a hexadecimal integer to decimal integer

For converting the hexadecimal fraction to its equivalent decimal, we would apply a similar method, except for negative powers for the base 16, since we proceed from left to right starting from the digit after the hexadecimal point of the given number. Consider for example, the hexadecimal number 0.2A.

Figure 1.18 shows the result of converting this to its equivalent decimal number. See to that the alphabet is converted to the corresponding digit of the hexadecimal system, before the conversion algorithm is applied. The equivalent value is found to be 0.4375.

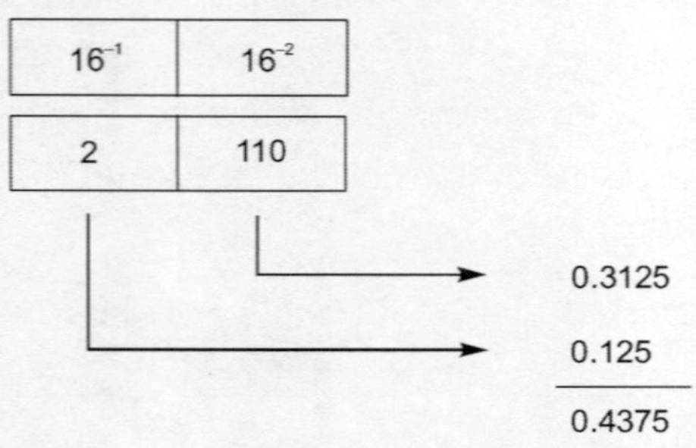

Figure 1.18 Converting a hexadecimal fraction to decimal fraction

1.8.13 Hexadecimal Integer/Fraction to Binary Integer/Fraction

Converting a hexadecimal integer to its binary equivalent is achieved by just reversing the idea discussed in the section 1.8.8. Each digit of the given hexadecimal number is written in their equivalent binary form in a four bit format and the final result is obtained by combining the results of the individual digits. Consider the hexadecimal number 4B and let us write its equivalent binary number by writing the binary equivalent of each of the digits of the number. Figure 1.19 shows how the result is obtained following this procedure. The alphabet B is first replaced with its numerical equivalent before the conversion. The result of the conversion is 1001011.

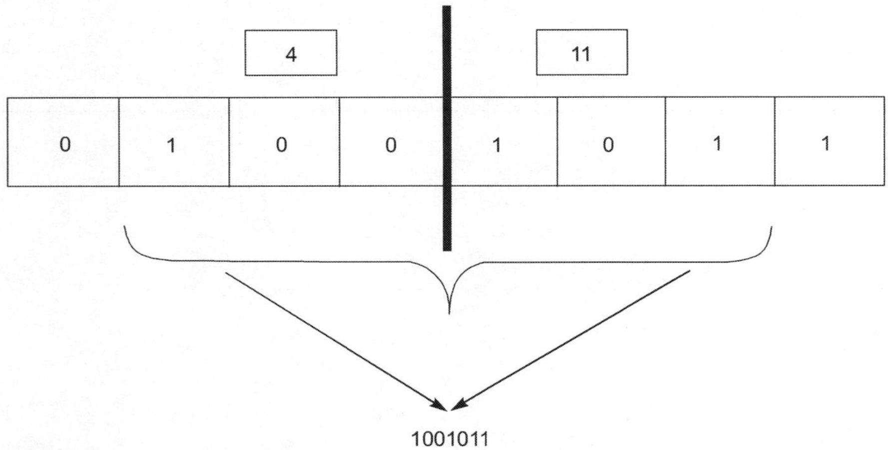

Figure 1.19 Converting a hexadecimal integer to binary integer

Similarly, when we are given a hexadecimal fraction, a similar kind of procedure is followed and the result is obtained. Consider for example, the hexadecimal fraction 0.AB. When we want to convert this to its equivalent binary fraction, we would convert each of the digits towards the right of the hexadecimal point to its binary format with four bits. We then combine the results to obtain the final result. The alphabets are converted to their numerical equivalents before the conversion process. Figure 1.20 shows how this conversion takes place and the result obtained is 0.10101011. When a mixed number is given for conversion, you have to carry out the conversion process individually and then combine the results to get the final result.

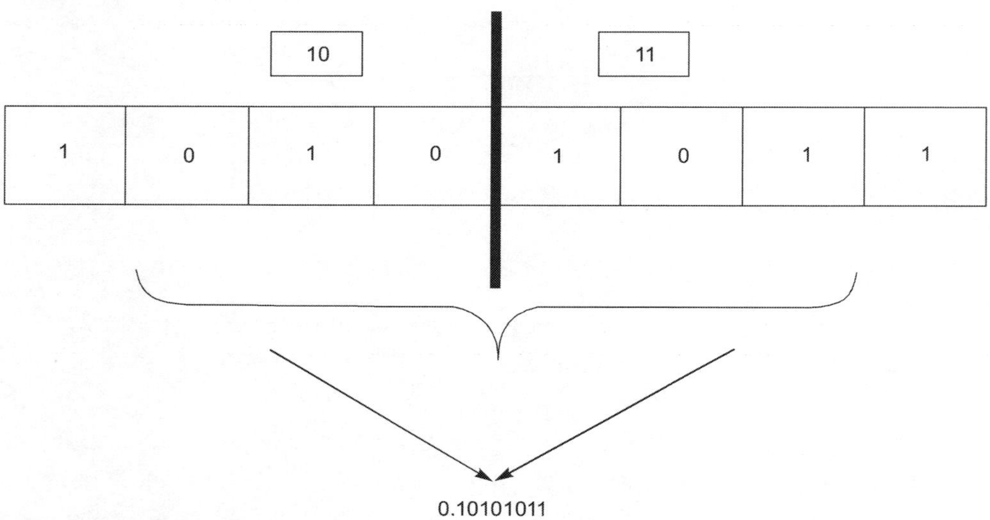

Figure 1.20 Converting a hexadecimal fraction to binary fraction

1.8.14 Hexadecimal Integer/Fraction to Octal Integer/Fraction

Conversion of hexadecimal integer to its octal is a two-step process. First convert the given hexadecimal number to its equivalent binary (Refer the section 1.8.13). Convert the binary number thus obtained to its equivalent octal (Refer the section 1.8.7). Hence the conversion is complete. Take for example, the hexadecimal number F6. The result of converting this to its octal equivalent would be 366.

If a hexadecimal fraction such as 0.F3 is given, the conversion process is simple. First convert the given fraction to its binary equivalent and then the binary fraction to its octal equivalent. (Refer the sections 1.8.13 and 1.8.7). I would request all the readers to check this with a few examples.

We have discussed all the possible inter-conversions between the number systems in use and I am sure all the readers would be thorough with the concepts discussed. I would advice all the readers to go through this section of this chapter to have a clear understanding of the number systems, since they are the base for learning any programming language.

1.9 ASCII CODE

If you are somewhat familiar with computers, then you know that all modern computers are "digital", i.e., internally they represent all data as numbers. In the very early days of computing (during 1940s), it became clear that computers could be used for more than just number crunching. They could be used to store and manipulate text. This could be done by simply representing different alphabetic letters by specific numbers. For example, the number 65 to represent the letter "A", 66 to represent "B", and so on. At first, there was no standard, and different ways of representing text as numbers developed.

Around the early 1960s an 8-bit byte was becoming the standard way that computer hardware was built, and you can store 128 different numbers in a 7-bit number. When all possible alphanumeric characters are counted (A to Z, lower and upper case, numeric digits 0 to 9, special characters like "%, *, /, ?" etc.) we would end up with a value of 90-plus. It was therefore decided to use 7 bits to store the new ASCII code, with the eighth bit being used as a parity bit to detect transmission errors.

American Standard Code for Information Interchange (ASCII) is a character encoding based on the English alphabets. ASCII codes represent text in computers, communications equipment, and other devices that work with text. Most modern character encodings, which support many more characters than did the original, have a historical basis in ASCII.

ASCII includes definitions for 128 characters: 33 are non-printing, mostly obsolete control characters that affect how text is processed; 94 are printable characters (excluding the space). The ASCII character encoding or a compatible extension is used on nearly all common computers, especially personal computers and workstations. Figure 1.21 represents the most commonly used 7-bit ASCII table and it has the ASCII values for the 128 characters supported and defined by ASCII encoding.

Even as these new standards are phased in, the 7-bit ASCII table continues to be the backbone of modern computing and data storage. It is one of the few real standards that all computers understand, and everything from e-mail to web browsing to document editing would not be possible without it. It is so ubiquitous that the terms "text file" and "ASCII file" have come to mean the same thing for most computer users.

ASCII	Hex	Symbol	ASCII	Hex	Symbol	ASCII	Hex	Symbol	ASCII	Hex	Symbol
0	0	NUL	16	10	DLE	32	20	(space)	48	30	0
1	1	SOH	17	11	DC1	33	21	!	49	31	1
2	2	STX	18	12	DC2	34	22	"	50	32	2
3	3	ETX	19	13	DC3	35	23	#	51	33	3
4	4	EOT	20	14	DC4	36	24	$	52	34	4
5	5	ENQ	21	15	NAK	37	25	%	53	35	5
6	6	ACK	22	16	SYN	38	26	&	54	36	6
7	7	BEL	23	17	ETB	39	27	'	55	37	7
8	8	BS	24	18	CAN	40	28	(	56	38	8
9	9	TAB	25	19	BM	41	29	)	57	39	9
10	A	LF	26	1A	SUB	42	2A	*	58	3A	:
11	B	VT	27	1B	ESC	43	2B	+	59	3B	;
12	C	FF	28	1C	FS	44	2C	,	60	3C	<
13	D	CR	29	1D	GS	45	2D	-	61	3D	=
14	E	SO	30	1E	RS	46	2E	.	62	3E	>
15	F	SI	31	1F	US	47	2F	/	63	3F	?

ASCII	Hex	Symbol	ASCII	Hex	Symbol	ASCII	Hex	Symbol	ASCII	Hex	Symbol
64	40	@	80	50	P	96	60	`	112	70	p
65	41	A	81	51	Q	97	61	a	113	71	q
66	42	B	82	52	R	98	62	b	114	72	r
67	43	C	83	53	S	99	63	c	115	73	s
68	44	D	84	54	T	100	64	d	116	74	t
69	45	E	85	55	U	101	65	e	117	75	u
70	46	F	86	56	V	102	66	f	118	76	v
71	47	G	87	57	W	103	67	g	119	77	w
72	48	H	88	58	X	104	68	h	120	78	x
73	49	I	89	59	Y	105	69	i	121	79	y
74	4A	J	90	5A	Z	106	6A	j	122	7A	z
75	4B	K	91	5B	[	107	6B	k	123	7B	{
76	4C	L	92	5C	\	108	6C	l	124	7C	|
77	4D	M	93	5D	]	109	6D	m	125	7D	}
78	4E	N	94	5E	^	110	6E	n	126	7E	~
79	4F	O	95	5F	_	111	6F	o	127	7F	□

Figure 1.21 ASCII coding

SUMMARY

I am sure that this Chapter would have been a good preparatory one for all those who had just started their affinity towards computers and programming languages. Especially those who are beginning their tryst with the computers would have surely felt comfortable with computing after learning the essentials.

To sum up, I would say that we started off with a very brief introduction to a computer program and significance of learning the same. We also learnt the various programming languages that evolved during different point of time and also their achievements and shortcomings. The main focus of this section was to introduce the beginners that programming languages are nothing new and they are in existence for decades and scores together. We also discussed the variety of programming languages that are used in different disciplines, both in developing applications and research. We categorized them as scripting languages and general purpose languages.

We discussed the different number systems and how they are useful in storing the information. We also discussed the inter-conversion of number systems with sufficient examples and figures. To conclude the Chapter, we had also learnt the ASCII character coding and its significance while programming. I am sure that the figures and illustrations that are given in this Chapter would be of profound use for the beginners and those aspiring to become programmers.

CRACK THE NUT

1. Convert $0100\ 1011_2$ to decimal.
2. Convert 173.34375_{10} to binary.
3. Convert the number 01111101.1001_2 to octal and to hexadecimal.
4. Convert 123.45_8 to binary.
5. Convert $C8.D3_{16}$ to binary.
6. Write down the binary equivalent of 2AC3.F1.
7. Convert 13_{10} to its equivalent octal and hexadecimal number systems.
8. Convert the number 13_8 to its equivalent hexadecimal system.

9. Identify the general equation for representing a number in any number system.
10. Convert the number 9F to the other three number systems.
11. Try to design a converter (if you are familiar with any basic level programming language) to convert one number system to another.
12. Learn how negative numbers are represented in Computing System (use the knowledge of binary, octal and hexadecimal numbers).
13. Try to learn about the different codes followed in computing (To quote a few, we have BCD, Gray, 8221, 2421, etc.)
14. Learn the basic arithmetic operations that you can perform with different number systems.
15. Learn about compliment method of storing numbers and explain their significance.
16. Learn about the IEEE method of representing the floating point numbers (real numbers).
17. Learn how octal and hexadecimal numbers are represented in C programming language while they are accessed through the functions scanf() and printf().
18. Give your suggestions and opinions on the applications of programming languages in various disciplines.
19. Get the list of different programming languages that evolved over a period of time and compare the efficiency of each language and explain their significance.
20. Explore the weighted and unweighted number systems in use.

2

ESSENTIALS OF C PROGRAMMING

2.1 INTRODUCTORY CONCEPTS

2.1.1 History of C

C, even after 38 years of its discovery, stands tall in the stream of Programming Languages. As you know, learning programming languages is very important, since you have to communicate with the computing system. There are umpteen programming languages in the world of computing and only a few are surviving and being used by the programmers and software developers now. One among them is C language. As mentioned earlier, it has its own place in the stream of programming languages and I can confidently say that it is the foundation for all the languages that were developed after its discovery. If you look at its history, it is quite interesting to note that it emerged from many other languages and surpassed the same to get ascendancy in the computing world. Let us learn the language starting with its history.

The C programming language and its direct descendants are by far the most popular programming languages used in the world today. This structurally tiny language originated at Bell Labs in order to write the first version of the Unix Operating System. In spite of its small beginnings, C has scaled up to run on powerful super-computers with gigabytes of memory. The language is extremely portable; it is possible to write a program in C for just about every platform in existence. Yet, if one takes a more in-depth look at C, one realizes that it is rather weak and has a small vocabulary. But, when we look at the efficiency and power of C, the weakness of C could easily be excused.

The history of C and that of the UNIX operating system are intertwined. You might almost say that C was invented and written for the purpose of writing the UNIX operating system. The seeds of UNIX began in 1965 with the completion of a project at MIT (Massachusetts Institute of Technology) called Project MAC, which was one of the first time-sharing computing systems. It allowed simultaneous use of the same computer, for at most thirty people, by using any of the 160 typewriter terminals placed around the campus and in the homes of faculty. It became so overloaded that MIT decided to embark on a more ambitious time-sharing system as a joint venture with General Electric and Bell Labs. This project was given the name MULTICS, which stands for MULTIplexed Information and Computing Service. The trio (Thompson, Ritchie and Kernighan) worked on this project for a few years, but in 1969, Bell Labs decided to discontinue the project MULTICS since it was too expensive and would take longer time to complete. At this point, an informal group of Bell Labs employees, led by Ken Thompson, started to investigate the alternatives.

Dennis Ritchie
(Father of C Language)

Thompson and his group started designing an operating system (UNIX) that would support coordinated teams of programmers in the development of products, and would simplify the dialogue between human and machine. Thompson decided that UNIX needed a high-level system language. A high-level language is one which abstracts the hardware away from the programming environment. However, he could not just pick one that already existed. He was restricted by the computing system that he was working, PDP-7, which had only 8 kilobytes of memory. This was such a problem that at the start of the project, he had to compile his program on the more powerful GE-635 machine and then transfer the program to the slower computer using paper tape. Thompson had had some previous experience with the BCPL (Basic Combined Programming Language) system language. But it was too slow and lacked the run-time support that Thompson thought was necessary. So, he took his experience of BCPL and created his own language which he called B. The language was not really a new language. More correctly, it was BCPL squeezed into 8 Kilobytes and filtered through Thompson's brain. B had some problems though, and small UNIX community felt limited by the restrictions in the language. It seemed that B needed some enhancements to satisfy the developers. The main enhancement that it needed was data types to properly handle the characters and floating point numbers. A data type is defined by two properties: a set of values and a set of operations. This means, a language with strong typing would not allow the programmer to do meaningless operations, for example, taking the square root of a string of characters.

One of Thompson's co-workers, Dennis Ritchie, in 1971, began extending B by first adding data types and continued to expand and rewrite the parts of language. In 1973, Ritchie thus completed the essentials of C and then rewrote the entire UNIX kernel using the language. This time he wrote it for the PDP-11, with its huge 24K memory. In 1978, Brian Kernighan and Ritchie published The C Programming Language, which has since then become somewhat of a Bible to C programmers, and is often called the "White Book" or "K&R".

Now that C was the main system language for the UNIX environment, its use quickly spread. As the popularity of UNIX grew, so did that of C. During the 1980s, the use of C language spread widely, and the compilers became available on nearly every machine architecture and operating system. As the language spread, it began to change slightly for each platform that it was written for, and at this point the first edition of K&R no longer fully described the language and it became clear that a standard version of C was needed. Therefore, in the summer of 1983, the American National Standards Institute (ANSI) established the X3J11 committee to define the C language. They took 6 years and finally defined ANSI C, in 1989.

C is now such a popular programming language that there have been several descendants from it that is in wide use. One such descendant Objective C was adopted by the NeXT Computer Company for their OpenStep operating system. Apple Computer recently bought out NeXT, and Objective C is now the main development language at Apple for its next generation operating system, code-named Rhapsody. Another descendant from C is Concurrent C, designed at Bell Labs by Dr. Narain Gehani and Dr. William D. Roome. This language is particularly well-tailored for parallel programming to write concurrent programs that can run efficiently on single computers, loosely coupled distributed computer networks, or on tightly coupled distributed multiprocessors. And of course there is the extremely popular C++, which is an Object Oriented extension of C.

How did C become so popular despite its deficiencies? C has many unique characteristic features that made it quite popular among the programmers.

2.1.2 Discovery of C—Is that Serendipity?

One can easily argue that C was written accidentally rather than being written intentionally. The reason is that, the intention of Ken Thompson was to enhance the performance of UNIX OS and he wanted to rewrite the entire kernel of UNIX. This persuaded Thompson to write a language and he started exploring various languages that existed and finally came out with B, which was later developed as C with the help of Dennis Ritchie and Brian Kernighan. So, the discovery of C is nothing but serendipity.

2.1.3 People Behind C

C was the brain child of many people, like Ken Thompson, Brian Kernighan and Dennis Ritchie. The contribution of Dennis Ritchie was so much that he is fondly called by the programmers as Father of C Language.

In 1978, Brian Kernighan and Dennis Ritchie published the first edition of *The C Programming Language*. This book, known to C programmers as K&R, served for many years as an informal specification of the language. The version of C that it describes is commonly referred to as K&R C. The second edition of the book covers the later ANSI C standard.

K&R introduced several language features:

- Standard I/O library
- Long integer data type
- Unsigned integer data type

Compound assignment operators =op (shorthand operators as we call it in our book) were changed to op= to remove the semantic ambiguity created by the construct i=–10, which had been interpreted as i =– 10 instead of the possibly intended i= –10.

Even after the publication of the 1989 C standard, for many years K&R C was still considered the "lowest common denominator" to which C programmers restricted themselves when maximum portability was desired, since many older compilers were still in use, and because carefully written K&R C code can be legal Standard C as well.

2.1.4 C as a Parent Language for Languages Emerged After It

Almost all the languages that were written after C, inherited many of its features. Take for example, C++, which is the so-called superset of C, has all the procedure-oriented features inherited from C. Pascal, another pretty language has most of the aspects taken from C, but it has its own style of programming. PERL (Practitioner Extract Report Language), which is a frequently used language in developing tools and algorithms, has many of its features directly inherited from C.

Even though C does not support object-oriented features or database ideas, the languages which are designed with object-oriented features and database aspects, take their fundamentals from C only.

2.1.5 Significant Features of C

- *It is powerful, flexible and imposes few constraints on a programmer.* This means, it is easy to learn and handle this language. A user need not struggle with the syntax and semantics of the language.
- *It provides faster program execution.* C's power and fast program execution comes from its ability to access low-level commands, similar to assembly level language, but with high level syntax. Its flexibility comes from the many ways the programmer has to accomplish the same tasks. C includes bitwise operators along with powerful pointer manipulation capabilities. This can be an advantage to an experienced programmer but disadvantage to a novice.
- *It provides modularity.* If you can divide your work into pieces and complete it, it won't be a burden to you. This is very often called as **Divide and Rule** concept. Sections of code can be stored in libraries for re-use in future programs. This structure of modularity also helps with C's portability and execution speed. The core C language leaves out many features included in the core of other languages. These functions are instead stored in the C Standard Library where they can be called on when needed. An example of this concept would be C's lack of built-in I/O capabilities. I/O functions tend to slow down program execution and also be machine-independent when running optimally. For these reasons, they are stored in a library separately from the C language and only included when necessary.
- *C is robust and transportable.* This means, it is quite easy for us to learn and easy to carry, install and work. It is not specific to any particular operating system and compatible with all the operating systems such as UNIX, DOS, etc. In simpler words, it is machine-dependent.
- *It is a programmer-friendly and free-form language.* C language is very flexible for the programmers as it does not impose any regulations for typing a program. The programmer is just required to open the editor and start typing the program. He/she can start at any part of the editor (an editor is also called as form and the editor normally used for typing a program in C is called as the Turbo Editor) and end anywhere. Editors of some of the other languages have a few restrictions. For instance, in the case of FORTRAN-77, the programmer has to start the coding of the program only from the 7th column of the editor. Similarly in COBOL, the coding has to be started from the 8th column of the editor. Regulations like these are not there in C. Hence, it is called as a free-form language.
- *It is a case-sensitive language.* C is a case-sensitive language. This means the uppercase and lowercase alphabets are treated differently when a program is written in C. So, be very careful when you start writing programs in C.

- *It is a middle-level language.* There are three categories of languages in all: High Level, Middle Level and Low Level. Those languages that use English like statements for writing the source code are put in High Level category, those which use binary digits (0 and 1) or mnemonics are put in the Low Level category and those that use both the high level and low level syntax are included in the category of Middle Level Languages. C fits in the middle level just because of its efficiency in handling both application-level programming and machine-level programming. You will probably be surprised if I say that you can make your computer monitor to burst by executing a program written in C. You can program your keyboard, mouse, etc. using C and make them work the way you like. That is the reason why C is regarded as the most powerful language among the stream of programming languages we have.

2.1.6 Importance of Learning a Language

Why do we learn languages? The fundamental difference between a human being and an animal is that the former knows to communicate and the latter does not know either to speak or to write. So, the reason behind learning a language is mainly to communicate with fellow human beings. Just as languages help to communicate with others, programming languages help to communicate with the Computing System. Not only that, we are able to come out with applications and interactive software, which make our task simple and easy.

2.2 BUILDING BLOCKS OF C

2.2.1 Methodology of Learning C

The next question in our minds will be How to learn a language? Different people adopt different methods to learn a language. Some languages could be learnt just by speaking, some could be by writing, some just by listening. But, almost all of the languages in this world fall under one domain, where we have to adopt a systematic procedure to learn and master it. Take for example; we would like to learn English. What is the procedure we normally follow? This is depicted in Figure 2.1.

Figure 2.1 Methodology of learning English

In Figure 2.2, we can see the methodology of learning C, which is very much similar to that of learning English.

Figure 2.2 Methodology of learning C

Grammar is a very important part of learning any language that human beings speak. Without that, the learning process would be incomplete. In learning English, we would probably learn the grammar after we learn how to form words with the help of the alphabets. It is not explicitly shown in the diagram since it is understood. There is a counterpart for grammar in C, which is nothing but syntax of the language. The characters, identifiers and keywords must be properly connected up in building a meaningful program. This stringing together of words to make meaningful statements is called the syntax. Also, we have to learn the data types and their qualifiers and operators that are available in the language, before we learn to make up the instructions. Let us start from the first block in the Figure 2.2.

2.2.2 The C Character Set

The character set of C language contains more characters as compared to English, which has just 26 alphabets. The following list gives the character set of the C language:

1. Uppercase alphabets — A to Z (ASCII—65 to 90)
2. Lowercase alphabets — a to z (ASCII—97 to 122)
3. Digits — 0 to 9 (ASCII—48 to 57)
4. Special symbols — @, $, #, $, ^, &, *,~, !, etc.
5. Mathematical symbols — +, –, <, >, =, /, etc.
6. Punctuation symbols — ?, ., ,, :, ;, “, etc.
7. Other symbols
   a. Symbols which we normally do not have on the keyboard, but still may use in programming (like playing card symbols—spade, clubs, diamond, etc.)
   b. Symbols that have no proper notations, but still on the keyboard (like space bar, backspace, enter, arrow keys)

We see that the character set is quite big and totally there are 256 characters in all, that could be used in C. We have to keep in mind that the language is case-sensitive and hence uppercase and lowercase alphabets are treated differently. Besides the first three categories of the characters listed above, we see the corresponding ASCII codes. What is this ASCII? ASCII stands for the American Standard Code for Information Interchange. It is a standardization established by the Americans for the users of the Computing System. Since the computers are used worldwide, there must be some standard system of representing the notations that we have on and off our keyboard. Languages like C, C++ use this ASCII coding system which could code only 256 (2^8). Since Java is a language user over the Internet, more characters are to be coded and hence the UNICODE (UNIversal CODE) system is preferred. It is a good thing if you remember the ASCII values of at least the Uppercase and Lowercase alphabets and Digits, since we will be later using them in string manipulation programs. Even if you cannot remember them, do not bother too much, since we always have options to get to know them.

When you make a keystroke 'A' from the keyboard, C language will interpret that as a number, more appropriately saying, as a binary number. The ASCII equivalent of 'A' is 65 and this 65 is converted to its binary equivalent, which is 1000001 (I would advice you to go through the chapter on Number Systems once again, in case you have any clarifications on conversion process). Thus, each one of the 256 characters will be coded in the binary format. Remember that each code is unique to itself.

Now that we are familiar with the character set of the language, we are ready to move on to the next block of the learning process, namely keywords and identifiers.

2.2.3 C Tokens—Keywords and Identifiers

Keywords Keywords are those, which the language designers have defined for a special purpose and shall be used only for that. As a user one cannot change the definition of a keyword. C keeps a small set of keywords for its own use. There are 32 keywords that, when combined with the formal C syntax, form the C language as defined by ANSI C. Some keywords were used along with the existing list and were used rarely in programming. To name a few asm, entry, fortran, etc. Look at the Figure 2.3 below to know the keywords. You can notice that all the keywords are in lowercase and none of them should be used for naming an identifier while writing/editing programs. The C compiler will flash with error message when we try to use any of these 32 keywords for naming an identifier. In simpler words, keywords are predefined for the compiler and cannot be redefined by the programmers.

auto	double	int	struct
break	else	long	switch
case	enum	register	typedef
char	extern	return	union
const	float	short	unsigned
continue	for	signed	void
default	goto	sizeof	volatile
do	if	static	while

Figure 2.3 Keywords in C

Identifiers Identifier is a technical term used to mean "name". An identifier in C could be the name of a variable, a constant, a function, a structure, etc. In simpler terms, they are nothing but labels. Identifier as the word itself suggests, an identifier would identify something for the compiler. It identifies what type of value is stored in a variable or a constant or what name a function or a structure has in a program. It is very important that we shall name all the variables and frequently used constants (like pi = 3.14) to make a prior intimation to the C compiler. Otherwise the compiler might find it difficult to recognize a variable or a constant or a function defined in a program. This is very similar to giving name to a new born baby in our homes.

2.2.4 Rules for Naming an Identifier

Before we give a name to the identifier that we use in a C program, we must know a set of rules that have to be followed, rather very strictly. These are given below:

1. The first character of the identifier must be a letter of alphabets (remember C is case-sensitive and hence there are 52 alphabets in C—26 lowercase and an equal number of uppercase) No other character should be used as the first character while naming a variable. For example, the name **single** for an identifier is acceptable, whereas the name **1single** is not acceptable, since the first character is a digit. Also, the name **single** and the name **SINGLE** are treated completely differently by the compiler since they are of two different cases (the former being a lowercase name and the latter being an uppercase name).

2. There should not be any embedded space within an identifier. For example, the name reg no for an identifier is not acceptable, since it has a space in between the words reg and no. The C compiler does not accept a space. In fact, it will not accept any other character (other than digit) in between two words of an identifier. If you wish to distinguish two words of an identifier, C has a provision. It allows the programmers to use an Underscore (_) between two words of an identifier. In other words, an underscore is regarded as the 53rd alphabet in C language. So, the above name reg no could be written as reg_no.
3. When you name an identifier in C, you are allowed to use digits within it. For example, the name mem_2_scr is acceptable. One can use digits even towards the end of an identifier's name. But no other character shall be used either within or towards the end of an identifier's name.
4. When you name an identifier, be sure that you are not using any of the keywords defined in the language. This point has been already discussed in keywords. The name of the variable shall be unique to itself.
5. This rule tells you about the length of an identifier's name. The identifier could contain a maximum of 32 characters including digits and underscores for its name. If this limit is exceeded, the same will not be notified by the compiler, since it is not a syntax error. If you exceed this limit, the compiler will simply consider only the first 32 characters of the identifier and will skip the remaining characters.
6. The last rule is not exactly a rule, but a convention to be followed while naming an identifier in C. The convention is Be Suggestive when you name an identifier. This means, you must have a valid reason for naming an identifier. For example, if I am writing a program to find the area of a circle, I can very well use names for the variables in my program as 'x' and 'y'. But when we have a look at this program at a later time, we cannot be sure what 'x' implies and what 'y' implies. In order to make a program easily understandable, we can be suggestive when we give names to the identifiers. In the above program, which I had mentioned, the variables could be given names like rad, area instead of x and y. So, we understand that the variable "rad" is going to store the value of radius and the variable "area" is going to store the area after it is calculated.

2.2.5 Data Types in C

Data types are an important feature of any programming language. Why do we have data types at all? Don't we know what type of input we are going to give to the computer or to the program that we have? It is not for our sake that we must learn the data types. It is necessary to know it so as to let the C compiler know what type of data we are giving as input and what type of data we want back in the output. This means whenever we use variables in our programs, we need to specify the type of value they will be holding throughout the execution of the program.

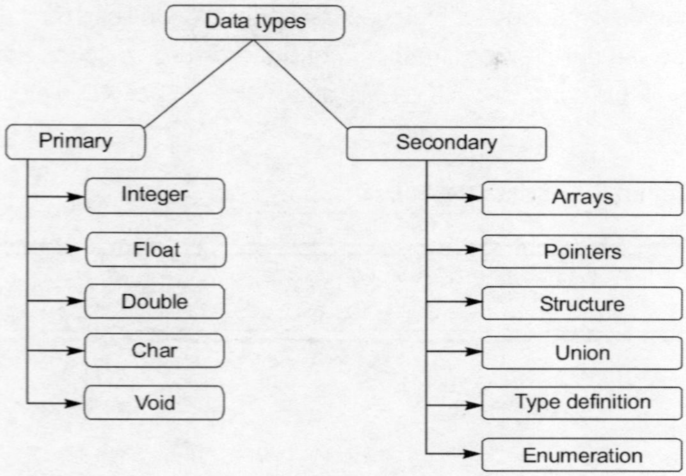

Figure 2.4 Data types in C

Data types are classified into two major categories namely, **primary** and **secondary**. The primary data types can again be classified into numerical data types and alphanumeric data types (character type). Numerical data types include integer, float (single-precision) and double (double-precision). Alphanumeric data type includes characters. Figure 2.4 shows the different data types available in the language.

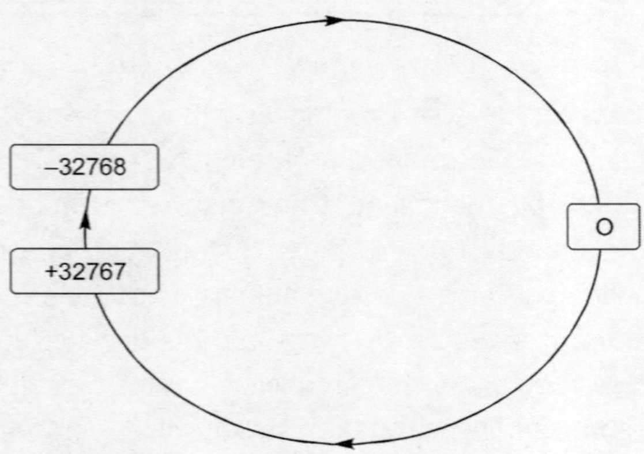

Figure 2.5 Integer data type representation

Data types are further classified into different subtypes based on their size and sign. Except the float data type, all the data types have their own qualifications. Integer has four subtypes, double has one and character has two. Put together there are seven numerical types and two character types. Each data type has its own size and range of values. Table 2.1 illustrates the different data types in the language with all necessary details. Do keep the table in memory, since the same will be required during programming. For the time being, do not bother too much about the last column of the table (format specifier). We shall come to that when discussing the input and output instructions.

Table 2.1 Detailed list of data types in C

Data type	Keyword used	Memory (bytes)	Range	Format specifier
Signed character	char	1	–128 – 0 – +127	%c
Unsigned char	char	1	0 to 255	%c
Signed short integer	int	2	–32768 – 0 – +32767	%d
Unsigned short integer	unsigned	2	0 to 65535	%u
Signed long integer	long	4	–2147483648 to +2147483647	%ld
Unsigned long integer	unsigned long	4	0 to 4294967295	%lu
Float	float	4	–3.4E38 to 3.4E+38	%f
Double	double	8	–1.7E308 to 1.7E+308	%lf
Long double	long double	10	–1.7E4932 to 1.1E+4932	%Lf
Void	void	–	–	–

Different data types have different ranges and hence we must be careful while we use variables pertaining to a particular data type. The C compiler will not check whether we have used the values within the range for a particular variable belonging to a particular data type. That is why we call C very often as a **Weakly Typed Language**. One can use any number of variables in a program. But all of them must be properly type declared using the **Type Declaration Instruction**. The compiler otherwise will report errors. What happens when we cross the limit for a particular variable of a specific type? For example, if a variable *x* is declared in the program as int, the compiler will treat that as 'signed short integer' (please see the table that the keyword used for a signed short integer is just 'int') and hence one can assign values only from –32768 to +32767. What will happen if we assign a value 32770 for *x* ? We are crossing the limit and we expect the C compiler to flash an error message. But it will not do so. Errors of this type are termed as logical errors, since they arise out of wrong thinking. The compiler will notify only the syntax errors in a program. So, for the above situation, what will be the result? Imagine the integer data type's range to be a circle, as shown in Figure 2.5. We see that after the upper limit value 32767, we have the negative extreme –32768. Since, 32770 is three values more than the actual limit, three will be added with the negative limit and the result will be checked again for availability within the range. If it lies within the range, then that value will be the one for the value which we assigned. In the above case the value which will be recognized by the compiler will be –32766. In the Table 2.1, the columns size and range are purely architecture-dependent. The values indicated are the most common ones found in 32-bit systems.

Thus, the compiler's response will be quite different and we have to be very careful when we program in C. You can just imagine the same kind of Circle represented in Figure 2.5 for representing the ranges of all the data types and try to get a clear idea of them. Do not ever memorize all the columns of the above table, as you will be practicing them in all your programs and they will be registered in your mind.

2.2.6 Essential Operators in C

Why do we have operators in programming? Obviously the answer is to perform operations. What are the various operations you can perform? Basically we can do addition, subtraction, multiplication and division. Some of you even would be familiar with modulus (remainder out of division). But basically only two operations are there in Mathematics, namely addition and subtraction. Multiplication is nothing but repeated addition and division is nothing but repeated subtraction. But C language has plenty of operators. In fact, it is very rich language with regards to the operators. No other language has such a large number of operators. All the operators are designated by symbols (combination of one or two characters). Let us look out what are the different operators available in this language.

Before you could learn the different types of operators in C, I suggest, you be familiar with the jargon involved. Consider an expression like c = a + b. In this expression, a and b are called independent variables and c is called the **dependent** variable. Mathematically we may write the above expression as c = f (a, b) and we read it as c is a function of a and b. We see the above expression has two sides a **Left Hand Side (LHS)** and a **Right Hand Side (RHS)**. Any expression basically will have two sides, as indicated above. In these kinds of expressions, the RHS will be evaluated first and the result will be assigned to the LHS. In the above expression, the variables a, b and c are called as **operands** and = and + are called as **operators**. In other words, any expression will involve operators and operands. Basically, an operator could be one of two kinds, namely **unary** and **binary**. Any operator that needs only one operand for performing the operation is called a unary operator (una means one) and that which needs at least two operands is called a binary operator (bi means two). Let us now explore the different categories of operators in C.

Category of Operators

1. Arithmetic
2. Relational
3. Logical
4. Bitwise
5. Assignment
6. Shorthand
7. Modification
8. Pointer
9. Structure and Union
10. Function
11. Array
12. Sizeof
13. Ternary
14. Typecast

Let us start with arithmetic operators. The arithmetic operators in C are listed below:

1.	Addition	+ (plus)
2.	Subtraction	– (minus)
3.	Multiplication	* (asterisk)
4.	Division	/ (forward slash)
5.	Remainder	% (percentage)
6.	Unary Minus	– (minus)
7.	Parenthesis	( )

It is quite simple to understand the way the addition, subtraction and multiplication operators work, since they occur in the basic mathematics, which we have learnt. It is important to understand how the division and the remainder operators work.

For example, consider the following expression: X = 1/2. What do you think the value in X will be? I think everyone will say 0.5. It is not. The value stored in X will be 0. How does this happen? When you work out the same in your calculators or any other computing devices, you will get 0.5 and when you work out this in C, it gives you a zero. Why? The reasoning is quite interesting and deviates from the basic arithmetic you have learnt in schools. We have a numerator (1) and a denominator (2) in the above expression and we see that both are integers. Integers are nothing but whole numbers, which do not have a fractional part. 1 is an integer, whereas 1.0 is not. Even if the fractional part is a zero, we don't consider it as an integer. When we do arithmetic operations with integers as operands, we tend to get only integers and it is applicable to all the arithmetic operators. So, we can form a tabular column like the one shown in the Figure 2.6 for the basic arithmetic operations (+, –, * and /).

As said earlier, the division operation has to be looked into very carefully, as C does a bit differently from the basic division process that we do with pencil and paper.

The next operator to be discussed in this category is Unary Minus (–). It carries the same sign as the Binary Minus, except for the fact that it works completely differently. We can say in more simple words that Minus Operator plays a dual role in C. As far as the jargon is concerned, you know what Unary and Binary are. Lot of questions does arise in our minds. Why do we have a Unary Minus, when there is a Binary Minus? Why don't we have Unary Plus? or Is there anything called Unary Plus? The answers for these questions are very simple and easy to understand. What do you do when you want to express a Positive number (an integer or a real number)? Yes, you are right.We simply express the number without having a need to prefix the 'plus' (+) operator. For example, '8' is a positive number and we do not say '+8' under normal circumstances. But, what about negative numbers, say, if we want to express a negative '8'? Obviously, we have to prefix a negative sign to the number, that is, we have to express it as '–8'. This 'prefixed minus' sign is called the Unary Minus. It is a must that all the negative numbers shall be prefixed with this minus.

Operations	Operand-I type	Operand-II type	Result type
Addition (+), Subtraction (–), Multiplication (*), Division (/)	Integer	Integer	Integer
	Integer	Float	Float
	Float	Integer	Float
	Float	Float	Float

Figure 2.6 Results of the basic arithmetic operations with data types

The next operator in the list is the Modulus Operator (designated by a % symbol). Most probably, this operator would have been used by you in your higher classes, but in a different way. We term that mathematically as a "remainder operator". What this operator does is that, it finds out the remainder of the two integer values. Of course, it divides one number by the other (like the division operator), but leaves the remainder as the answer (unlike the division operator, which leaves the quotient). Then, what could be the answer for 8%5? Yes, the answer is 3, since upon dividing 8 by 5, the remainder that we get is 3. What could be the answer for 5%8? If your answer is 0, then it is wrong. Then, what else is the answer? It is 5. Wondering how!!! Work out this division with a pencil and a paper or just ask yourselves, what could be the quotient on dividing 5 by 8? Or how many times do 8 go in 5 exactly? Obviously 8 will not go in 5 or simply saying, 8 goes 0 times in 5. So the quotient you obtain is 0 and then to get the remainder, you multiply the divisor with the quotient and the product obtained is subtracted from the dividend. (In division process, the numerator is termed the dividend and the denominator is termed the divisor. Normally to get the remainder of the division, we first multiply the quotient with the divisor and subtract the product obtained, from the dividend). In the above example, when we multiply the quotient (which is 0) with the divisor, we tend to get 0 as the product and when we subtract the product, that is, 0 from the dividend, which is 5, we get 5. This is the justification for 5% 8 = 5.

I can give you a blind way of understanding this. For any modulus division of the format m%n, the answer will be 'm' provided (i) m < n and (ii) m, n > 0. But, don't just keep this blind idea in your mind, since you may be required to justify in the way I have done above. Now, try the following: –5%8, 8% – 5 and –5%–8. What do you think the answers will be for the above expressions? If your answers are –5, 3 and –5, they are right. But, how? Again apply the same idea of division. When you evaluate the number of times 8 goes in –5, it will obviously be a 0. Hence, the remainder is –5. When you evaluate the number of times –5 goes in 8, it will be –1 and when this quotient is multiplied with the divisor (–5) we get 5 and when this product is subtracted from the dividend, we get 3. And finally when you evaluate the number of times –8 goes in –5 it will again be a 0 and hence the remainder is –5. Can you conclude anything from the above discussion, with regard to the % operator? Yes, the sign of the result of % division is always decided from that of the numerator of the expression. So we can have a table for deciding the sign of the result of the % division as shown in the Figure 2.7.

%	+	–
+	+	+
–	–	–

Figure 2.7 Operation and results with % operator

What could be the result of the expression 5.85 % 9.4? Don't take too much pain in solving this expression, since % operator is applicable only to integer and its associated data types and not to any other data type. Thus, the above expression will give rise to an error and hence be ruled out.

The last one in the list of arithmetic operators is parenthesis ( ). This set of parenthesis also plays a different role with regard to the context in which it is used. When it is used in an arithmetic expression (an arithmetic expression is the one which has arithmetic operators alone operating on the operands), it shall be considered as an arithmetic operator. In any complex expression, the expression within the innermost parenthesis shall be evaluated first and then the one in the next innermost and so on. For example, consider the following arithmetic expression: a+b/(c*(d+e)). In this, we have an expression 'd+e' within parenthesis and another expression "c*(d+e)" also within paranthesis. The former expression is within the innermost parenthesis and the latter within the next innermost (or probably the outermost) parenthesis. So, when the entire expression is evaluated, "d+e" is evaluated first and the result is multiplied with 'c'. This is the importance of using a set of parenthesis in an expression. We can give priority to a particular part of the whole expression thereby overriding the normal precedence rules. It is right time for me to explain the precedence rules associated with the arithmetic operators. The following list provides the rankings or priority to the arithmetic operators.

1. ( ), – (Unary Minus)
2. *, / and %
3. + and – (Binary Minus)

This is otherwise called the hierarchy of operators and all the expressions we use in C language are governed by such hierarchy rules. As we go on with this discussion, I will let you know the hierarchy of all the operators pertaining to their category.

There comes a question in our minds. If any expression contains both '*' and '/', which will be preferred first by the C compiler? Preference will be given to the operator that comes first from left. For example, consider the following expression: c = 4*6/8%3–4+9. What will be the result of the above expression after you apply the above precedence rules? It is 5. How? When the C compiler starts to evaluate the expression on the Right Hand Side, it traces it from the left and when it does so, the first operator encountered is '*' and hence 4*6 will be evaluated first, whose result is 24 and this intermediate product is now divided by 8, since '/' appears next to '*'. So, we get a quotient of 3. This in turn is divided by 3 and the remainder of this division happens to be 0. The next operator encountered is '–' and when 4 is subtracted from 0, we get –4 and finally we have '+' using which we add –4 and 9 and get the result as +5.

What is the importance of knowing hierarchy? Why do we want to implement in languages? Let us understand the reason with an example.

Try to evaluate the following expression: c = 5/9*(f–32). This is the formula to convert the temperature in Fahrenheit to Centigrade. Using the hierarchy of arithmetic operators we get zero as the result, for any input for the temperature in Fahrenheit. This happens because according to the hierarchy, f–32 will be first evaluated, which will give a value and then 5/9 will be evaluated, which will give a zero. When Zero is multiplied with any value, obviously the result is Zero. Thus, even though the expression given above is mathematically correct, it will not give the right result. Some modification should be done in the above expression to get the correct answer—change the expression to c = 5.0/9*(f–32) or c = 5/9.0*(f–32) or c = 5.0/9.0*(f–32)—it will work.

Such is the importance of hierarchy of operators in C and it has to be followed when you are writing a program or developing an application in C.

The next category of operators to be discussed is the Logical Operators. The logical operators in C are NOT, AND and OR, designated by the symbols, ! (exclamation mark), && (double ampersand) and | | (double pipe) respectively. These symbols accept the inputs and process them in the same way as the arithmetic operators. But the way they treat their operands is different. For a better understanding of how the operands will be treated, you should know what **Boolean Values** are. In digital electronics, there are two numbers normally used, 0 and 1. We had discussed about these two numbers earlier as binary digits and now we will call them as Boolean Values. In C language, any non-zero number is treated by the compiler as a '1' (stands for true) and any zero as a '0' (false). All the three logical operators accept the inputs and internally convert them to either '1' or '0' based on their values. For example, consider the following logical expression (any expression that has logical operators in it): c = a&&b, where the values of a and b are 20 and 5 respectively. Since 20 and 5 are non-zero values, they are internally converted to '1' and hence the above expression reduces from c = 20&&5 to c = 1 && 1. Thus, we see that regardless of the values we give, they are either treated as a '1' or a '0'. Out of the three logical operators, NOT is a Unary operator and the other two are binary operators. Thus, it is not possible to have an expression like c = a!b.

The logical operators give the result also in Boolean format. This means the result of the operation will either be a '0' or a '1'. Let us consider first the operator AND (&&). Using 'double ampersand' for AND is a must. We may commit a typing mistake by just typing a single ampersand. It will lead to a different meaning and a different answer and this type of problems will not be notified to the users or the programmers. Hence, the programmers and the learners must be very careful when they use this operator. In digital electronics, this operation is achieved with the logical devices called **GATES**. There are two categories of gates called primary and secondary. The primary gates are AND, OR and NOT. They work based on the combination of the inputs given to them. For AND and OR gates we can give any number of inputs and for a NOT gate, we can give only one input. Let us consider for simplicity that we give two inputs for AND gate. There are obviously four different combinations. They are (0,0), (0,1), (1,0) and (1,1). They are formulated inside a table called the **truth table**. This table for a two-input AND gate is shown in the Figure 2.8.

Input a	Input b	Output c
0	0	0
0	1	0
1	0	0
1	1	1

Figure 2.8 Truth table of a two-input AND gate

Thus we can now generalize an AND operation and define it more formally in terms of C. An AND operation is defined as that which gives a true output if and only if all its inputs are true. See that we have used only two inputs for justifying this general statement. This holds good for any number of inputs. Also, AND logic is often called as Boolean Multiplication and is very useful for Boolean Algebra.

An OR gate works in a way just opposite to that of an AND gate in digital electronics. The truth table of a two-input OR gate is shown in the Figure 2.9.

Input a	Input b	Output c
0	0	0
0	1	1
1	0	1
1	1	1

Figure 2.9 Truth table of a two-input OR gate

Now we can define the OR operation from the outputs generated in the table as follows: An OR operation is defined as that which gives a true output if at least one of its inputs is true.

A NOT gate is a Unary gate and hence there can be only one input for it and it produces an opposite output for an input. For example if the input is true, it produces a false output and vice versa.

Thus a NOT operation is defined as that which produces a true output when the input is false and vice versa.

Thus, when you have an AND operation throughout an expression, the expression will be checked till a first '0' is encountered. When you have an OR operation throughout an expression, it will be checked till a first '1' is encountered. Suppose that there are 100 inputs in an expression and throughout we have an AND operator, then the expression will finally yield a '0' if the expression contains a zero either in the beginning, in the middle or at the end. The evaluation of the expression will stop once the first '0' is encountered. If in the same expression, we have an OR operator, the result will be a '1' if the expression contains a '1' in the beginning, in the middle or end. The evaluation will stop once the first '1' is encountered. This is the way the logical AND and OR operators work. One has to be very careful while performing operations with AND and OR operators, especially when they both occur in a logical expression. The hierarchy of logical operators is as given below:

1. NOT (since Unary)
2. AND
3. OR

The next category of operators is relational operators. They are otherwise called comparison operators. They are six in number and they are:

i. >= (greater than or equal to)
ii. > (greater than),
iii. <= (less than or equal to)
iv. < (less than)
v. == (equal to) and
vi. != (not equal to)

As the name suggests, they are used for comparing two inputs and all of them are binary operators. Probably, these operators would be familiar to you since they are available in the basic mathematics itself (only a few symbols are new to you). These operators accept the input and return either a true or a false (i.e., a '1' or a '0'). For example, if we have an expression X = 10>=3, it evaluates to true, since '10' is greater than '3'. Thus, there will be a '1' stored in 'X' after the evaluation. If we have, say, X = 10<3, the expression will evaluate to false and hence a '0' will be stored in 'X'. The operators '==' and '!=' are called otherwise as 'equality' and 'inequality' operators respectively. The main difference between the logical and relational operators is that, the logical operators accept the input in Boolean form and return the result also in Boolean form, whereas, the relational operators accept the input in Numerical form (Arabic Numerals) and return the result in Boolean form.

Relational operators are simple and easy to work with and they carry the basic meaning as in mathematics. We have to be very careful when we use '==' (double equal to or equality operator) in relational expressions, since we have another operator '=' (single equal to), which stands for 'assignment' operator. We will get unanticipated results when we use one in place of another. Before going to the discussion of the above difference, let us know about the assignment operator. An assignment operator is the one which is used in almost all the expressions, whether relational or logical or arithmetic. The whole expression will be called an assignment expression, which has two sides, a Left Hand Side (LHS) and a Right Hand Side (RHS). Usually, the RHS will be the first to be evaluated and the result will be assigned to the LHS. Also, in any assignment expression, the LHS will always have only one variable (not even a constant is allowed) and the RHS may have any number of variables or constants clubbed together by the valid operators of the language. For example, C=A+B+6 is a valid expression. But the expression 3=R+T is not valid, since we have a constant on the LHS. Also, the expression A+B=C is not valid since there are two variables on the LHS. Now let me come to the point of using an assignment operator in the place of an equality operator. Consider the expression C=A+B. Here we add the values of 'A' and 'B' and assign the result to 'C'. Consider the expression C==A+B. Here we are adding 'A' with 'B' and comparing the result with 'C'; that is, we are checking whether the sum of A and B and the value of C are the same. If so, the above result will yield a '1' and if not, we will get a '0'. Assuming the value of A as 8, the value of B as 3 and the value of 'C' as 11, the second expression will yield a '1', whereas the first will yield a '11' (will overwrite any previous value available in 'C'). Thus, one has to be very careful with the assignment operator and the equality operator.

The next category of operators is bitwise operators. The bitwise operators are a special category in C language, since no other programming language prior to C supports this kind of operators. In fact, it is the bitwise operators in C which have made the language so popular. There are six bitwise operators in C, namely one's complement operator (~), bitwise logical AND (&), bitwise logical OR (|), exclusive OR (EX-OR / X-OR - ^), shift left (<<) and shift right (>>). Among these operators, one's compliment operator is a unary operator and all the others are binary. When we use a bitwise operator in an expression, the operands in the expression are converted to their binary format and then the expression is evaluated. Let us first have a look at the binary bitwise operators and then go to '~'.

Consider the bitwise logical AND. It is designated by the symbol '&'. For a logical AND operator we use a '&&'. We have to be very careful again. For example, consider the expression C=4&&5

and another expression C=4&5. I would like to project the common mistake that is being made while typing a program. One may type '&' instead of '&&' and vice versa. This sort of error will not be notified to the users since it is a logical error. The result of both these expressions will vary and hence one has to take utmost care while using a '&' and '&&'. Let us see what happens to the above two expressions. The first one is a logical expression and the second one is a bitwise logical expression. In the first expression, both 4 and 5 will be converted to intermediate value '1' since they are non-zeroes and hence the expression will be reduced to C=1&1, whose result is a '1', according to the definition of AND operator. But when we look at the way the second expression is evaluated, it will show the difference between the two operators. In the second expression, both 4 and 5 are converted to their binary equivalents. Normally the conversion will be such that the binary format has 16 bits or 2 bytes of memory. The reason for taking 16 bits will be discussed when we do data types in C language. For discussion sake, let us confine ourselves to 3 bits. The binary of 4 is 100 and the binary of 5 is 101. When we use these two binary formats and operate them with '&', we have to consider each pair of the three pairs of bits available and perform the logical operation on them (see Figure 2.8). The result of the expression is 100, whose decimal equivalent is '4'. Thus, we get '4' in the expression C=4&5 and we get a '1' in the expression C=4&&5. The moral is: Use the operators appropriately and get the proper results.

Let us consider the bitwise logical OR. It is designated with the symbol '|'. For a logical OR operator we use a '||'. We have to be very careful again in using a logical OR and bitwise logical OR. Using one in place of the other will yield unwanted results. For example, consider the expression C=4||5 and another expression C=4|5. Again this may be a typing error as in the case of using a '&' in the place of '&&'. The error will not be notified by the compiler, since we have been using the right syntax only. The result of both these expressions will vary and hence one has to take utmost care while using a '|' and '||'. Let us see what happens to the results of the above two expressions. The first one is a logical expression and the second one is a bitwise logical expression. In the first expression, both 4 and 5 will be converted to intermediate value '1' since they are non-zeroes and hence the expression will be reduced as C=1||1, whose result is a '1', according to the definition of OR operator. But when we look at the way the second expression is evaluated, it will justify the difference between the two operators '|' and '||'. In the second expression, both 4 and 5 are converted to their binary equivalents.

The binary of 4 is 100 and the binary of 5 is 101. When we use these two binary formats and operate them with '|', we have to consider each pair of the three pairs of bits available and perform the logical operation on them (*see* Figure 2.9). We have the result of the expression as 101, whose decimal equivalent is '5'. Thus, we get '5' in the expression C=4|5 and we get a '1' in the expression C=4||5. Again, to get the correct results use the operators appropriately.

The next bitwise operator is Exclusive OR (otherwise termed as Ex-OR or X-OR) designated with the symbol '^' (caret). In the same way as we defined the logical AND, OR and NOT operators, we have to formally define an X-OR operator too. An X-OR operator could be defined as the one in which the output will be true if and only if we have ODD number of TRUE inputs. This could be justified with a truth table shown in the Figure 2.10. In this truth table, we have used just two inputs X and Y and so we have the outputs for four different combinations of the inputs. Had there been three inputs, there must be eight possible combinations. We could see from the table that only for odd number of TRUE inputs we get a TRUE output and for other combinations we get a FALSE output.

X	Y	Z = X^Y
0	0	0
0	1	1
1	0	1
1	1	0

Figure 2.10 Truth table of a two-input X-OR gate

I will just illustrate you how this operator functions when we write programs with an example. Consider, for example, the problem of swapping (interchanging) the values of two variables. For swapping, we may have umpteen solutions, like, we may go for using a temporary variable, we may just use the arithmetic operators, etc. Let us consider two variables X and Y and let their values be 2 and 5. When a temporary variable, say temp, is used, the following set of codes will result in swapping the values of two variables:

1. temp = X (2 transferred to temp)
2. X = Y (5 transferred to X)
3. Y = temp (2 transferred to Y)

The diagrammatic representation of the stepwise swapping process for easy understanding is given in Figure 2.11.

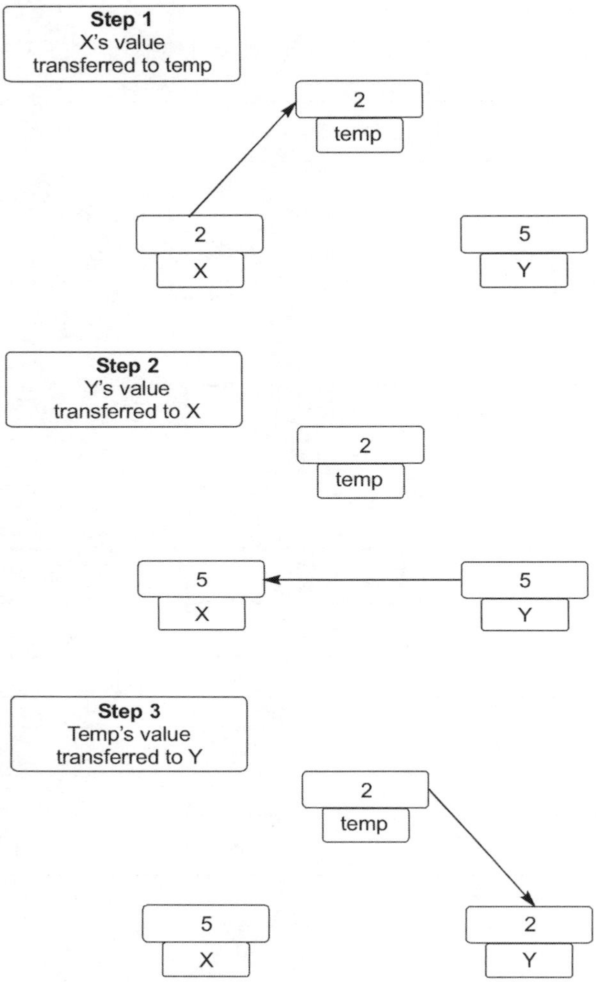

Figure 2.11 Swapping values using a temporary variable

The second method for swapping the variables values will not be using a temporary variable. Rather, we will be using the variables X and Y and use the basic arithmetic operations addition and subtraction on them and swap the values. The following are the steps:

1. X=X+Y (2+5=7, 7 replaces 2 in X)
2. Y=X–Y (7–5=2, 2 replaces 5 in Y)
3. X=X–Y (7–2=5, 5 replaces 7 in X)

At the end of these three steps, we see that X has 5 and Y has 2 and hence the values are swapped. This process is shown in the Figure 2.12.

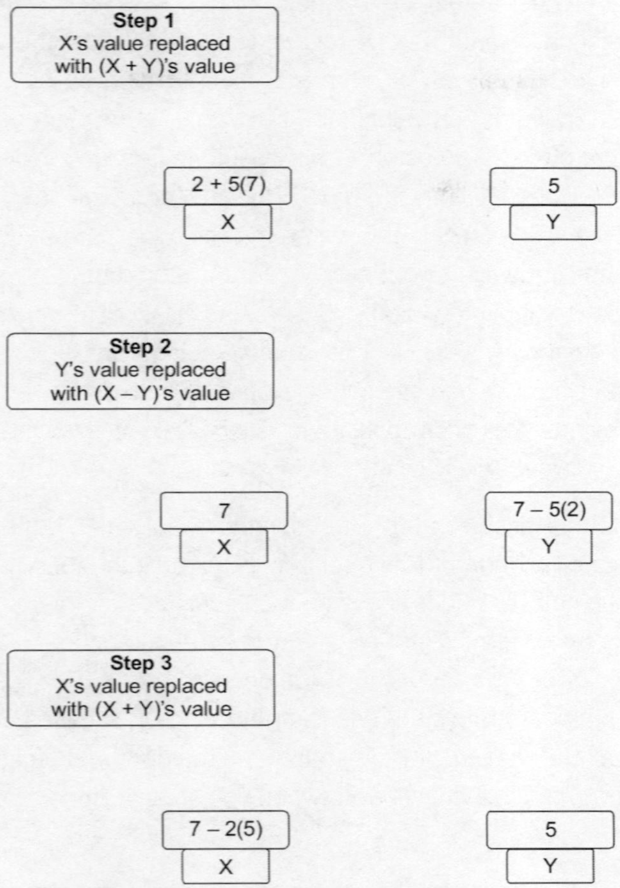

Figure 2.12 Swapping values without using a temporary variable

The same process can be accomplished with the X-OR operator and the steps are given below:

1. X = X^Y
2. Y = Y^X
3. X = X^Y

At the end of these three steps, the values would have swapped. Consider for example, the values of X and Y to be 2 and 5 respectively. So, to start with we will have 2 in 'X' and 5 in 'Y'. After the above three steps are performed, the value in 'X' will be 5 and that in 'Y' will be 2. Hence the swapping will take place. Consider the first step, where we have X = X ^ Y, that is X=2 ^ 5. Here the values 2 and 5 will be converted to their binary as usual and then operated at bit level. Hence, we have the intermediate expression as (considering three-bit pattern for the integers 2 and 5),

X = 010 ^ 101, which will yield an output 111 (refer Figure 2.10). The decimal equivalent of this binary is '7' and it is stored in 'X' and the previous value of 'X' is lost (that is 2 is lost). Consider now the second step, where we have Y = Y ^ X, that is Y = 5 ^ 7. The equivalent binary expression for the above is Y = 101 ^ 111, which will yield 010. The decimal equivalent of this result is '2' and it is stored in 'Y'. Hence the previous value of 'Y' is lost (that is, 5 is lost). Now after the execution of steps 1 and 2, we have, '7' in 'X' and '2' in 'Y'. Consider now the third step, where we have X = X ^ Y, that is X = 7 ^ 2. The equivalent binary expression for the same is X = 111 ^ 010, which will give 101. The decimal equivalent of this is '5' and it is stored in 'X'. Thus, we see that, at the end of the three steps, the value of 'X' is 5 and that of 'Y' is '2'. This means the initial values in 'X' and 'Y' have been swapped. Thus, we have another logic for swapping the values of two variables apart from those that we had already discussed. Where do we exactly use this operator? Since this is a bitwise operator, this finds application at low level where one can swap the information between two registers or two memory locations very easily.

The next operator for discussion is shift right operator (>>), which is also a binary operator like '&', "|' and '^'. The basic function of this operator is to shift the binary equivalent of the given number to its right by the number of bits specified. Consider, for example, the expression, a = 9>>2. Here, we mean that the binary equivalent of the number '9' has to be shifted to the right by two bits. Let us assume the 16-bit pattern for the above operation. In a 16-bit pattern, the leftmost bit is called as the Most Significant Bit (MSB) and the rightmost bit is called as the Least Significant Bit (LSB). So, when you count the bits, always count from the right and proceed to left and start the counting from '0' and not '1'. (As a convention, when you are learning C, do the counting from '0' and not from '1'.)

0	0	0	0	0	0	0	0	0	0	0	0	1	0	0	1

Figure 2.13 Binary pattern of '9' in the memory

Figure 2.13 shows the pattern (binary equivalent of '9') for 9 in the memory. The bits are shifted in the direction towards right (because of shift right operation). The first bit to be shifted will be the LSB. Since, we have shifted the LSB towards right side, we would be apparently losing that bit and hence that position will become vacant. To fill up the position, we shift the next LSB (the bit at the second position from right) to the vacant bit. Like this we will be shifting all the bits one by one till we reach the MSB.

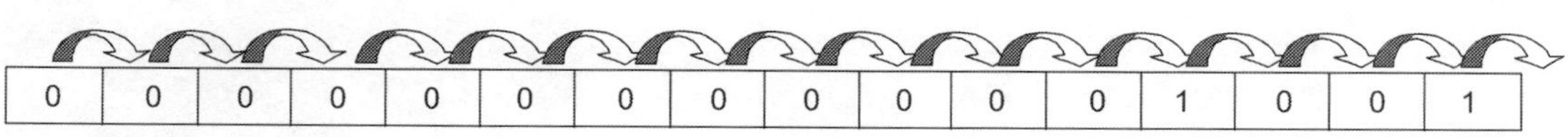

Figure 2.14 Methodology of right shift in the memory

	0	0	0	0	0	0	0	0	0	0	0	0	1	0	0

Figure 2.15 Status of binary '9' after shifting by one bit to right

This is indicated in the Figure 2.14. The resultant pattern after one complete shifting is shown in the Figure 2.15. We could see that the MSB becomes vacant at last. But, one cannot leave any of the bits vacant. Either it shall have a '0' or a '1'. So what could be the bit available in the vacant slot? Will it be occupied by the bit that is expelled from the LSB or will it be a different one? One has to keep in mind, while shifting a bit towards right or left that, the shift is not a rotate. This means, the bit that is lost at the rightmost end (LSB) is lost forever. The vacant position at the leftmost end (MSB) is filled with a '0' (this is called a leading zero). Thus we have a new pattern as shown in the Figure 2.16. We see from the pattern that, after shifting the given pattern (decimal '9') by one bit, the value of the pattern has changed. The decimal value of the above pattern is '4'. In the expression that we have considered, we have to shift the given pattern by two bits. The above process is illustrated for only one bit. That is in the above discussion, we have shifted the given pattern by one bit towards right. So, we have to repeat the entire process with one more bit and for this, we have to take the above pattern as the source (that is the new value '4'). If we carry out the same procedure with the above pattern, we will finally get the pattern which is shown in the Figure 2.17.

0	0	0	0	0	0	0	0	0	0	0	0	0	1	0	0

Figure 2.16 Pattern after including a leading zero during the right shift

0	0	0	0	0	0	0	0	0	0	0	0	0	0	1	0

Figure 2.17 Pattern after shifting the binary of '9' to the right by two bits

The decimal value of this pattern is '2'. Thus we have shifted the binary value of '9' by two bits towards right using the operator '>>' and obtained the result as '2'. This result is stored in the variable 'a'. The intermediate result in the process was '4'. Hence, we can conclude that upon shifting the binary equivalent of any decimal number towards right side by '1' bit, we actually divide the number by '2'. Thus, we can generalize the result as, if we shift the binary equivalent of a decimal number 'd' towards right by 'n' bits, we actually divide the number by 2^n.

So far we have been discussing the bitwise operators, which are all binary operators. The operator '~' (pronounce this operator as tilda), which is used for getting the 1's complement of the given number, is the only unary operator among the bitwise operators. To have an in-depth discussion on this operator, we shall actually know what a "complement" is. We have many number systems in use like Decimal (10 numerals—0 to 9), Binary (2 numerals 0 and 1), Octal (8 numerals—0 to 7) and Hexadecimal (16 numerals—0 to 9 and A to F). Each number system has two types of complements called as "radix complement" and "radix-minus-one complement". For example, the decimal number system has 10's complement (radix complement) and 9's complement (radix-minus-one complement), the binary number system has 2's component (radix complement) and 1's complement (radix-minus-one complement) and so on. Consider a decimal number '43'. The 9's complement of this number could be found out by subtracting each of the digits from '9'. On doing this, we get '56', which is the 9's complement of '43'. To find out the 10's complement, we have to first find the 9's complement and add '1' to the result. For the number '43', 9's complement is 56 and the 10's complement will be '57'. Similarly, when we consider a binary number like 11011, the 2's complement could be obtained by subtracting each of the digits from '1'. The 2's complement could be obtained by adding one to the result of the above process. On subtracting each of the digits of 11011 from '1' we get, 00100 and adding '1' to it we get, 00101.

The operator '~' does the job of finding the 1's complement for the given number after converting it to bits. For example, consider the expressions a = 0 and b = ~a. The value in 'a' is '0' and it has to be converted to its binary equivalent in 16-bit pattern. (We are taking 16-bit pattern because we consider '0' as an integer and normally an integer will take 2 bytes—16 bits, in memory.) The 16-bit binary equivalent of '0' is shown in the Figure 2.18 (memory picture for 'a'). When we evaluate the expression b=~a, the 16-bit pattern of 'b' will be the one shown in Figure 2.19. Can you predict what will be the value of 'b'?

0	0	0	0	0	0	0	0	0	0	0	0	0	0	0	0

Figure 2.18 Binary equivalent of '0'

1	1	1	1	1	1	1	1	1	1	1	1	1	1	1	1

Figure 2.19 Pattern after the 1's complement of '0'

If you had followed a procedure that had been taught to you in converting a number from binary to decimal for getting the result of the above pattern, then you are wrong. If you had followed that method, then your answer would be 65535 and C does not yield this result. Instead, C will do the calculation in a different way. The decimal equivalent of the second 16-bit pattern noted above would be '–1'. Wondering how??? Let me justify the answer.

In subjects like Digital Electronics or in Computer System Architecture, you would have come across the concept of encoding and decoding and, most appropriately, we can call it as **sign-magnitude** representation of a number. In sign-magnitude representation, the leftmost bit, which is called the **Most Significant Bit (MSB)** is used for indicating the sign of the number.

If it is set as '0', the sign of the number stored would be positive and if it is set as '1', the sign of the number stored would be negative. C uses exactly this idea in giving the result to you. It will store all the negative numbers in an encoded format (usually the format used is 2's complement approach). When we apply the '~' operator to the first 16-bit pattern, all the zeroes in it are converted to ones, due to the fact that the operator is a 1's compliment operator and it has to do like that.

Keeping the above logic in mind, let us try to analyse how the decimal equivalent of the second 16-bit could be '-1' and not '65535' as you would have calculated. From the second 16-bit pattern (Figure 2.19), we notice that the MSB is '1' and hence the number stored in the pattern is surely a negative number. But, we have to decode the pattern, since it is encoded in 2's complement format. In order to decode that, we have to find out its 2's complement. For finding 2's complement of a given binary pattern, we have to first find the 1's complement and then add a '1' to the Least Significant Bit (LSB) of the result. The 1's complement of the second 16-bit pattern is shown in Figure 2.20 (note that it is the same as the pattern in the Figure 2.18).

0	0	0	0	0	0	0	0	0	0	0	0	0	0	0	0

Figure 2.20 1's compliment of the pattern in the Figure 2.19

When we add a '1' to the LSB of the pattern shown in the Figure 2.20, we will be getting the pattern shown below in the Figure 2.21.

0	0	0	0	0	0	0	0	0	0	0	0	0	0	0	1

Figure 2.21 2's compliment (by adding a '1' to the LSB of Figure 2.20)

The decimal equivalent of the pattern in Figure 2.21 would be '1'. So, we decide the sign of the resultant pattern from Figure 2.19 and the magnitude from Figure 2.21, which together will give the result as '–1'. Many results in C language will not comply with that of the users of the language and one has to be thorough in the concepts of the language, to be able to work like a C compiler. With this operator '~', we have come to an end of the discussion related to the bitwise operators.

The next operator is Assignment Operator designated by the symbol "=". Note that we have already discussed "==" operator (called "equality" operator) in the Relational Operators category and one has to be extremely careful when they use '=' and '==' operators, since they both yield completely different results. For example consider the expressions, a=b and a==b, where the variables 'a' and 'b' have the values 3 and 4 respectively. In the first case, the value of 'b' is transferred to 'a' and hence we will lose the value already existing in 'a' and hence in the end, we will have the value '3' in both 'a' and 'b'. When we look at the second expression, it is obvious that we are comparing the values of 'a' and 'b' and checking whether they are equal. In our case, they are not equal and hence we will get a 'false' value returned from the evaluation (refer to the earlier discussion of relational operators). Thus, the result of the second expression would be a '0' (Zero). The beginners in C often commit the mistake of confusing the operators which are alike. In C, different operator symbols will be used for different contexts and one must be clear about using the symbol in the right context.

The next category of operators is Shorthand Operators. As the name suggests, they are really used in programming for cutting short an expression that may be lengthy. For example, consider the expression, a=a+b, where we have two variables 'a' and 'b'. In the expression, we are trying to update the value of 'a' by adding the value of 'b' with already existing value of 'a'. When we are doing so, instead of saying a=a+b, we can very well say 'a+=b' and we notice that the expression is cut short because of '+='. Thus, whenever we update a variable's value, we can use a shorthand operator. The following are some of the shorthand operators: –=, *=, /=, %=. A shorthand operator enables us to cut short the expression without losing the meaning. Some operators do not have their shorthand form, e.g. ">=", "<=", etc. since they already have two symbols in their notation.

We have so far discussed all the essential operators, which are important for basic programming in C. Let us temporarily stop the process of learning the operators here and we will learn more operators in the forthcoming sections, whenever one is needed. Let us move to the next building block of C language, namely, I/O instructions.

2.2.7 C Instructions/Statements

Type declaration instructions We are aware of the data types available in C (please refer 2.2.5). Once we start to write programs in C, we must follow certain regulations and structure. The very first part of any C program would be (not necessarily in all the programs) the declaration of variables and their corresponding types. This section is called as type declaration part. For example, consider the following C statements (As of now do not worry about the syntax of the statement since we will deal with the same later):

```
int a,b=10;
float c;
```

In the first statement, we have used two variables a and b and both of them are of type integer and in the second statement we have used another variable c, which is of type float. This section of the program informs the compiler in advance that there are going to be three variables in the program with the names a, b and c. All the variables that would be used in the program shall be necessarily declared in the type declaration itself and not anywhere else. This means, if we are using a variable in the 100th line of a program containing 500 lines, we cannot be in a mindset to declare that variable in 99th line of the program, since the compiler would flash that as an error.

This issue, however, has been fixed in C++, where the user is free to declare the variable at the place it is used for the first time. The style of type declaration that is used in C enhances the structure of the program and the program does not look clumsy.

When the variables are declared, the C compiler does three things. The following steps illustrate these three things.

1. Identifies memory location for each variable declared
2. Associates name for each memory location (as used by the user)
3. Assigns value for each variable (if it is given by the user, that would be used or a garbage value would be used)

For example, in the first statement above, both a and b are declared as integers and b is given a value of 10 by the user while it is declared. This way of giving an initial value while declaring the variable is called as initialization during declaration. Let us execute the above steps for the variables a and b. In the case of a, the compiler will first reserve a location in memory for the variable, say, 1000th byte of memory. Then it associates the name, 'a', with that memory location. Finally, it assigns a value to that variable. In this case, the value would be garbage, since the user has not given any value. This garbage value would be within the range of the corresponding data type and it would be surely decided by the compiler and not by the user. Also, the garbage value for a variable would be different for different users for the same kind of program that may be executed simultaneously in two systems. In this case, 'a' is an integer and more precisely, a short signed integer and hence the value assigned for it would be within the range of –32767 – 0 – +32767 (please refer the Table 2.1). In the second case, the variable 'b' is declared as integer and also given a value of 10. The first two steps of the above procedure is executed for 'b' as it was for 'a'. The third step alone is quite different for 'b' and it would be given a value of 10 instead of being given a garbage value, since it has been initialized as it is declared.

I/O instructions Input and output instructions or the functions are the heart-throb of any language, since the programmer would be able to communicate with the computer only through these instructions. This is exactly the counterpart of framing sentences in learning English (as depicted in the Figure 2.2 earlier). There are many input and output instructions in C, of which the user has necessarily got himself/herself familiar with two important ones, namely scanf() and printf(). In C language, the instructions (what we term as commands in some other programming languages) are mostly in the form of functions. We will see more about the functions in the Chapter 8. But, for the beginners of the language, let me tell you that functions are nothing but that do some specific task assigned to them and they must follow the syntax rules defined by the language. C language is full of functions (any function name in C would be followed by a set of parentheses, that may or may not be empty and it is a very easy identity for them) and hence it is very often called as a compiled language or a procedure-oriented language.

The two important and essential input and output functions, as I had quoted earlier are scanf() and printf(). Let us discuss how and why we should use these functions.

The function scanf() is used to get the input from the input device (normally a keyboard in case of C, since C is a Command Based Language). The expansion for the word "scanf" is "scan format". It has the following method of usage, formally called as the prototype (or model).

```
int scanf(const char *format, ...);
```

Its prototype is available in the header file called stdio.h. It returns the number of input fields processed successfully. It processes input according to the format and places the results in the memory locations pointed to by the arguments. (stdio.h is nothing but a standard input and output header, containing prototypes of all the input and output functions. We will see more about a header file later in the chapter 2). The more general model of scanf() is as given below:

```
scanf("format specifier(s)", (list of) address(es) of the argument(s));
```

As an example, consider the following statement which uses a scanf().

```
scanf("%d%d",&a,&b);
```

In the above statement, we see that we had used "%d". Now, you can just go back to Table 2.1, where we had used one column called "format specifier". All the data types in C have their own format specifiers and one has to be very careful in using them while programming, especially in the functions scanf() and printf(). In the above example, I have used two %d format specifiers inside the function scanf() and this means I am trying to input two short signed integers. This information is easily recognized by the C compiler, since we had used the format specifier %d. Had we used two "%f" inside the function scanf() in the above example, then the compiler would expect two float values from the user. So, the only way to let the compiler know that we would be giving a float input is by providing proper format specifier.

Now, let us analyse the job of the function scanf(). This function is just doing the job of a postman. A postman will go through the postal address in the envelope and drop the letter in that address. Similarly the function scanf() gets a specified value from the input device (keyboard) and drops that value in the address or memory location of the corresponding variable.

We have seen in the previous section that when a variable is declared, three things happen at the compiler level. These steps invariably would have happened for the variables a and b used here. They both would be having their own memory locations and have a garbage value.

Inside the function scanf(), we see that we are using the arguments a and b along with a "&" (Ampersand) symbol. We had already used this symbol when we discussed the bitwise operators. The symbol "&" is a bitwise logical AND operator and it has its own meaning as discussed. But the same symbol is used here and it has a different meaning here. The function of "&" here is to fetch the "address of" the argument that follows it. So, it is called as "address of" operator. In the above case, the addresses of a and b are being fetched by the "&" operator and in the addresses, the values provided for them by the user are dropped. Thus the garbage values are replaced by the new values given by the user. Let me remind you that as long as the values for variables are not provided by the user, they would continue to have garbage values in the program.

Let me conclude this discussion by saying that the function scanf() instructs the compiler to fetch the value pertaining to a data type from the input device (keyboard) and drop the same in the address located by "&" for that corresponding variable.

Remarks

1. Another important thing to be noted here is that the values that are scanned are dropped according to the order specified in the list of arguments. In the above case, the first value scanned would be dropped in the address of 'a' and the second would be dropped in that of 'b' and so on. In more simpler words, when we use more than one format specifier in the function scanf(), the first one will associate itself with the first argument and its address in the list, the second one with the second argument and its address and so on.
2. In the above example, we could see that we have used the format specifiers inside the function scanf() without providing any space in between them. No special character, not even a comma, is supplied in between them. What would be the result had we used the same function scanf() in the following way?

```
scanf("%d,%d",&a,&b);
```

If you closely see the above version of the function scanf(), we have provided a comma (,) in between the two format specifiers "%d". This question would surely test the reasoning power of a C programmer. Normally the answer for this question, even by a professional programmer in C, would be, "an error". But it is not so. The above version of the function scanf() is absolutely permitted and error-free. I would justify this for sure.

When we execute this code of the program, the compiler would expect two integer values from the user. Normally when the user is keying in more than one value from the input device (keyboard), he/she would separate the two values or the two inputs with a space or a tab or an enter key. This argument goes by the common sense that when we write words, we separate them using a blank space. These three separators are called as the delimiters and they are invariably the default delimiters used by anyone who writes a program in C or in any other programming language. In the above newer version of the function scanf(), we see that the format specifiers are separated by a comma and hence when we give the two integer values from the input device (keyboard), we have to use a comma in between them. This means we have to overlook the three default delimiters discussed above. The delimiter in this case would be a comma. For that matter, any character could be used other than a comma and the same character shall be used as the delimiter when we execute the program by giving the inputs. We can even have a scanf() like the one below:

```
scanf("%d ^ %d*%d",&a,&b,&c);
```

In the above version of the function scanf(), we see that the first two format specifiers are separated with a '^' and the last two are separated with a '*'. This means, when we give inputs we have to use a '^' between the first two and use a '*' between the last two. I hope you would be satisfied with my argument. Just try this when you write programs later.

Another important I/O instruction is the function printf(). This function is used to print the values or the processed results or any message the user wants to print in the output device (normally a visual display unit, the screen in our case or a printer). Thus the function printf() would try to retrieve values from memory when the user wants to print some variable's value or just print a message as given by the user. The expansion for the word "printf" is "print format". It has the following method of usage.

```
int printf(const char *format, ...);
```

Its prototype is also available in the header file stdio.h. The function printf() formats a variable number of arguments according to the format, sending the output to stdout (output device). It returns the number of bytes output. In the event of error, it returns EOF (end of file). The more general model of scanf() is as given below:

```
printf("control string with or without format specifier (s)", <list of arguments>);
```

For example, one can have a function printf() like the one shown below:

```
printf("The values of a and b are %d,%d",a,b);
```

In the above case, we see that the format used satisfies the general format given above. Thus the result of the above printf() would be "The value of a and b are 5 and 20" (had we given 5 and 10 respectively for a and b). Thus we see that first %d in the expression associates with the first argument in the list and the second %d with second argument in the list. Thus we can generalize this result by saying that if there are 'n' format specifiers in the function printf(), there must be 'n' corresponding arguments and always the format specifiers will be associated with the arguments in a particular order (as stated above) and never in a random manner.

Now, a big question would arise in our minds. It is "what would happen if there is a mismatch, that is, either there are more format specifiers than the actual number of arguments or otherwise?" Let me tell you what will happen.

Let us consider the first case, where we have more number of format specifiers than the actual number of arguments as in the following example:

```
printf("%d,%d,%d",a,b);
```

In this case, we have three format specifiers and only two arguments. For the first two format specifiers, the compiler would associate a and b. But what will happen to the third format specifier inside the function printf() above. Since the third format specifier could not find any variable for getting itself associated, it prints a garbage value for it (the value would be within the range of the short-signed integer and is randomly decided by the compiler and not by the user). So the output could be like the one given below:

```
10, 20, 2515
```

(You as a user may get a different value for the third value here in this output.)

Let us consider the second case, where we have the number of format specifiers less than that of the arguments. In this case, the output generated would be 10, 20 (since there is no format specifier in the control string, the compiler does not bother to print a value).

In the general format of the function printf(), we have stated that the presence of the format specifier is optional. This means we can have another version of the function printf() like the following:

```
printf("Hello, how are you?");
```

You may see that the above version of the function printf() does not have any format specifiers and hence any arguments. This means the presence of format specifier necessitates the presence of an argument. If there is a format specifier and there is no argument, then we would get a garbage value in the output and if there is no format specifier and there is an argument, we would not get anything in the output. And as a last case, we may have the function printf() without any proper control string or even a format specifier. It is absolutely valid and would work without any problems. Thus, the following are the different versions of the function printf() a programmer might use.

1. printf(); — This would not print anything and is of little use, but syntactically valid.

2. printf("%d",a); — This would print the value of 'a' by getting the same from the memory, but you can see that there is no control string accompanying.

3. printf("The value of a is %d",a); — This would also print the value of 'a' but along with the control string given by the user.

4. printf(""); — This would not print anything and of little use in the program, but syntactically valid.

5. printf("%d,%d"); — This would print two garbage values for the format specifiers belonging to the type short-signed integer, since there is no argument accompanying them.

6. printf("%d,%d",a): — This would print the value of 'a' and a garbage value, since there is only one argument and two format specifiers and the first format specifier will associate itself with the first argument.

7. printf("%d",a,b); — This would print the value of 'a' alone, since there is only one format specifier, even though there are two arguments.

8. printf("%d"); — This would print a garbage value as there is no argument accompanying the format specifier.

9. printf("",a); — This would not print anything, since there is no format specifier and only an argument.

When we discuss the function printf(), it will be apt to discuss the **escape sequences**. Escape sequences are those which are highly useful when formatting our programs. The escape sequences which are very frequently used are:

\n Newline character

\t Horizontal tab

\a Audible alert

\v Vertical tab

\b Backspace

\r Carriage return

\f Form feed

\\ Backslash

\" Double quote

\' Single quote

Any character that is accompanying the backslash will be printed along with the control string in the format specifier. The escape sequences are specially designed for the function printf(). Normally they are not used in the function scanf() and even if we use them, it does not serve any purpose.

The newline character (\n) when used in the function printf(), would generate a new line with the cursor blinking at the beginning of the line. This character works just like the carriage handle of a typewriter, which when pulled and moved towards the left, generates a new line and positions itself so that the user could start typing from the beginning of the new line. Assume that the function printf() is written as printf("hello\nworld\n"). The output would look like the screenshot shown in the Figure 2.22.

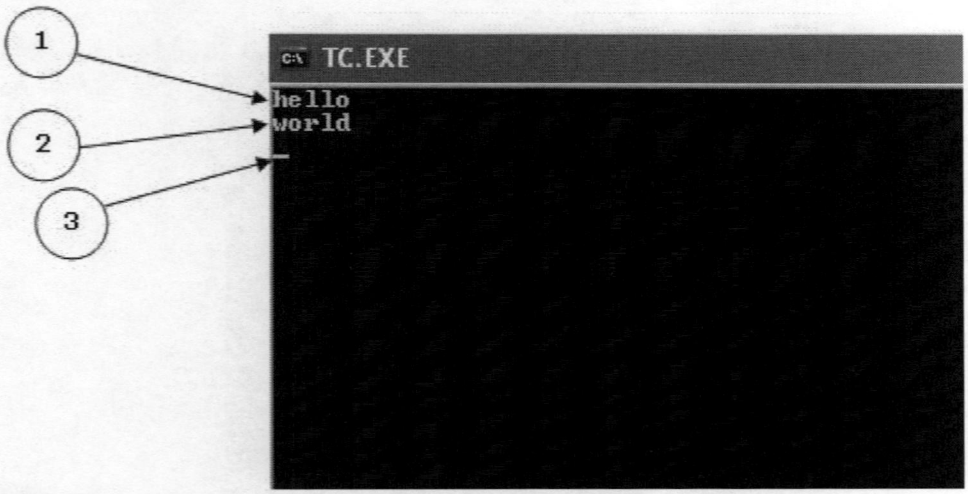

Figure 2.22 The newline character (\n) in action

In the Figure 2.22, we have three indicators 1, 2 and 3. The indicator 1 denotes the first word in the control string of the function printf(), 2 denotes the second word and see that these two words are available one after another, since there is a "\n" in between them in the control string. Indicator 3 denotes the position of the cursor after printing the words inside the control string of the function printf(). The cursor is positioned at the third line of the output since there is a "\n" after the word "world" in the control string of the function printf(). Now you would probably know what must be the coding to generate an output that is expected as in the Figure 2.23. Yes, you are right. The coding shall be printf("hello\n\nworld\n"). There are two "\n" characters between the words "hello" and "world", so that a blank line is left between them in the output. There is only one "\n" character after the word "world" and hence the cursor is positioned in the line immediately after the word "world" in the output.

Figure 2.23 The newline character (\n) in action in a different manner

The horizontal tab character (\t) when used in the function printf(), would generate a tab space (normally a tab space corresponds to 8 blank spaces, which can be modified by the programmers). This is shown in Figure 2.24.

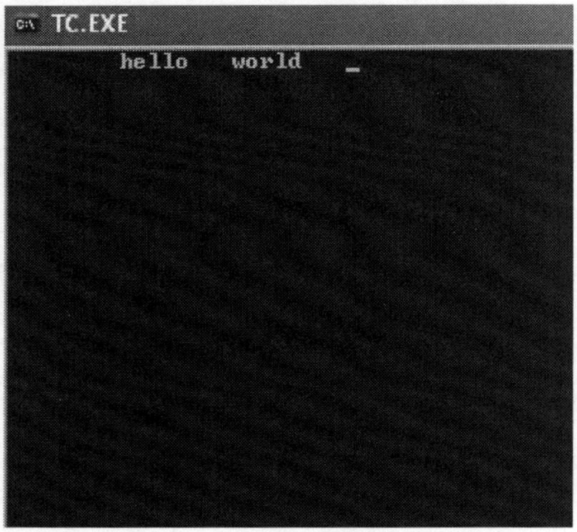

Figure 2.24 The tab space character (\t) in action

We can see from the Figure 2.24 that there are three tab spaces used in the function printf(), one before the word "hello", one after that and one after the word "world". That is the function printf() would look like printf("\thello\tworld\t")

The audible alert character (\a) would generate a beep sound (provided there is an in-built speaker) along with the other contents of the function printf(). This cannot be exemplified using a figure.

The vertical tab character (\v) generates a tab space vertically, unlike the horizontal tab. In most of the recent compilers, this would not work properly.

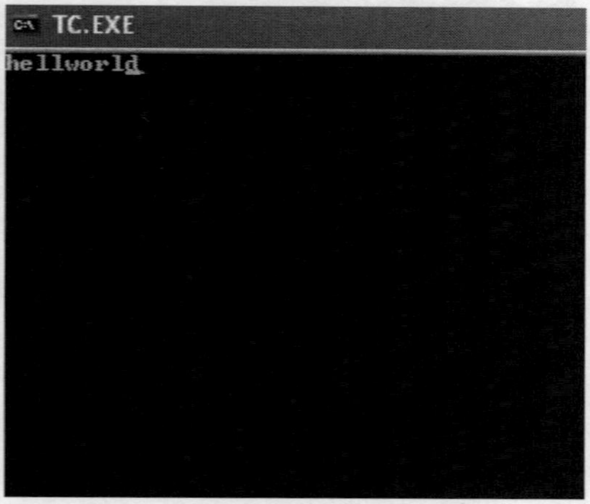

Figure 2.25 The backspace space character (\b) in action

The backspace character (\b) would make the cursor to position itself at the last character of the output generated by the function printf(). If there is any character left out from being printed in the control string of the function printf(), this character will erase the last character of the previous word and would start printing the first character of the next word. The Figure 2.25 shows a typical '\b' character in action. The following is the way the function printf() would have been used for getting the above output.

```
printf("hello\bworld\b");
```

We had used a '\b' character after the word "hello" and there is one more word "world" following "hello" in the control string of the function printf(). So, after printing the first word, the cursor would move one space backwards and start printing the second word and finally would position itself at the last character of the second word.

The carriage return character (\r) would place the cursor at the beginning point of the same line, unlike the '\n' character. Consider the version of the function printf() as printf("hello world\r"). The output of this is shown in the Figure 2.26.

Figure 2.26 The carriage return character (\r) in action

Since we had used a carriage return character (\r) after the word "world" in the control string, we see that the cursor is positioned at the beginning of the same line. What would happen if the function printf() looks like printf("hello\rworld");

The output would be like the one shown in the Figure 2.27. The justification for the output could be given by you.

Figure 2.27 The carriage return character (\r) in action in a different manner

The form feed character (\f) would be used while working with a printer, and if we want to print a page or an output, we can send the same to the printer and programmatically, we can tell the compiler when there shall be a page break and when there shall not be while a program or its output is getting printed.

Any character that is used along with '\' will be printed in the output. For example, let us assume that we had used "\\", then we can anticipate a '\' at the output. Also, if we want to print a character like '$' in the output, then we may include a '$' along with a '\'. This is shown in Figure 2.28.

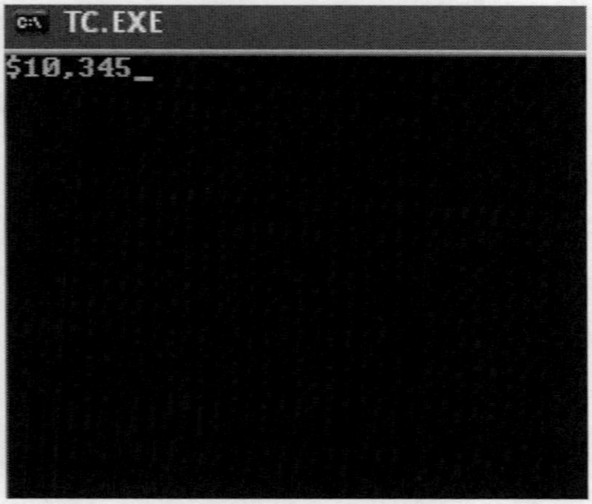

Figure 2.28 $ character is printed along with the output

The escape sequences, I hope, are now clear to you. You will probably have to work more with these sequences to get to know the real implications of these.

Arithmetic instructions Arithmetic instructions will involve the arithmetic operators. For example, consider the following expression: c = a + b. This is a very simple example for the arithmetic instruction, where we have the operator '+' between the operands a and b. In basic mathematical principles, we would express the same instruction as c = f(a,b) and this would be read as "c is a function of a and b". Thus, a and b are independent variables and 'c' is a dependent variable, whose value depends on the values of a and b.

The following is also an example of an arithmetic instruction: c = a+b/d%e. Here, we notice that apart from the operator '+', we have the operators '\' and '%'. Since we had already learnt about the hierarchy of arithmetic operators, you must know the order in which the above arithmetic instruction would be executed. All the arithmetic instructions have two sides, a Left Hand Side (LHS) and a Right Hand Side (RHS). Always the LHS will have a variable and that too only one variable. A constant is not allowed on the LHS. The RHS of the instruction will have any number of variables and constants. If we write the expression c=a+b as a+b=c, it will not work in programming. Even though a+b=c serves the purpose of the operation, it is not accepted since it is not an assignment statement but an equation.

Control instructions Control instructions are used in a program to control the flow of the program. The control instructions are of two types in general. They are conditional control and loop control. Conditional controls work on the basis of conditions and loop controls work on the basis of loop cycles. We will discuss more about the conditional and loop control instructions later in this chapter.

2.2.8 The Function main()

We know that C language is that which has many built-in functions that are categorized into different headers and included as and when required in a program. We had seen two such functions earlier when we discussed the I/O instructions, namely printf() and scanf(). The function main() is one of the functions in C and the most important of all the available functions.

When a program begins running, the system (operating system) calls the function main, which marks the entry point of the program. Every program must have one function named main(). No other function in the program can be called main(). The function main() has one of two forms:

int main (void) *block_statement*

int main ( ) *block_statement*

You will probably come to know about the blocks later when we introduce you to programming ideas. Now, take for granted that the function main() has two forms of usage as represented above.

As said earlier, without the function main(), no C program would compile or execute. The name of the function itself implies that it is the core function in any program.

2.2.9 Conditional Branching

There are three conditional control instructions available in C language and they are listed below:

if

if-else

if-else-if

Let us have a detailed discussion on all these controls.

if This is the simplest condition of the three. The following is the syntax of the **if** condition:

```
if (<expression[s]>
{
Statement-1;
Statement-2;
........................
........................
Statement-n;
}
```

→ **Block of Statement (s)**

The expression in the above syntax may be the one that has a single condition or multiple conditions clubbed together with the logical operators. Following the expression, we see that we have a block of statements. In C programming, when any set of statements is included within the curly braces ({ and }), we could say that a block is created. The advantage of having a block is that we can execute a set of statements in a single go, upon meeting the condition(s) in the expression. The statements inside a block will all be executed or will all be skipped together. It is not possible (it is not sensible too) that we can select a particular statement from the block and execute or skip it (unless and otherwise there is a nesting of *if*). Also, one cannot execute statements inside a block at random. The block statements are always executed in the order of their appearance inside the block. The block following the expression inside *if* is also called as the **body** of the *if* statement. The body will be executed if the return value of the expression is non-zero (true) (we know that the logical and relational operators and the expressions involving them will return a Boolean zero or a Boolean one as the result). Thus when the return value is true, the block following it will be executed and when it is false, the block will be simply ignored. The following fragment exemplifies the usage of *if* statement in programming code.

```
if (x>0)
{
printf("Good Morning");
printf("Have a Nice Day");
}
```

Thus, if the input for the variable 'x' is greater than 'zero', then the messages inside the function printf() will be executed in that order and the block will be terminated after that. If the input value for x is less than 'zero' or 'zero' itself, then the entire block will be skipped.

The simplest form of the above *if* is as given below:

```
if (<expression[s]>) Statement;
```

Here, we see that there is only one statement and not a block as in the syntax given already. When the expression returns a true, the only statement available will be executed, otherwise, we will not get any output. When there is only one statement after the expression, then the set of curly braces is optional. This means that even though the curly braces are provided, the expression will just look lengthy and the curly braces are redundant. The following piece of code exemplifies this idea.

```
if (x>0)
printf("Good Morning");
```

We see that only one statement is executed if the expression 'x>0' returns a non-zero value. But see that in both the cases, the set of parentheses is essential and one cannot skip them. For example, the following piece of code is invalid.

```
if (x>0)
printf("Good Morning");
```

The 'if' statement that is explained above has very least applications in real life programming, the reason behind this being no alternative option available once the expression inside the set of parentheses returns a "zero" (false). Thus, the designers of the language have introduced another conditional control, which is an advanced version of the *if* statement. This is discussed in the following section. Figure 2.29 is the representation of *if* structure in the form of a flowchart.

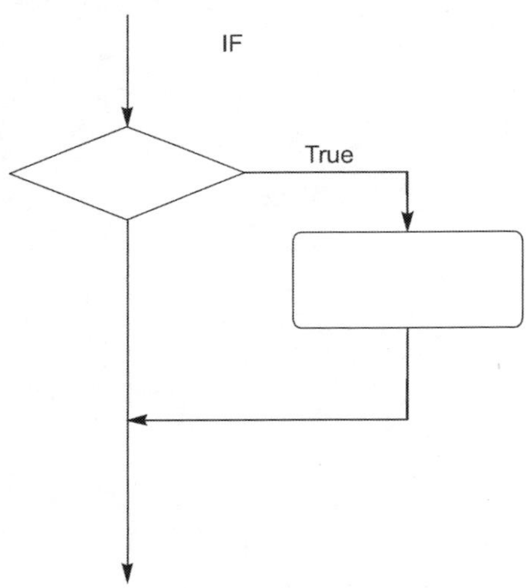

Figure 2.29 Flowchart representation of *if* structure

if-else The following is the general syntax of *if-else* structure:

```
if (<expression[s]>)
{
Statement-1;
Statement-2;
........................
........................
Statement-n;
}
else
{
Statement-1;
Statement-2;
........................
........................
Statement-n;
}
```

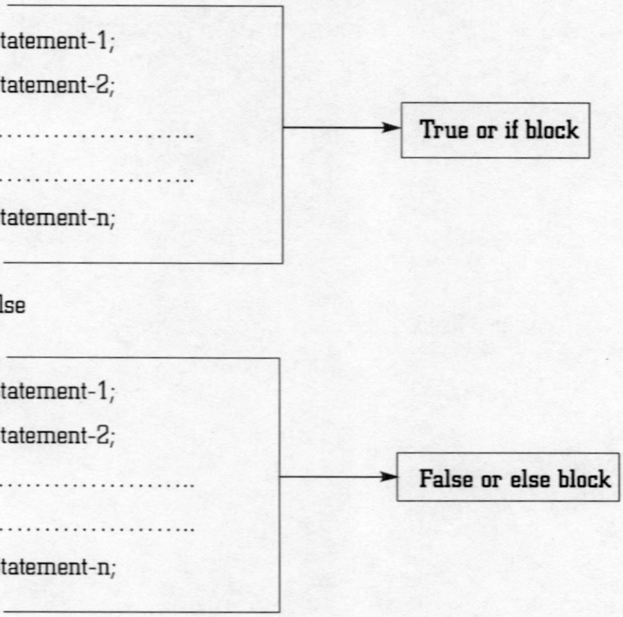

We see from the general syntax that there are two alternatives for the expression given within the set of parentheses. This means, one of them will always be executed. The expression inside the parentheses may return a true (non-zero) or a false (zero). If it returns a true, then the block following *if* will be executed and hence it is called a true or if block. If the expression returns a false, then the block following *else* will be executed and hence it is called a false or else block. In no case, will both the blocks be executed or be skipped. The following piece of code is an example for an *if-else* fragment used in programming.

```
if (x>0)
{
printf(“Good Morning”);
printf(“You have entered a positive value”);
}
else
{
printf(“Good Evening”);
printf(“You have entered a negative value”);
}
```

Thus, if we enter a positive value for 'x', the block following *if* would be executed, otherwise that following *else* would be executed. As in the previous case, the set of parentheses is essential and one cannot skip it.

The following are the alternative forms of the *if-else* structure used in programming.

Form – 1

```
if (<expression[s]>)
{
Statement-1;
Statement-2;
.......................
.......................
Statement-n;
}
else
Statement;
```

True or if block

Form – 2

```
if (<expression[s]>)
Statement;
else
{
Statement-1;
Statement-2;
.......................
.......................
Statement-n;
}
```

False or else block

Form – 3

```
if (<expression[s]>)
Statement;
else
Statement;
```

In the first form, we see that there are multiple statements for execution in the true block and there is only one statement in the false block. Since there is only one statement following the *else* statement, we had skipped using the set of curly braces (as it would be redundant, as discussed earlier in the previous section). The following is the example code for this.

```
if (x>0)
{
printf("Good Morning");
printf("You have entered a positive value");
}
else
printf("You have entered a negative value");
```

In the second form, we have multiple statements for execution in the false block and only one statement in the true block. Again, the usage of a set of curly braces after the *if* is redundant and hence is skipped. The following fragment exemplifies this.

```
if (x>0)
printf("You have entered a positive value");
else
{
printf("Good Evening");
printf("You have entered a negative value");
}
```

In the last form, we have only statement after *if* and one after *else*. This is the simplest way in which we can use the *if-else* structure in our coding. The following is a piece of coding that tells us this.

```
if (x>0)
printf("You have entered a positive value");
else
printf("You have entered a negative value");
```

While programming, all the three forms of *if-else* would be used frequently. Figure 2.30 shows the flowchart representation of *if-else* structure.

At this juncture, I would like to introduce you a new operator. Recall that you had one category of operator called **ternary operator**. The ternary operator is called so because it has three sections (or arguments). The following is the general syntax of a ternary operator.

Argument-1? Argument-2 : Argument-3;

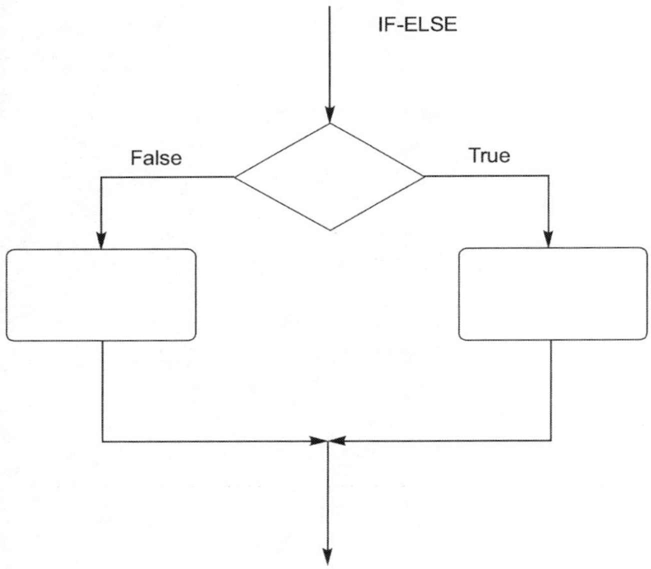

Figure 2.30 Flowchart representation of *if-else* structure

Normally, argument-1 would be an expression or a combination of expressions. Argument-2 is the statement that would be executed upon the expression returning a true (non-zero) value and the argument-3 is the statement that would be executed upon the expression returning a false (zero) value. Now, I believe you would have understood the reason why I introduced this operator at this point. Yes, you are right. This operator is just a replacement for the third form of the *if-else* structure. Just go back and see the general syntax of the third form of the *if-else* structure and it will just be the same idea that we had discussed.

Assume that we have the following piece of fragment in some part of our program.

```
if (x>0)
printf("You have entered a positive value");
else
printf("You have entered a negative value");
```

The above piece of coding could be replaced with the ternary expression, whose general syntax is given above. The following will be the replaced code for the piece of coding given above.

```
x>0? printf("You have entered a positive value") : printf("You have entered a negative value");
```

It is obvious from the expression above that the simplest form of the *if-else* could be replaced with a ternary operator and we call it a ternary expression. Another advantage of using a ternary expression instead of the simplest *if-else* is that we need not use the set of parentheses for the expression, as we do in the *if-else* structure. Even if we use them, it is just redundant and its presence or absence will not be a big issue with respect to the coding.

Consider the following piece of coding:

```
x>9 && x>0? printf("Good") : printf("Bad");
```

Suppose that we enter a value of '10' for 'x', then the output for the above piece of coding would be 'Good'. We notice that even if we have two expressions clubbed with an AND operation, we could skip the usage of the parentheses. This is the advantage of ternary operator over the simple *if-else* structure.

Another wonder that we can do with a ternary operator is that we can nest them to any number of levels and execute. For example, consider the piece of coding given in Figure 2.31.

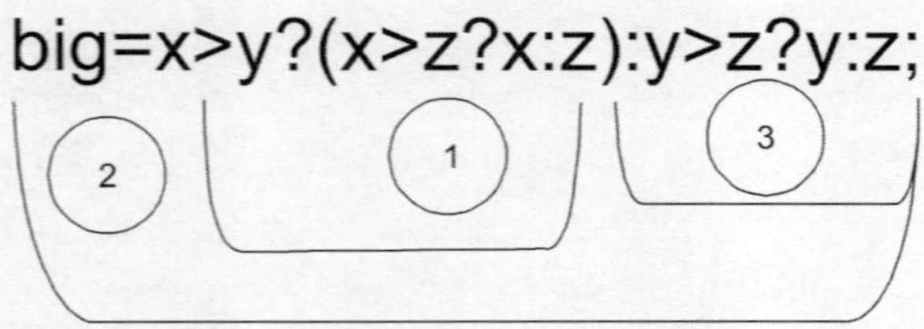

Figure 2.31 Nesting of ternary operator

In the above piece of coding, we see that the ternary operator is used thrice and this is what we call as nesting. We have nested the ternary operator up to three levels (the indicators 1, 2, and 3 show the three ternary expressions). When we give values for x, y and z, they will be tested accordingly and the result (the biggest of three values) will be stored in the variable 'big'.

To begin with, the ternary expression indicated by 2 will be in action. The first argument for the expression indicated by 2 will be the condition x>y, the second argument will be that indicated by 1 (another ternary expression by itself) and the third argument will be that indicated by 3 (yet another ternary expression by itself). Let us consider the following cases for the above nested ternary expression.

When the condition x>y returns a false, obviously it means that the reverse condition y>x is true. Now the control comes to the third argument (indicated by 3) of 2. Here the condition y>z is checked and if it is true, then the biggest value will be the one in 'y'. Otherwise the biggest value will be the one in 'z'.

When the condition x>y returns a true, then the control would go to the second argument (indicated by 1) of 2. There the condition x>z is checked and if it returns a true, then the control will store the value of 'x' in the variable 'big' and that will be the biggest of the three. If the condition returns a false, then the value of 'z' will be stored in the variable 'big' and hence that will be the biggest of the three.

As an example, consider the values for x, y and z as –5, 4 and 98 respectively. To start with, the value '–5' will be compared with '4' and obviously –5 is less than 4 and hence the condition will return a false (Boolean zero). So, the control will skip the second argument of the ternary expression indicated by 2 and will go to the third argument indicated by 3, which is another ternary expression by itself. Here, 4 and 98 are compared (remember that already –5 and 4 are compared in the first argument and it has been confirmed that 4 is bigger among them) and again it will return a Boolean Zero, i.e., a false. This means, the control would now go to the third argument of the ternary expression indicated by 3 and there we have the value of z and it is nothing but '98' and it will now be assigned to the variable 'big'. This way, you may shuffle the values and test the validity and the way the above expression works.

There is a shortcoming with the ternary expressions, though it has many advantages. There can be only one argument after the question mark (?) and after the colon (:). This means, we cannot create blocks (*if* block or *else* block) after the question mark or the colon. In other words, only the simplest form of the *if-else* could be replaced with a ternary expression and not the other forms that are discussed earlier.

if-else-if The first conditional structure (*if*) has no optional statement or block to be executed, except the block that follows the *if* statement, when the conditional expression returns a false. The second conditional structure (*if-else*) has two blocks (one following the *if* block and one following the *else* block) and hence there is always an alternative block that is executed even when the condition returns a false. If we need more than one condition to be tested and hence execute different blocks according to the one that returns a true, then we are provided with another structure in C and it is *if-else-if*. Here the conditional expressions are tested one after another and the corresponding block of code is executed and the rest are skipped. The following is the syntax of the *if-else-if* structure.

```
if (<expression-1>)
{
Statement-1;
Statement-2;
........................          Block-1
Statement-n;
}
else
if (<expression-2>)
{
Statement-1;
Statement-2;
........................          Block-2
Statement-n;
}
else
if (<expression-3>)
{
Statement-1;
Statement-2;
........................          Block-3
Statement-n;
}
else
if (<expression-4>)
{
Statement-1;
Statement-2;
........................          Block-4
........................
Statement-n;
}
```

```
.......................
.......................
else
if (<expression-n>)
{
Statement-1;
Statement-2;                  ──► Block-n
.......................
Statement-n;
}
else
{
Statement-1;
Statement-2;                  ──► Default Block
.......................
Statement-n;
}
```

We could see from the syntax that there are 'n' different expressions and corresponding blocks for each of them to get executed. The control starts evaluating the expression from the very first *if* and proceeds according to the return value of that. That is, if it gets a return value of false for the first conditional expression, then it will skip the first block and proceed to evaluate the second conditional expression and if that returns a true, then it will execute the block following that. When all the conditions in a particular *if-else-if* structure fail, then the default block available as the last one will be executed and it is purely optional to have such a block. But it will be sensible to provide a default block in such cases, as the user will not know the outcome if all the conditional expressions fail. But what happens then? Do you think the control will skip the entire structure and come to the subsequent part of the program? The answer is "no". The control will check all the remaining conditions, regardless of whether a particular condition returns true and the corresponding block gets executed. In this aspect, the compiler is said to waste the execution time and it is built in such a way.

For example, consider the following fragment of a program that contains *if-else-if* structure, which checks whether the given character is a vowel or a consonant.

```
if(a=='a' || a=='A')
printf("Apple");                              ──────► Block-1
else
if(a=='e' || a=='E')
printf("England");                            ──────► Block-2
else
if(a=='i' || a=='I')
printf("Ice Cream");                          ──────► Block-3
else
if(a=='o' || a=='O')
printf("Owl");                                ──────► Block-4
else
if(a=='u' || a=='U')
printf("Umbrella");                           ──────► Block-5
else
printf("The character is not vowel");         ──────► Default Block
```

When the program gets executed, if the conditional expression 2 returns a true, then the control will execute the block corresponding to that, but still will check the remaining conditions for their return values, before it leaves the entire structure to the subsequent part of the program. We see that there is a default statement following all the formal conditional expressions. If the user inputs a consonant, none of the available conditional expressions will return true and hence the default statement (The character is not a vowel) would be the output and the user could understand that the input he/she has given is processed and an outcome is given. If this default statement has not been provided, then the user must be unaware of what is happening around if he/she has supplied a consonant as the input.

The following Figure 2.32 represents the *if-else-if* structure in the form of a flowchart. The structure can be extended beyond the specified number of levels in the Figure 2.32, as required for the program. The *if-else-if* structure is also termed as *if-else-if* chain.

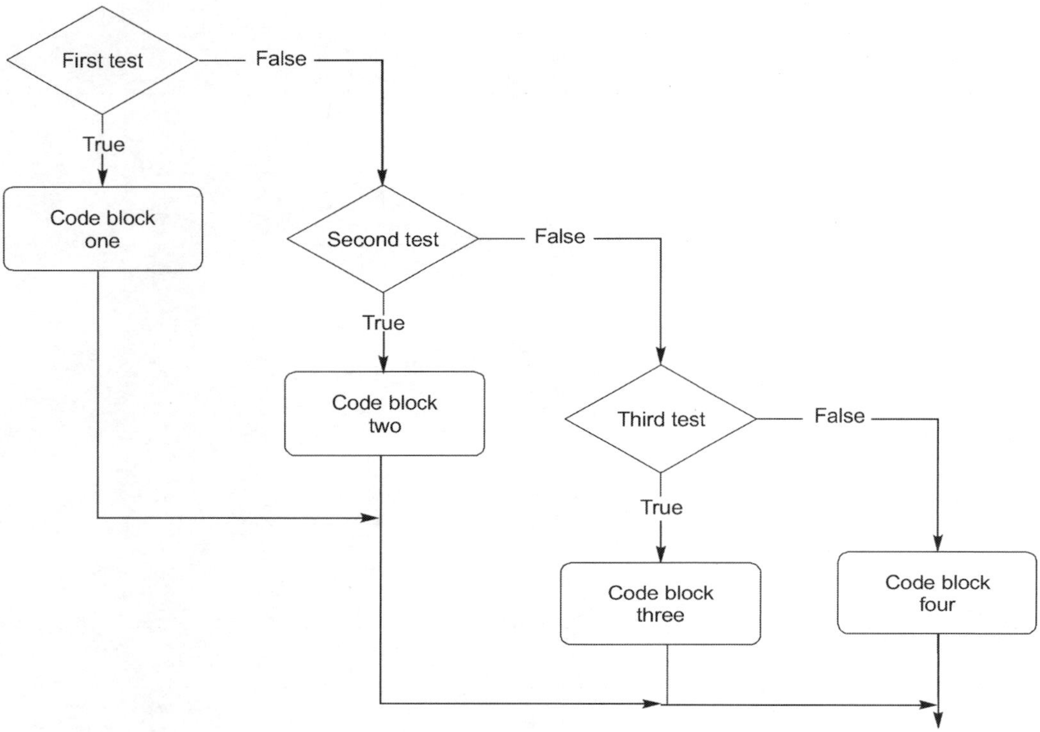

Figure 2.32 Flowchart representation of *if-else-if* structure

When we discussed the syntax of *if-else* structure, we noted that there are three different forms of the structure that could be used in programming. Similar to that, we have umpteen forms for the *if-else-if* chain, since one block or the other would be provided with only one statement and hence we can skip the curly braces for that block. As an example, two of the alternate forms of *if-else-if* structure are provided below. Based on this, you may derive many similar forms by applying a similar logic.

In the example provided above, we had used the simplest form of the structure, where each block contains only one statement and hence we could skip the curly braces that are provided in the general syntax.

```
if (<expression-1>)
Statement-1;                    ————► Block-1
else
if (<expression-2>)
```

```
{
Statement-1;
Statement-2;
.......................
Statement-n;
}                          ──► Block-2
.......................
else
if (<expression-n>)
Statement-1;               ──► Block-3
else
{
Statement-1;
Statement-2;
.......................
Statement-n;
}                          ──► Default Block
```

We see from the above alternate form of *if-else-if* that the first and the nth conditional expressions are provided with only one statement following them and the second conditional expression and the default block are provided with more than one statement for execution. Hence, we could skip the curly braces, when we have only one statement following the conditional expression. Even if we provide curly braces, we will be just lengthening our code and they will be redundant.

We had enough information about the conditional structures and their different forms of execution and usage. We will now discuss another control structure in C language, which is Case Control Structure.

2.2.10 Multiway Branching—Case Control Structure

The case control structure (otherwise called as switch-case control structure) is a multiway branching structure, which may be termed as an alternate to the *if-else-if* chain. Similar to *if-else-if* chain, one can have many options in case control structure. It is otherwise a far better and tidy way of representing the *if-else-if* chain. The control will execute the option based on the match it does before getting in the switch-case block. The following is the syntax for the case control structure.

```
switch (expression)
      {
case item-1:
      Statement-1;
      Statement-2;
      ......................
      Statement-n;
      break;
case item-2:
      Statement-1;
      Statement-2;
      ......................
      Statement-n;
      break;
      ......................
      ......................
case item-n:
      Statement-1;
      Statement-2;
      ......................
      Statement-n;
      break;
default:
      Statement-1;
      Statement-2;
      ......................
      Statement-n;
  }
```

Case item

Case item

Case item

Default case

Switch-case block

In each of the cases above, the **case item** can be only a constant and shall not be a variable. Even the constant that is used must be a numerical or a character constant and not anything else (like a float or its associated data types). This is in contrast with the *if-else-if* structure, where one can use any type of constant or a variable. We will later do a comparative study of the case control structure and *if-else-if* chain, so as to better analyse which to use when.

In order to terminate the *switch* control once a particular *case item* is executed, we supply a *break*, so that the control comes out of the *switch-case* block. Consider for example, the following fragment of a program, wherein we print the given single digit numbers in words, using case control structure.

```
switch(m) {
case 0: printf("Zero",m);  break;
case 1: printf("One",m); break;
case 2: printf("Two",m); break;
case 3: printf("Three",m); break;
case 4: printf("Four",m); break;
case 5: printf("Five",m); break;
case 6: printf("Six",m); break;
case 7: printf("Seven",m); break;
case 8: printf("Eight",m); break;
case 9: printf("Nine",m); break;
default: printf("\nThe number %d is not a single digit number",m); }
```

For a given value of 'm' (which would be scanned using the proper scanf() in the program), the individual cases inside the *switch-case* block is matched and the output is generated accordingly. If the given value does not match with any of the cases inside, then the default case would be executed and here we could see an analogy between the default block in the *if-else-if* structure and the default case inside the *switch-case* block. Both are provided for the same purpose: in the former structure, to provide the user with some output when none of the conditional expressions return a true and in the latter structure, when none of the cases match with the input given by the user.

We had used the *break* statement in all our cases (except the default) for obvious reasons that are mentioned earlier. What will happen if we do not use *break* for a particular case item? What will be the outcome if we use *break* for the default case?

If we do not use *break* statement for a particular case item, then the statements related to that case along with the subsequent case will be executed. For example, consider the following program fragment that prints the single digit numbers in words.

```
switch(m) {      case 1: printf("One"); break;
          case 2: printf("Two");
          case 3: printf("Three");break;
default: printf("The number is not a single digit number");
}
```

We see that there is no *break* used in the second case item. Suppose that we enter a value of '2' for 'm' from the keyboard, we expect only the output "Two" to be printed. But along with that, we get the output "Three" also, since according to the control, the block of the case item starts at 'case 2' and ends at the *break* statement of the case item 'case 3' and hence whatever statements available after the case item '3' till the next *break* statement will be executed. We understand that *break* statement is the one that acts as a 'terminator' for a given case item. What do you think the output will be if there is no *break* statement in the cases 2 and 3? The output will have both "Two" and "Three". So, it is concluded that *break* statement is essential for terminating the execution of the *switch-case* block and bringing the control outside it.

In the above program fragment, the variable 'm' can be of any type related to integers or characters. Float and its associated data types are not allowed within the switch as well as in the case items.

Can you guess how we can rewrite the program fragment that we had written already for illustrating the *if-else-if* structure? (The program using which we have printed a word for the given vowel). The following would be the corresponding program fragment.

```
switch(m) {
case 'A': case 'a': printf("Apple"); break;
case 'E': case 'e': printf("Elephant"); break;
case 'I': case 'i': printf("Ice"); break;
case 'O': case 'o': printf("Owl"); break;
case 'U': case 'u': printf("Umbrella"); break;
default: printf("The character is not a vowel");}
```

We see that, unlike the logic that is used in *if-else-if*, we have not used the logical OR operator for checking the nature of the alphabet that we enter. The reason is that the *switch-case* structure does not allow the usage of operations of any kind in the case. The logic, however, is achieved in a different way.

Let us list out the advantages of case control structure over *if-else-if* structure. See that the advantages also give a fair idea on how the case control differs from *if-else-if* structure.

1. This structure is useful in case the user wants to terminate the control upon matching a particular case item and bring the control outside the *switch-case* block, unlike the *if-else-if* structure. Remember that in *if-else-if* structure, there is no way to bring the control outside the structure, even when a particular conditional expression returns a true value and the block following it is executed. In case control, it is achieved through the *break* statement.

2. When the statements of a particular case item are written, we do not enclose them within curly braces, unlike the *if-else-if* structure. It will be an error if we do so in *if-else-if* structure. But, in case control, the presence of the colon (:) denotes the beginning of case block and the *break* statement denotes the ending of a block. So, there is no need for enclosing the set of statements of a case item within curly braces. This is true for any number of statements for a particular case item.

3. The *if-else-if* structure, when professionally written, needs indentation of the program lines. Indentation will make the program to have a clear look and the users can easily identify the sections of the program. For example, consider the following fragment of a program written using case control structure.

```
switch(m) {        case 1: printf("One"); break;
      case 2: printf("Two"); break;
      case 3: printf("Three");break;
      default: printf("The number is not a single digit number"); }
```

The same program, when written using *if-else-if* looks like the one below.

```
if (m==1)
            printf("One");
                        else
                              if(m==2)
                                          printf("Two");.....
```

The structure needs indentation to be done in each line of the code as shown above. If we proceed with this sort of indentation, then the coding will creep towards the right side of the screen and will not be fitting within the working area of the editor. But the case control does not require indentation to be done for the blocks following the case items, since the provision of colon (:) and the *break* statement does the job of identifying the correct block and it is matched with the correct case item.

4. We need not use the set of curly braces to create a block of statements, as in *if-else-if* structure, since the presence of colon (:) and *break* satisfies this one too.

5. We can use all sort of operations inside the case control as in the *if-else-if* structure, but with a caution. We can perform the operations only with the values pertaining to integer and its associated data types.

6. Another advantage of the case control structure is that, one can use any order for the case items. That is the programmer need not follow any particular logical order or a sequence. For example, consider the following fragment:

```
switch(a)
{
case 3: printf("Three");break;
case 2: printf("Two"); break;
case 1: printf("One"); break;
default: printf("Bye");
}
```

The output for the above program would be fine, even if the order of the cases is jumbled. Let us now slightly modify the above fragment and present the same as shown below:

```
switch(a)
{
default: printf("Bye");
case 3: printf("Three");break;
case 2: printf("Two"); break;
case 1: printf("One"); break;
}
```

What do you think will happen? Since we have told that the cases could be in any order, in the above fragment, the default's position is also jumbled unlike the earlier version of this fragment. The output for this will be quite different from that of the earlier one. If we give a value of 1 or 2 or 3, then the output will be fine. But if we give any values other than the above, we are supposed to get the result "Bye", and of course we will get it, but along with that we will also get "Three". This is because, the default is not terminated with a 'break' statement and hence the printf() statements associated with both the default case and the succeeding case are executed. Thus the position of the default case could also be changed, provided we give a *break* statement following the default case.

There are many disadvantages of case control structure, which makes the usage of *if-else-if* structure inevitable. The following are some of the disadvantages and shortcomings with case control structure.

1. We cannot use data types other than integer and its associated types and character type constants (only unsigned character) inside the *switch-case* block for the case items. For example, the following case would be invalid and hence will get you an error message.

   ```
   case 1.4: printf("Good"); break;
   ```

What will happen if the variable is declared as a float and the case item is an integer? Consider the following example to know about this situation more clearly.

```
float m;
scanf("%f',&m);
switch(m)
{
case 1: printf("One"); break;
case 2: printf("Two"); break;
default: printf("Good"); break; }
```

In this case, if the given input is '1.5', what will happen to the output of the program? Will any one of the cases be executed or will we get an error? The answer is quite simple. The rule is that we shall not use constants other than integer (and its types) and character in the case items and we are never restricted from declaring a variable of any type and passing it to the switch statement. This means that when the above value (1.5) is given for 'm', it will be accepting and matching the value (1.5) with all the cases and it is matching with case 1 (the value that is given by us will be truncated to an integer value) and hence that case will be executed. When we give values like 3.6, the default case will be executed.

2. What will happen if we have cases like the one given below?

```
case 1+2: printf("Good"); break;
case 2>1: printf("Big"); break;
case 'a' && 'b': printf("AND"); break;
case 'a' & 'b': printf("Hello"); break;
```

Again, the answer is very simple to understand. The operation will be performed and the result will be assigned for each case item. In the above three cases, the case items have the results as 3, 1 (Boolean), 1 and 96 respectively. These values can be realized by applying the suitable operations to all of them, since we knew all the operators used in them. This proves the fact that any kind of operation is possible with the case items. What about the following one then?

```
case a&b: printf("True");break;
```

where 'a' and 'b' are variables of type integer. This type of case is not allowed, due to the reason that a case item cannot have anything other than constants.

3. We cannot use the statement *continue* (we will discuss about this statement and its usage at a later stage when we discuss loops) inside the case control structure, unless there is a loop used within a particular case item. Since the loops allow the usage of the statements *break* and *continue*, the programmer will be tempted to use both of these in case control also. But they must know that *continue* is not allowed in case as a stand-alone statement.

4. Even though we are allowed to use character constants for a case item, we cannot use strings (which are a collection of two or more characters). For example, the following case item is invalid.

```
case '>=': printf("This is greater than or equal to operator"); break;
```

This means, any operator that contains more than one character for designating it cannot be used as the case item. A few other examples are the operators like ?:, <=, ==, ++, —, >>, << and so on. But do remember they can be used along with the numerical and character constants in the case item.

We will be using a lot of ideas that are learnt in this section of the chapter later while learning advanced concepts in the language.

2.2.11 Unconditional Branching—goto Statement

goto is a statement found in almost many computer programming languages. It is a combination of the English words go and to. When used and executed, it causes an unconditional transfer of control (a "jump") to another statement in the same program. The statement is specified using some kind of label, which may be an identifier or a line number depending on the language in use. At the machine code level a *goto* is a form of branch or jump statement.

In some languages, the functionality of *goto* may be present without explicit use of the keyword *goto*, such as where a *break* or *continue* keyword may be present (we have seen the usage in *break* in the previous section).

Different programming languages impose different restrictions with respect to the jump location of a *goto* statement. For example, in the C programming language it is not allowed to jump to a label contained within another function. For the control to unconditionally jump, the label must be provided in the same program and that too in the body of the same function. The following illustration explains the function of *goto* statement.

```
scanf("%d",&opt);
goto aa;
switch(opt) {
```

```
case 2<=2: printf("Three");break;
default: printf("good"); }
aa: printf("You are here");
```

If we give a value for the variable 'opt', say, 2, the function scanf() will accept the value and it will not pass the control to the switch statement, since there is a *goto* statement available immediately after the function scanf(). So, the control will abruptly transfer to some part of the same function (in this case within the function main()) and execute the coding in that.

Why one shall not use goto statement? The *goto* statement has been the target of much continued criticism and debate, with the primary negative claim being that use of *goto* results in unreadable and generally difficult-to-maintain type of coding called as a **spaghetti code**. Spaghetti code is a pejorative term for source code which has a complex and tangled control structure, especially one using many *goto*'s, exceptions, threads, or other "unstructured" branching constructs. It is named such because program flow tends to look like a bowl of spaghetti, i.e., twisted and tangled. Structured programming greatly decreased the incidence of spaghetti code, and is widely regarded as one of the most important advances in the history of programming languages. C, is a structured programming language and the usage of statements like *goto* will surely spoil its structure. Java, which is one of the programming languages, that uses many of the syntax from C, does not use *goto* statement. Professional programmers in C, very rarely or do not at all use this programming structure. Probably, in our book, this would be the last place, where you will be seeing the usage of this statement.

2.2.12 Loop Control Structures

Loops are used to repeat a block of code. Being able to have your program repeatedly execute a block of code is one of the most basic but useful tasks in programming. You could have seen many programs or websites that produce extremely complex output (such as a message board) are really only executing a single task many times. (They may be executing a small number of tasks, but in principle, to produce a list of messages only requires repeating the operation of reading in some data and displaying it.) Now, think about what this means: a loop lets you write a very simple statement to produce a significantly greater result simply by repetition. In C programming, loops are important and unavoidable.

In C we have three kinds of loops, namely, *while*, *do-while* and *for*. Before proceeding, make sure that you have understood the concepts of C's true and false (Boolean Values—1 and 0), because it will be necessary when working with loops. All the loops in C have **Loop Counter**, which keeps track of the number of cycles of a loop. A loop counter is used only for this purpose and it does not server any other purpose in program coding. In some programs, one of the variables used in the program for calculation itself will be used as a loop counter, where it has to play a dual role. The loop counters are normally used in the conditional expressions that appear in the loop fragment.

While Let us start from the **while** loop and understand the way it functions. The syntax of the *while* loop is given below:

```
Initialization of Loop Counter;

while (<conditional expression [s]>)   ──► Loop Header
{
Statement-1;
Statement-2;
........................               ──► Loop Body
Statement-n;
Loop counter modification;
}
```

Basically, there are **four** sections in any loop. They are initialization, conditional expression, loop body or set of loop statements, modification. If you have a closer look at the above syntax, you may notice that all the said sections are available in it. The general procedure for executing the above structure in programs is stated below.

1. Initialization of Loop Counter
2. Checking the Conditional Expression
3. Executing the body of the loop
4. Modifying the Loop Counter

The execution of the loop begins with the first step, proceeds to the second step and based on the return value (true or false) of the expression, the control proceeds to the third step. Once the third step is executed, it goes to the next step and modifies the loop counter and then goes to the second step where the conditional expression will be checked again. Thus, we see that the first step of the loop is executed only once and the remaining loop cycles make the control to move with the other three steps. Once the conditional expression returns a **false**, the loop will terminate at that step. So, to conclude, the control starts at the first step and terminates at the second step (unless there is a *goto* statement somewhere inside the loop). For example, let us suppose that we are writing a program to print numbers from 1 to 100 on the output screen. The following would be the program fragment for achieving that using the *while* loop.

```
x=1;
while(x<=100)
{
printf("%d\t",x);
x=x+1;
}
```

We see that the above fragment complies with the general syntax in all aspects and will be executed based on the four steps that we had discussed above. So, it begins with the initialization, where the variable 'x' is initialized as '1'. Then comes the conditional expression, 'x< = 100'. This expression will return a true value as long as x's value does not exceed 100. Since, now, the value of 'x' is '1', the control will get into the body of the loop and execute the same (step 3). When it is executing it we could understand that '1' is getting printed on the output screen. Then, we see that the value of 'x' is incremented by '1' (since we want the output to contain numbers from 1 to 100 in steps of 1). After incrementing the value of 'x' by '1', the control will proceed to check the value of 'x' (conditional expression) and proceed to execute the loop based on the return value of the expression.

Thus the loop is executed 100 times and when the value of 'x' becomes 101, we can see that it goes beyond the limit of 100 and hence the loop will be terminated. Figure 2.33 gives the screenshot of the output that is generated for the above piece of coding.

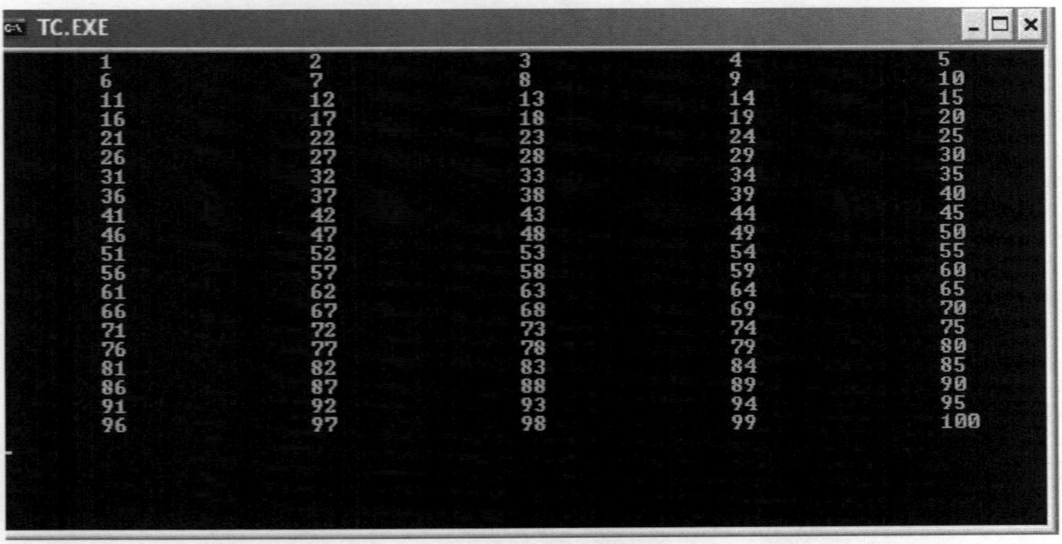

Figure 2.33 *while* loop in action—printing numbers from 1 to 100

See that we had used one of the escape sequences '\t' in order to get an output, in which we have enough space between the numbers 1 to 100. What would you do in case the program requires you to print numbers from 100 to 1? Just initialize the value of 'x' as 100 and decrement the value of x in steps of 1 as you proceed and while doing so, check whether the value of 'x' is not going below '1', i.e., when the value of x becomes 0, the loop will terminate. Thus, working with the **while** loop is quite simple.

do-while This is another loop control, like the *while*. This serves the same purpose as the *while* serves in programming. But the fundamentals of *do-while* are much the same as *while* loop the following is the syntax of *do-while* loop.

```
do                                        --> Loop Header
{
Statement-1;
Statement-2;
……………………                                --> Loop Body
Statement-n;
Loop counter modification;
}
while (<conditional expression [s]>);     --> Loop Footer
```

The general procedure for executing the above structure in programs is stated below.

1. Initialization of Loop Counter
2. Executing the body of the loop
3. Modifying the Loop Counter
4. Checking the Conditional Expression

Can you identify any difference between *while* and *do-while* loops, on comparing the syntaxes of *while* and *do-while*. Not too many know? Of course, there are two differences, if you closely observe. The primary difference is that the conditional expression in while loop is checked at the entry of the loop and the same is tested here at the exit of the loop. The second difference is due to the first one. Since the condition is checked at the entry of the loop, the loop body will not be executed at all, once the expression returns a false. But, in the case of *do-while* loop, the body of the loop is executed first and only then the condition is checked and hence, even when the expression is going to return a false, the loop body would have been executed at least once. As an example, consider the following program fragment, whose outputs are shown as screenshots in the Figures 2.34 and 2.35.

Program Fragment-1

```
int x=0;
while(x!=0)
{
printf("%d",x);
}
printf("Nothing inside the loop is printed");
```

Figure 2.34 Output of the Program Fragment-1—*do-while* loop in action

Program Fragment-2

```
int x=0;
{
printf("%d",x);
}while(x!=0);
printf("\n\nThe value of 'x' is printed");
```

Figure 2.35 Output of the Program Fragment-2—*do-while* loop in action

We could see that the initial value of the variable x in both the cases is one and the same. The conditional expression also checks for the inequality of the variable x with zero in both the cases. But in the case of *while*, since the conditional expression is checked at the entry itself and returns a false, the control will never get inside the loop and execute the loop body. But in the case of *do-while*, since the condition is checked at the exit, the body of the loop is executed once. We find that the conditional expression however, returns a false in this case too. But still we are able to get the output value of x, unlike the earlier case. Thus we can conclude that the *do-while* loop will execute at least once in its lifetime, regardless of whether the condition is true or false.

This loop is normally used in creating menu-driven programs, where we would like to have the menu of items to do on the screen, regardless of whether we are executing a particular menu or not. But remember the fact that, whatever can be done with the *while* structure, can be done with the *do-while* structure too.

for This is the most versatile and handy-to-use loop in C language. The syntax of the loop is also very simple that, the user would not forget to provide any particular component when he/she is writing a program. In the other two loops, we see that the initialization is done outside the loop structure and the modification is done inside the loop body. The *for* loop has such a syntax that all come under one line. The following is the syntax of the loop.

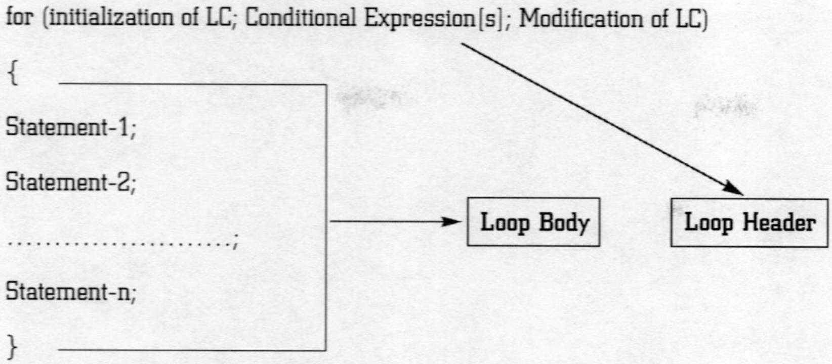

We notice from the header of the loop that the three essential components are available under one roof and the user may not forget to supply them when the code is written. Also, we see that the three components of the loop header, namely, the initialization of loop counter, conditional expression and modification of loop counter are separated with a semicolon symbol, which is also essential. Also, if the loop body has only one statement, then the set of curly braces that are provided in the program are not needed and the loop will still work fine and also be error-free.

The general procedure for executing the above structure in programs is stated below.

1. Initialization of Loop Counter
2. Checking the Conditional Expression
3. Executing the body of the loop
4. Modifying the Loop Counter

As an illustration for the working of *for* loop, consider the following program fragment. We are printing the numbers from 1 to 100, as we did using the *while* loop earlier.

```
int x;
for(x=1;x<=100;x=x+1)
printf("\t%d\t",x);
```

The output of the program is shown in the Figure 2.36.

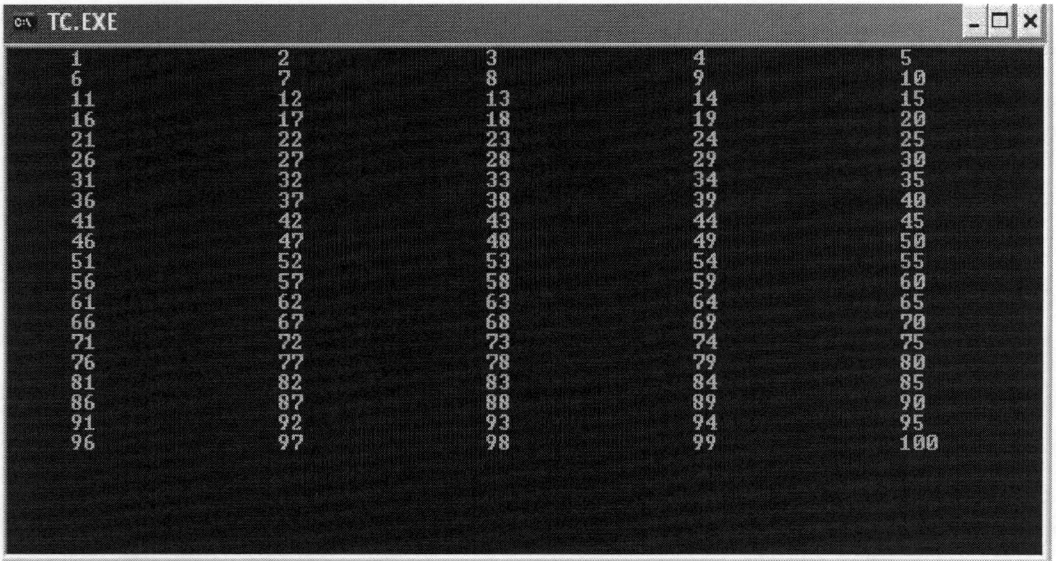

Figure 2.36 *for* loop in action—printing numbers from 1 to 100

There are many forms of *for* loop, which are called as **odd loops**. These are just like the different structures for the conditional controls we had discussed earlier (*if*, *if-else* and *if-else-if*). The syntaxes of the various odd loops are given below.

Odd loop-1

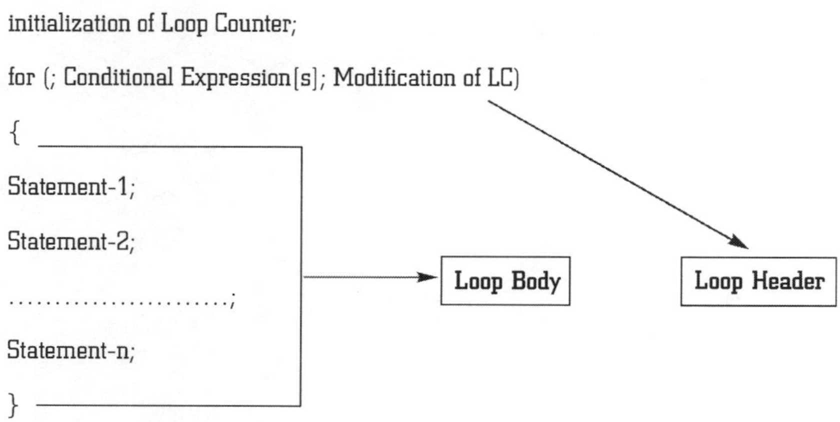

Here, we see that the initialization of the loop counter is done outside the loop, like the way we do it in the *while* and the *do-while* loops. But still the two semicolons are necessary within the set of parentheses and the first column is left blank.

Odd loop-2

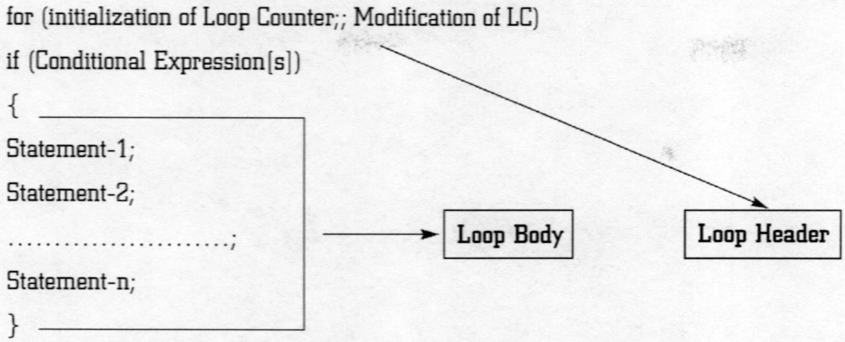

In this structure, we see that the conditional expression is used within the *if* structure and the column corresponding to it is left blank and the two semicolons are still there.

Odd loop-3

In the third form of the odd loop, we can see that the modification of the loop counter is done within the body of the loop and the corresponding column is left blank in the structure. Still we could see that the two semicolons are present.

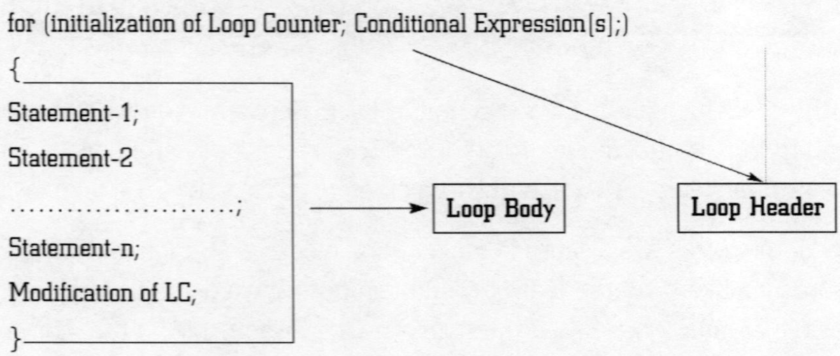

Odd loop-4

In the structure shown below, we could see that all the three components are provided outside the structure. But, this kind of odd loop is of rare usage in programming and it does not sound sensible if we use this kind of *for* loop, since this structure defeats the basic purpose of the loop that all the components shall come under one roof.

To conclude, do know that there are four versions of *for* loop (called odd loops) and we use them very rarely in our programs, as they do not comply with the purpose of the *for* loop.

```
initialization of Loop Counter;
for (;;)  ──────────────────────────►  Loop Header
if (Conditional Expression[s])
{
Statement-1;
Statement-2
........................;             ──►  Loop Body
Statement-n;
Modification of LC;
}
```

We will be using the three loops that we have discussed so far, in our programming later and you will see and appreciate how much these loops would be useful to you in performing repeated operations.

SUMMARY

After having grasped the fundamentals of C programming in this chapter, you will probably be confident of programming. In this chapter, we had started with the history and development of C language and the people who were instrumental for its emergence as a powerful language.

Following the introductory ideas, we discussed the building blocks of C language, in which we had learnt the essential ideas like character set of the language, keywords and identifiers, data types, a few category of operators, types of instructions, the importance of the function main(), conditional and unconditional branching, multiway branching and loop control structures. We had seen in detail the way in which the ternary expression is used and also learnt that they can be nested at any number of levels.

We also discussed how to use the functions printf() and scanf() effectively along with the escape sequences. The different data types and the kinds of operators available in the language would have very clearly justified the powerfulness and the efficiency of the language. The intricacies of the case control structure and how it is advantageous over the *if-else-if* chain is also discussed in this chapter and above all, the reason why the professional C programmers avoid using *goto* is also learnt in detail.

I am sure that you would have got more ideas from this chapter and I shall tell you that preserve all those you have learnt in this chapter and you will need them very often in the forthcoming chapters. There is no wonder why C is called as **God's Programming Language**.

CRACK THE NUT

1. Find out from the following the valid and invalid names for an identifier. Also, justify why some of the names given are invalid
    1. Long
    2. printf
    3. 43F
    4. F_9
    5. value$sum
    6. exit flag
    7. 3lotsofmoney
    8. _valid_flag
    9. Jerry7
    10. _foo42
2. There is a variable 'a' declared as integer in a program. Then, what will be the result of the following expression. Justify your result. Also, try parenthesizing the expression in various ways and analyze the result. What do you conclude?

    a = 300*300/300
3. Evaluate the following expressions with the help of the precedence rules
    1. 5*5/5+4*6%4–6+8
    2. 7–4+6*7+8–4*2/12%6/3
4. Evaluate the following logical expressions
    1. a = 10&&4
    2. b = 5&&3
    3. c = 56||4

4. c = 67||0
5. c = 9&&0
6. a = 0&&4
7. x = 34&&23&&0
8. x = 33||0||4
9. x = 1&&0||4
10. x = 3||5||6

5. Evaluate the following relational expressions
   1. X = 10>5
   2. Y = 4<=4
   3. X = 3>=3
   4. X = 4<3>–2
   5. Y = 10>6<7

6. What would happen if the binary equivalent of '9' is shifted towards left by two bits? i.e., consider the expression a = 9<<2 and find out the result.

Identify the outputs for the following programs:

7.
```
main()
{
float me = 1.1;
double you = 1.1;
if(me= =you)
    printf("Good Morning");
else
    printf("Good Evening");
}
```

8.
```
main()
{
int i=-1,j=-1,k=0,l=2,m;
m=i++ && j++ && k++ ||l++;
```

```
printf("%d %d %d %d %d",i,j,k,l,m);
}
```

9. main()

```
{
    int i=3;
    switch(i)
        {
            default:printf("zero");
            case 1: printf("one");
                break;
            case 2:printf("two");
                break;
            case 3: printf("three");
                break;
        }
}
```

10. main()

```
{
    printf("%x",-1<<4);
}
```

11. main()

```
{
int c=- -2;
printf("c=%d",c);
}
```

12. main()

```
{
int i=10;
i=!i>14;
```

```
    printf ("i=%d",i);
    }
13. main()
    {
    printf("\nab");
    printf("\bsi");
    printf("\rha");
    }
14. main()
    {
    int i=5;
    printf("%d%d%d%d%d%d",i++,i—,++i,—i,i);
    }
15. main()
    {
    int i=400,j=300;
    printf("%d..%d");
    }
16. main()
    {
        int i=5;
        printf("%d",i++ + ++i);
    }
17. main()
    {
        int i=5;
        printf("%d",i+++++i);
    }
```

18.
```
main()
{
int i=1,j=2;
switch(i)
    {
    case 1:  printf("GOOD");
        break;
    case j:  printf("BAD");
        break;
    }
}
```

19.
```
main()
{
int i;
printf("%d",scanf("%d",&i));  // value 10 is given as input here
}
```

20.
```
main()
{
int i=0;
for(;i++;printf("%d",i)) ;
        printf("%d",i);
}
```

3 BASIC PROGRAMMING TECHNIQUES

3.1 INSTALLING THE C COMPILER

The installation of the C compiler is quite simple. Just open Google in the Internet Explorer or Netscape and type Turbo C in the keyword area. You will get umpteen options for downloading the different versions. You may download the version, which you may wish to.

Carefully go through the **read me** document of the version that you are downloading and extract the files in the corresponding drive of your computer and now you are ready for installation of the compiler.

Follow the steps given in the read me document and do the installation in your local hard drive. Once the compiler is installed, it is ready for usage and you can start writing programs.

3.2 EDITING, SAVING, COMPILING AND EXECUTING A "C" PROGRAM

Anyone who works with a computer and developing a script (whether it is a document or a presentation) need to have a good platform to work with. For example, we schedule our day-to-day activities, make it as a record in our diaries or a scribbling pad and then execute the same. Similarly when we learn and write programs in any programming language, we must script the programs in an editor and then execute the program to check its validity. Before we start using any editor, we must know the background details of it and learn to be convenient with it. This chapter deals with how to make use of the editor for typing the C programs and all related aspects. I have given a simple program in C and a line by line explanation is also provided for the syntax used. I have also elaborated on how the program is saved, compiled and executed. At the end of the chapter, the most important concept in C language called Debugging is being dealt with using a simple example and explanation.

3.2.1 What is an Editor and What is Editing?

An editor is a program where the user types and does all the formatting and editing processes. A notepad is a typical editor used in a Windows mode. When we work in a DOS mode, we use DOS editor and all that we can do in a Windows editor, could be done in DOS editor (options like save, copy, paste, exit, etc. are possible). Another famous editor still used by many of those who work in Windows mode is a WordPad. Those who develop web pages use the default editors like notepad or dedicated editors like FrontPage Express or Microsoft FrontPage. Different programming languages use dedicated editors for editing the programs typed using their syntax. For all the programming languages, one can very well use a notepad or even a DOS editor. But still the very purpose of having a dedicated editor would not be solved, if they are not used by the programmers.

As for example, programming languages like FORTRAN and COBOL use a very famous editor called a Norton Editor, shortly called as NE. Java uses a different editor called Kawa. Visual Basic has its own editor (built-in with the package), UNIX uses vi editor for typing either C programs or shell scripts. Likewise, C uses a dedicated editor called Turbo Editor, shortly called as TC. Figure 3.1 shows how a TC editor would look like.

Editing is a very cumbersome process, if it is done professionally, especially in languages like FORTRAN or COBOL. But in languages like C, C++ or Pascal, the process of editing is very easy and any individual could do the editing work, without having much expertise on the editor.

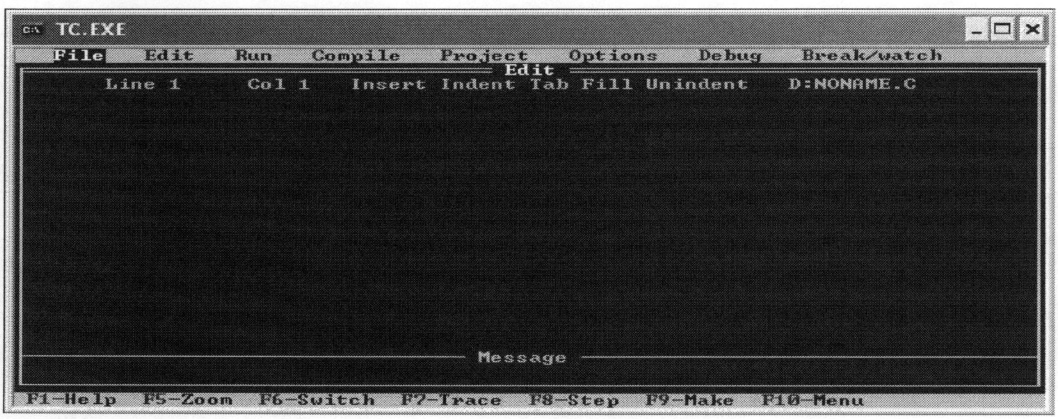

Figure 3.1 A Turbo Editor or TC

When a user is editing with a word processor, with FrontPage Express or with Microsoft Word, he/she must be an expert in using them, since they have such intricate designs and tools for the user to learn and to use. Even, when the user is editing with the help of DOS editor, he/she must know about the options available with the editor, so that no difficulty is faced during the process of editing. When a programmer is using a Norton Editor for editing a FORTRAN or a COBOL program, he/she must know the regulations that these languages impose on the user in using them.

When one uses a notepad editor in Windows, there is absolutely no expertise needed, since it is an easy-to-use editor. Similar to that, when we are using a TC, no difficulties arise during the process of editing a C program. In Chapter 2, we had been discussing the significant features of C and if you can recall, one such feature is that C is a free-form language. There we had mentioned that it is very easy to use a Turbo Editor. We had also mentioned a point that if the user once opens the editor, he/she can start typing the program from any point of the editor (whether fitting or not fitting within the visible part of the screen). Figure 3.2 gives a sample program fragment typed in the editor and Figure 3.3 shows the same program typed in a different style and these two figures imply the point that the Turbo Editor could be used without any difficulty. The output of both the styles will be one and the same, as shown in Figure 3.4.

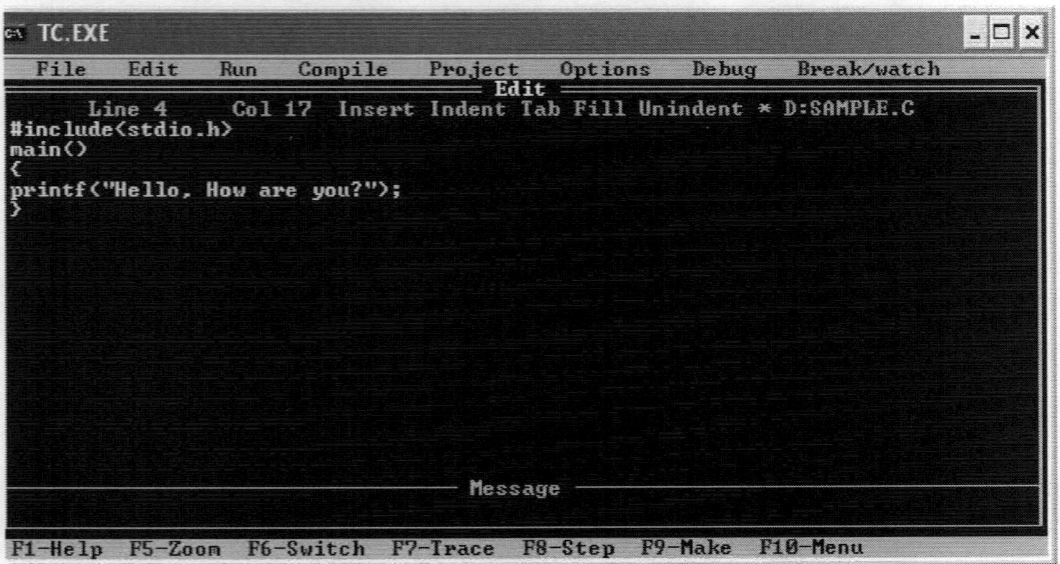

Figure 3.2 A simple program written in TC Editor

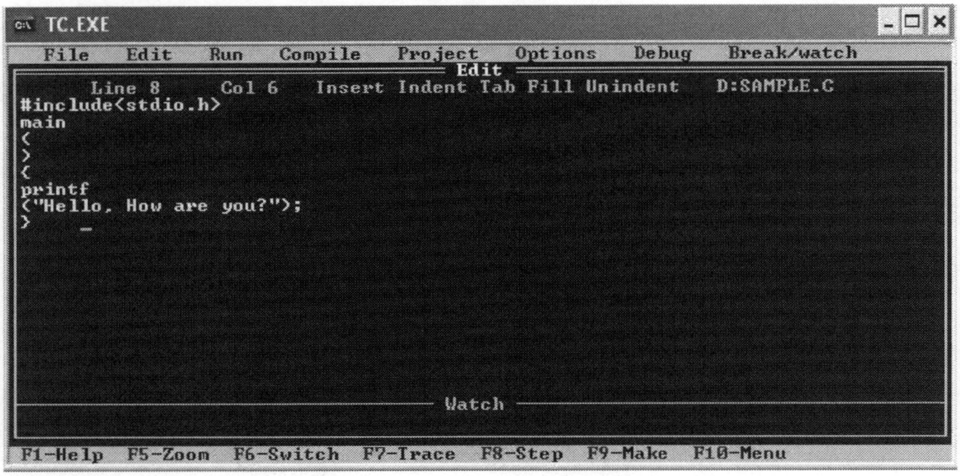

Figure 3.3 Program in Figure 3.2 written in a different style

Figure 3.4 Output for the programs in Figure 3.2 and Figure 3.3

Now that we know the way a TC editor looks and is used, we must know the different parts of the editor. As usual, the editor has the menu bar that contains options like file, edit, etc. When we would like to use a particular menu item, we must use the hotkey combination (as we do in DOS or in Windows). Remember that the roll of mouse is very minimal in TC editor. We can even say that mouse could not be used at the level of TC, since it is totally command based. So, to activate the option File from the menu, we have to use the key combination Alt+F from the keyboard. Once we use this combination, the corresponding menu pops down as shown in Figure 3.5.

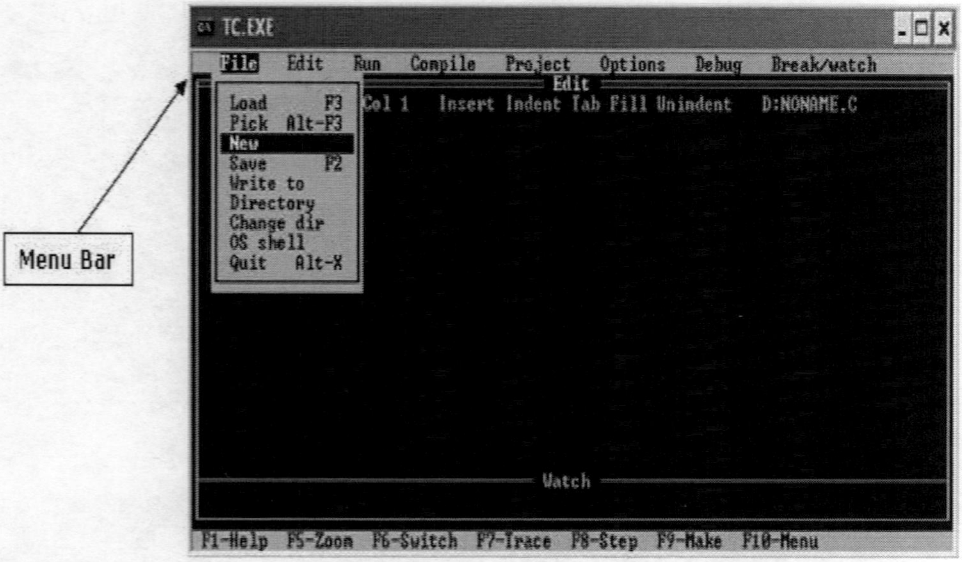

Figure 3.5 Menu bar and the file option in the menu bar

Normally, the Edit option will not work and will be disabled. But still we could perform the editing operations like cut-paste and cut-copy, as we do with other editors. Turbo Editor inherits these functions from its predecessor, the WordStar editor. When one wants to copy and paste a particular block of code to some other part, the following is the procedure.

Procedure for copy-paste and cut-paste of a block of code

1. Place the cursor at the beginning point of the code which you want to copy and holding the control key down, press the keys 'k' and 'b'.
2. Take and place the cursor after the last character of the end of the block of code that you want to copy and holding the control key down, press the key 'k' twice.
3. On completing the above two steps, you will get a highlighted block as shown in the Figure 3.6. Now the block is ready to be copied or moved (cut and paste) to another part of the editor. Figure 3.7 shows how it is pasted and Figure 3.8 shows how it is moved (cut and paste). For pasting it to another location of the editor, take the cursor to that particular position and by holding down the control key down, press the keys 'k' and 'c'. For moving (cut and paste) to another location from the present location, take the cursor to the specified location and by holding the control key down, press the keys 'k' and 'v'.

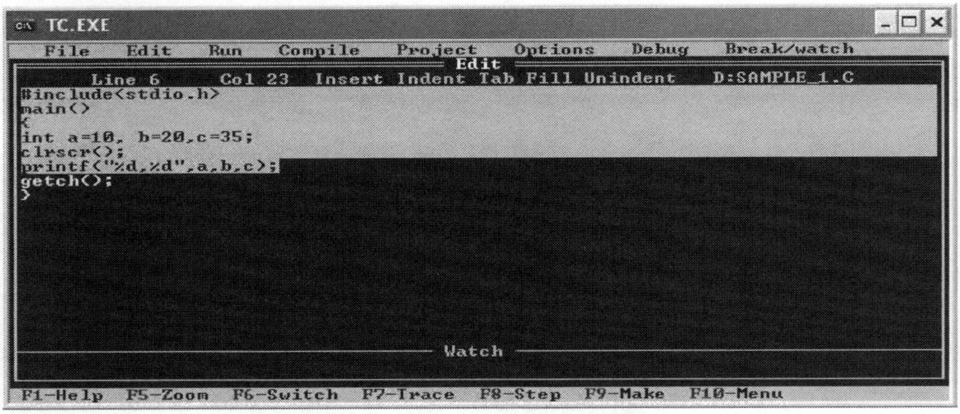

Figure 3.6 Block of code to be copied or moved to another location

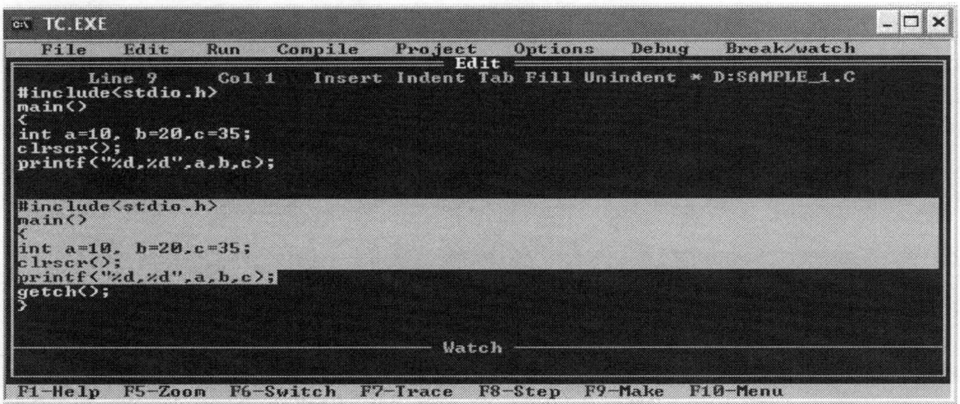

Figure 3.7 Block of code copied to another location in the Editor

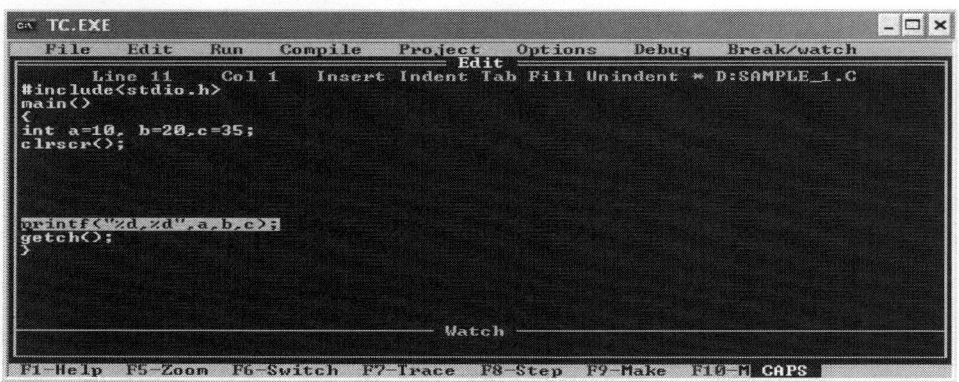

Figure 3.8 Block of code moved to another location in the Editor

If you need more help on the commands and operations that could be performed in the Turbo Editor, you may press the key **F1** by keeping the cursor on the editing part of the screen. All the cursor movement commands could be viewed by further pressing **Page Down/Page Up** keys. This is shown in the Figure 3.9.

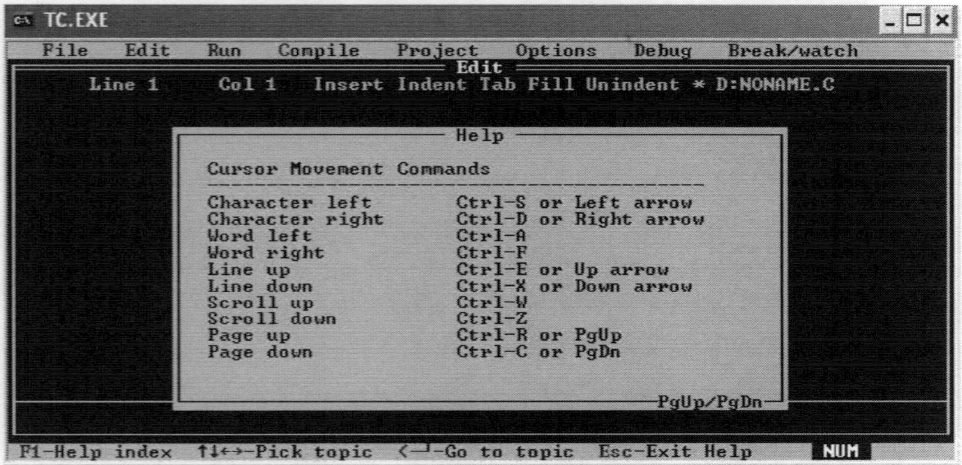

Figure 3.9 Cursor movement commands

If we want to have help regarding the **File** option in the menu bar, then use the combination **Alt+F** and then press **F1**. You will get all the information with respect to **File**. This is represented in the Figure 3.10.

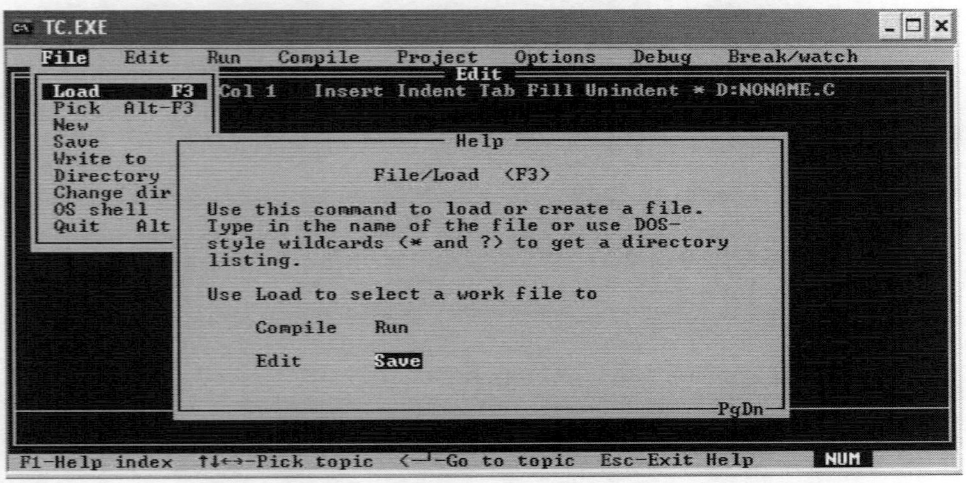

Figure 3.10 Help on File option

The next option available after the **File** is **Edit** and normally it will be disabled as said earlier and still we can get the help on it by highlighting the option using the key combination **Alt+E** or **F10** and then pressing **F1**. Figure 3.11 represents the help corresponding to the **Edit** option.

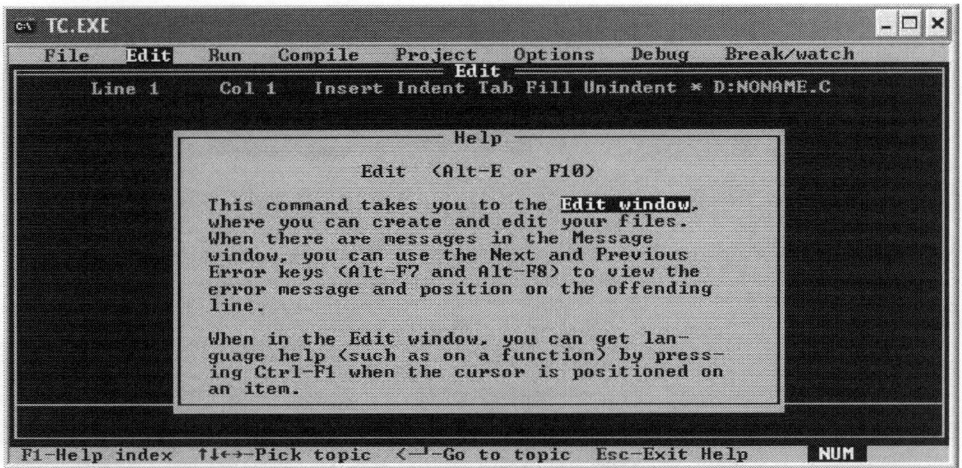

Figure 3.11 Help on Edit option

The same way, you can get the help for all the options on the Menu Bar and know the way they work. As you do programs, you will be better knowing them, since most of them will be used by us during the different contexts of a program.

3.2.2 Editing and Saving Your Work

Create shortcut on your desktop for the C application file, so that you can always open the editor for typing your work from the desktop and you don't have to go to the drive for opening the editor each and every time.

Once the editor is opened, you can start typing the program coding in it and you do not have to follow any rules (as mentioned earlier, C is a free-form language). Let us type the following simple program in that.

```
#include<stdio.h>
main()
{
printf("Welcome to the world of C Programming");
}
```

Figure 3.12 shows the way it is typed in the editor and you can see that the file is not named (the default name noname.c could be seen at the right corner of the editor screen).

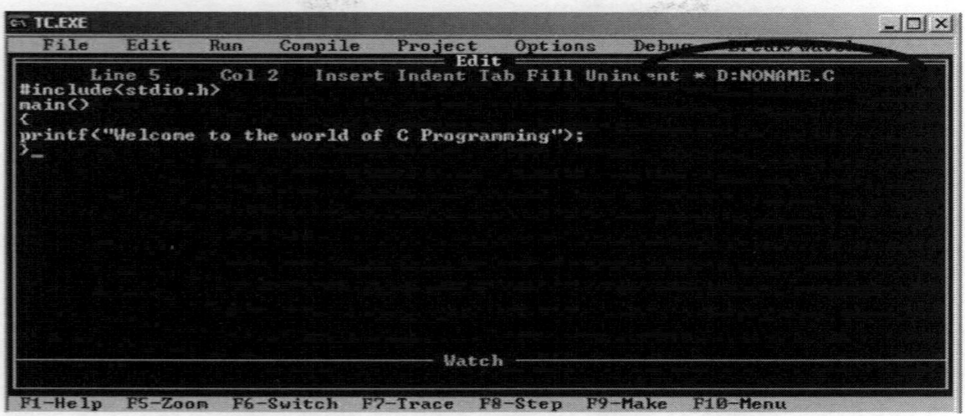

Figure 3.12 A program typed in the Editor

In order to name the file, we have to save our work. For saving the work, we have two options. We can use the key combination **Alt+F** and get the **File** menu to pop down and from it we can choose the option **Save**. You can see that there is a shortcut key assigned beside the option Save in the File menu, which is F2. So, we can directly access F2 instead of the other way around. When we choose either of the above methods, we will get a small window (a Save dialog box) that will expect a file name to be typed. Figure 3.13 represents the appearance of a dialog box for saving the program that is typed. Just start typing the name of the program in the dialog box and it is not necessary even to give the extension. Normally, C programs are saved with an extension .c and the case (uppercase or lowercase) is not a big deal in saving the file.

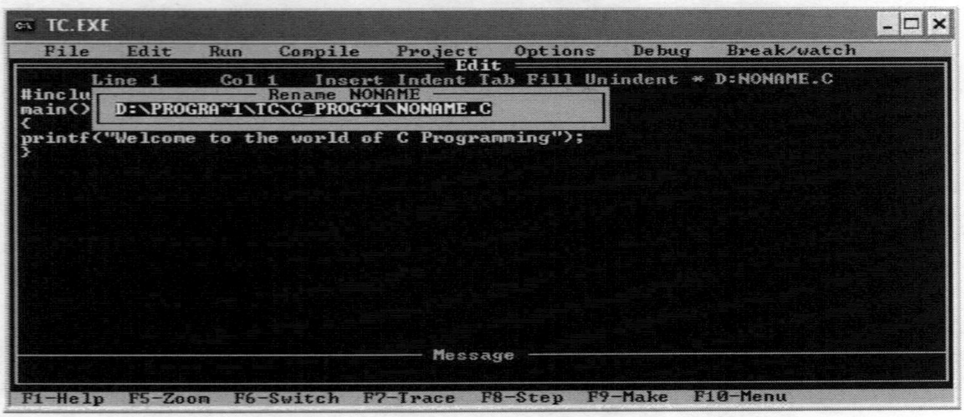

Figure 3.13 Save dialog box (obtained by using the key F2)

In Figure 3.14, we can see that the file is named as **sample** (skipping the extension) and in Figure 3.15, we see that the name **noname.c** is changed to **sample.c**.

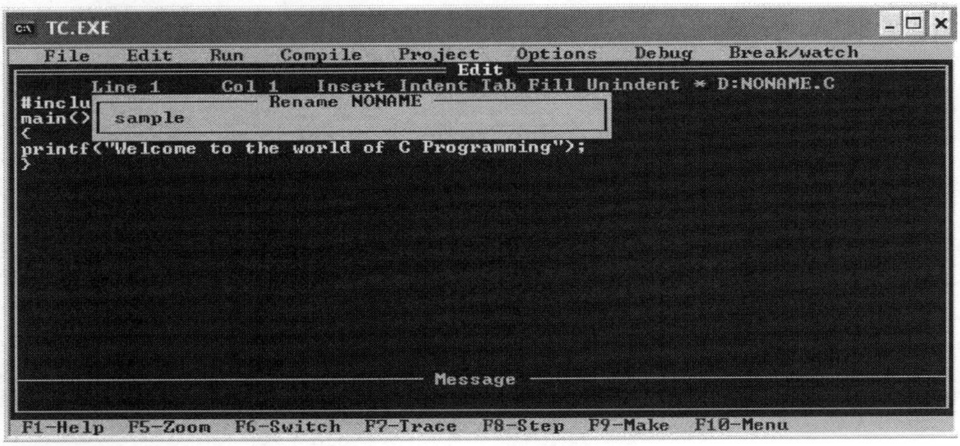

Figure 3.14 File name typed in save dialog box without the extension

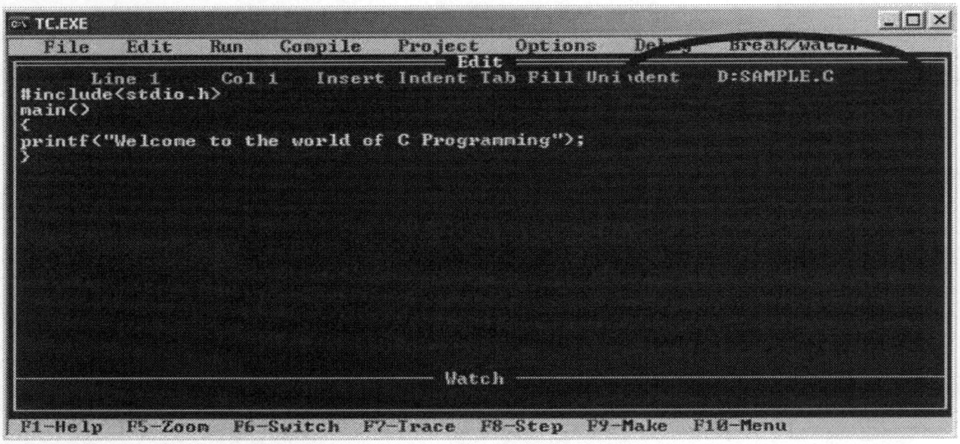

Figure 3.15 New name (sample) given for the file typed in the Editor

Each time we make any alterations in the program, we just have to update them using the key **F2** and the change will be updated properly in the corresponding file. Now that you have saved your work, it is time to compile and trace the errors in the program. We will learn that in the following section.

3.2.3 How to Compile a C Program?—The C Compilation Model

The process of compiling a C program is a very essential part of programming, which allows the user to proceed further with the validation. The compilation process will generate the executable or .exe format of the corresponding .c file. Without the proper compilation of C programs, it is not possible to execute and check the validity of the programs. On compilation, the user actually checks for any syntax or linker errors and fixes the same during the process. Compilation has three sub-stages, namely, compiling, linking and making, all done one after another. The final stage, **making**, would create the .exe format of the C file.

For compiling the C program, use the key Alt+F9 and a window will appear showing the status of the compilation process. Figure 3.16 represents the process of compilation. We see that the compilation is success and this implies there is no syntax error in the program and we are ready to link our program. For linking and making .exe file, we use the key F9 alone and the corresponding screenshots are shown in Figures 3.17 and 3.18.

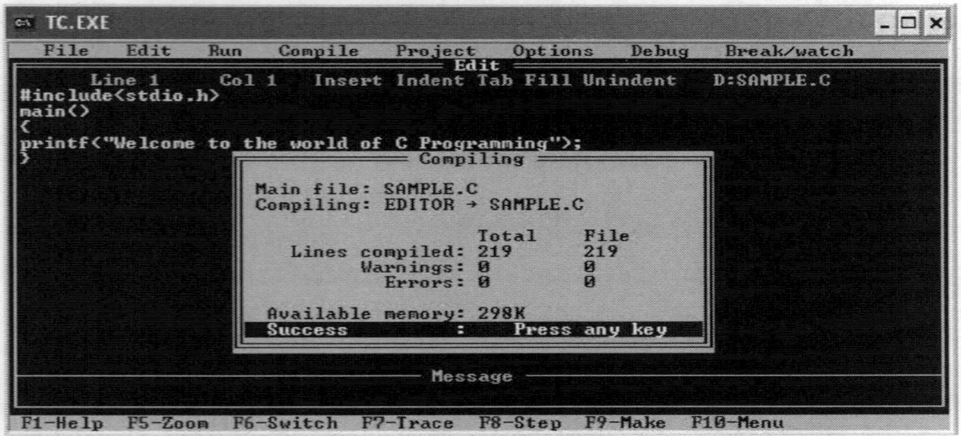

Figure 3.16 Compilation stage

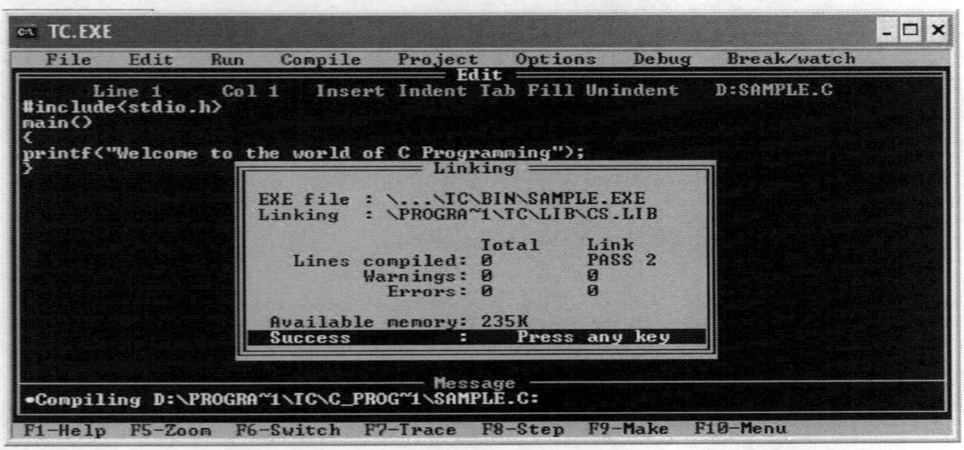

Figure 3.17 Linking stage

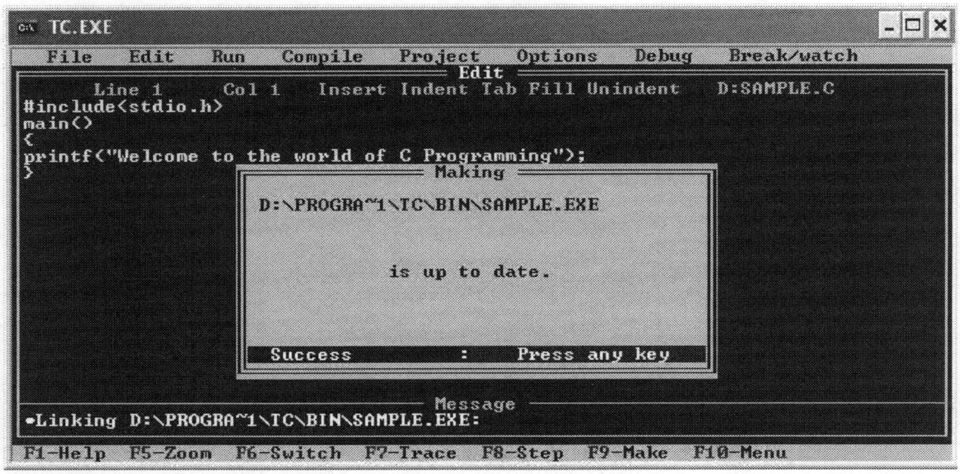

Figure 3.18 Making stage (making .exe file)

If the program is free from any syntax errors or linker errors, the above three stages of the compilation process would be success as shown in the figures above. Suppose the program has syntax errors or linker errors, it will be flashed in the space allocated for that down the editor (the space titled Message). Figure 3.19 flashes that a syntax error has occurred in the program and Figure 3.20 shows the details of the error. Figure 3.21 shows a linker error that might creep into your programs.

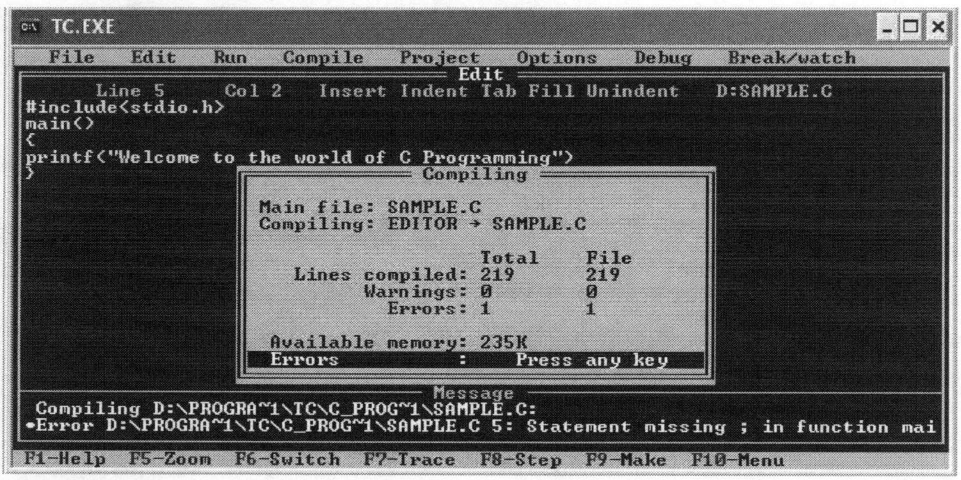

Figure 3.19 Error message flashed due to syntax

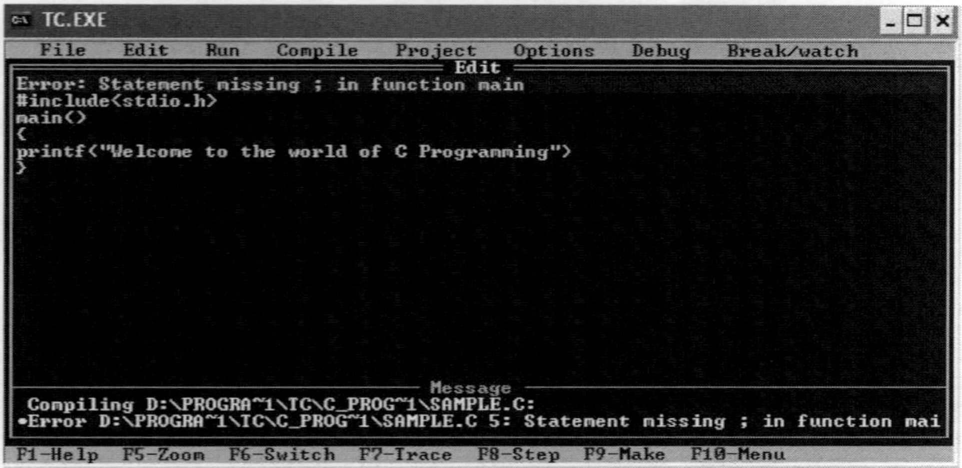

Figure 3.20 Details of the syntax error

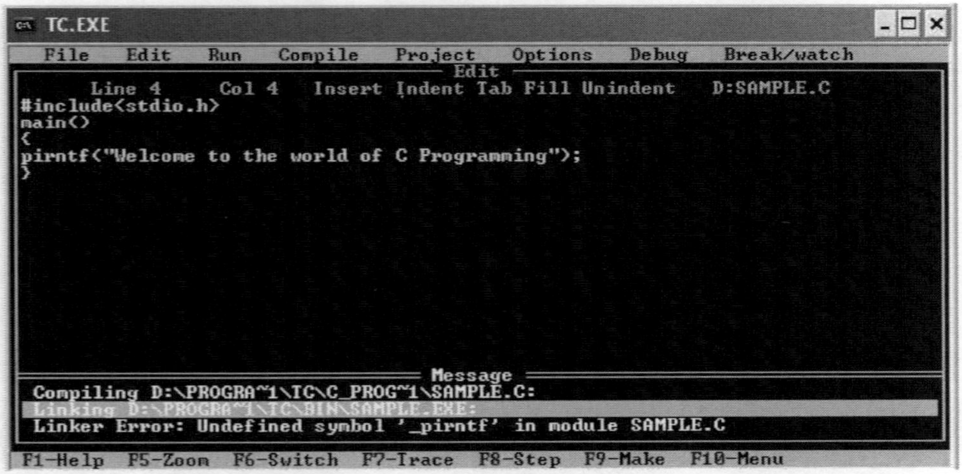

Figure 3.21 Linker error in the program

It is the programmer's responsibility to take care that no logical errors are appearing in the program, since only syntax errors and linker errors will be noticed and flashed by the compiler. For example, consider that in one of our programs, we have divided a given number by 'zero'. Since divide by zero is not allowed, we would anticipate an error while we run the program. But that will not happen, since it is a logical error and is due to the programmer. So it will be skipped from being traced by the compiler.

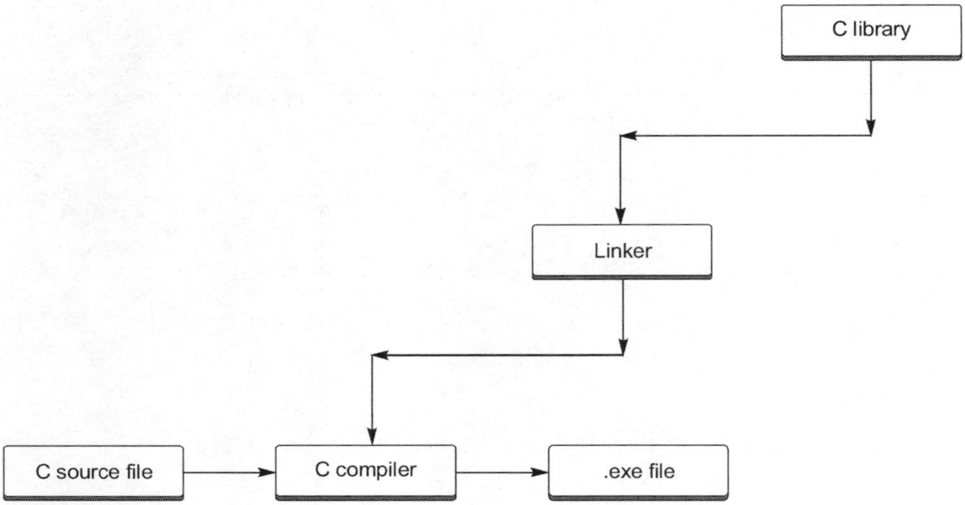

Figure 3.22 C compilation model

The C compilation model is depicted in the Figure 3.22, where the compilation process is indicated as a two-stage process. All the syntax errors are notified by the compiler and the other type of errors are reported to the compiler by the linker. The linker communicates with the library, if there is any issue in the program related to the library functions or built-in functions and reports to the compiler. Thus, even when the program is error-free with respect to the syntax, the linker might have reported some error due to any unrecognized functions or function names. The error shown in the Figure 3.21 is one such error, where the spelling for the function printf() is typed wrongly and hence the linker is unable to recognize the word, as it won't be available in stdio.h or in any other headers.

Also, if there is any error in a program, do fix all the errors and try to analyse why the errors come in the program. If you can do this, you will be surely a master in Programming. One good practice is to keep a database of errors with you and this way, you can easily trace the errors and fix them easily.

3.2.4 How to Execute a C Program?

Executing a C program depends on whether you get a successful compilation and hence .exe file. If we are able to get the executable form of the source code, we are ready to execute our program. Consider the example program that we have been discussing so far (the file named sample.c) and look at the Figure 3.18. We can notice that the making of .exe file is successful and the user could validate his/her program. For executing the file, use the key combination Ctrl+F9. You will be provided with another screen, which is called as the **user screen**. So far, we have been working with the editor and the corresponding screen is called as the **editing screen**. In the user screen, we can give the inputs for the program and check the validity of the program logic. In our example, we need not give any inputs, since we are just displaying a message. When the program is bug-free, we will get the message in the user screen. Figure 3.23 shows the way the message appears in the user screen. But in actual practice, when you press the key combination **Ctrl+F9**, you will not be able to view the output in the user screen properly, since the control will be available only for a moment in the user screen and immediately comes back to the editing screen. In order to view the output, the user has to use the key combination **Alt+F5**. The users might find it very difficult to use the said key combination every time they run a program and shall also want the control to remain in the user screen as long as they wish.

The question is whether we can achieve this idea? The answer is yes. C has console I/O functions for achieving this, which we will discuss in the following section of this chapter.

Figure 3.23 User screen displaying the output

3.3 USING CONSOLE I/O (INPUT/OUTPUT) FUNCTIONS

As said in the previous section, C provides console I/O functions if the user wants to view the output and make a delay in getting the control back to the editing screen from the user screen. One such function is **getch()**. The function is expanded as **get character**. This means the user, after viewing the output, has the option to make the control remain in the user screen itself and after entering a **character** from the keyboard, the control would be taken to the editing screen. The character entered may be any character available in the keyboard. Normally the function **getch()**

would be used before we close the program while writing the coding, in the penultimate line of the program. The above program would appear as shown below.

```
#include<stdio.h>
main()
{
printf("Welcome to the world of C Programming");
getch();
}
```

When you run the above program, the status of the user screen would be like the one shown in the Figure 3.24 in which you could see that the control is waiting for the user's input character (the function **getch()** works here) to get back to the editing screen.

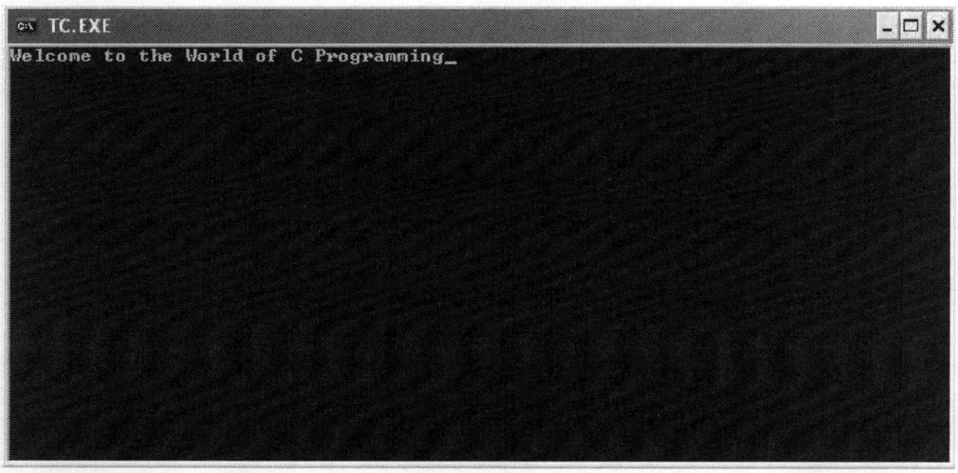

Figure 3.24 Function getch() in action

There are other versions of the function **getch()** in usage. They are the functions **getchar()** and **getche()**. We will be dealing with these two functions in detail later in our Chapters. These functions could also be used instead of **getch()** in our program to make a delay in getting the control from the user screen to the editing screen. But they will have a different effect when used, giving rise to the same kind of result.

For example, when we use the function **getchar()** instead of **getch()**, then the character that is typed by the user to get the control from the user screen to the editing screen will also be displayed along with the message. Figure 3.25 proves this fact. We see that an extra character (e) is displayed with the message.

This means after typing the character, the user has to press the enter key or any similar key to get the control back to the editing screen. This would be an additional work for the user of the program and so the programmer would naturally prefer using the function **getch()** instead of the function **getchar()**. When the function **getche()** is used in place of the functions **getch()** or **getchar()**, the result would be the same. But the way in which **getche()** behaves is quite different from that of the function **getchar()**. When we use the function **getche()**, that will display the character also, that is typed along with the message displayed, but on typing the character, it will get the control back to the editing screen and one need not press any other keys like enter or escape from the keyboard, unlike that in using the function **getchar()**. But, try to understand the fact that both **getchar()** and **getche()** echo the character that is read from the keyboard, whereas the function **getch()** will just read the character and would not echo that on the screen.

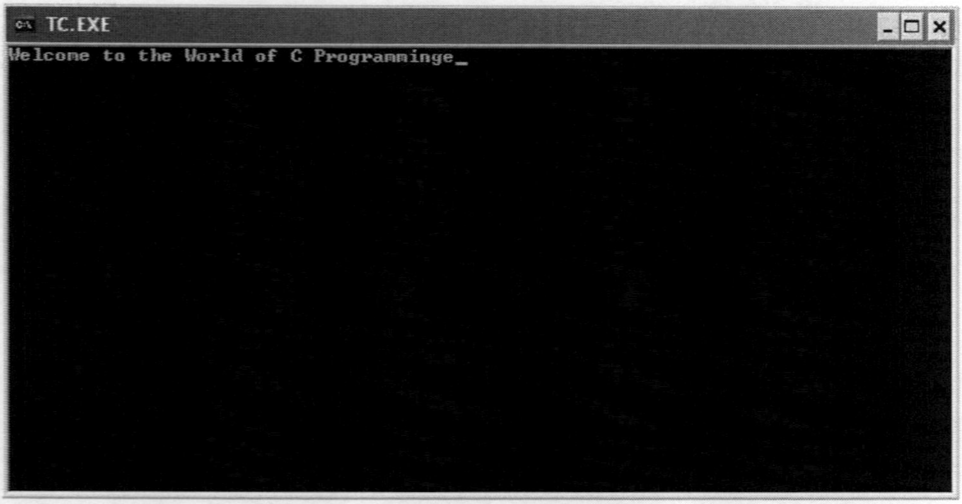

Figure 3.25 Function getchar() in action

All these functions are grouped under the category **conio.h** (header file corresponding to the console input and output functions). There is one more function in the header **conio.h** that is used quite frequently in all our programs and that is the function **clrscr()**. This function clears the screen during each run of a program. For example, consider the program, which we have written above. In that we have not used the function **clrscr()** and hence, each time the program is executed, we will have a junk of the same kind of output in the user screen and this is true even if we run two different programs, one after another. The Figure 3.26 represents a user screen in which we have the outputs of two different programs and this has resulted since we have not used the function **clrscr()** in either of the program codes.

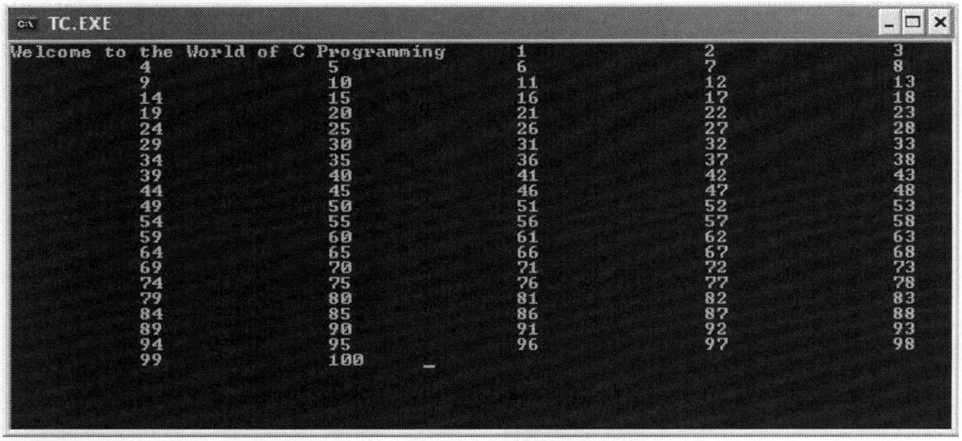

Figure 3.26 Outputs of two programs, where clrscr() is not used

Figure 3.26 has the outputs of the program showing the message and the numbers from 1 to 100 printed. When we open another program and execute it, we will get the third output along with the first two. This surely will make the user screen to have a clumsy look, which any professional C programmer would not like to have. The usage of the function **clrscr()** immediately after the **variable declaration section** (and not before that) of the program would solve this issue and each time a program is executed, the earlier output would have been cleared and the output of the program presently executed would be available in the user screen. Figure 3.27 shows the way the function **clrscr()** is used in the coding of the program. In this program, we have used a variable named 'x' and we had used the function **clrscr()** after that. If there is no variable declared, then the function could be used straightaway in the beginning without any reservations.

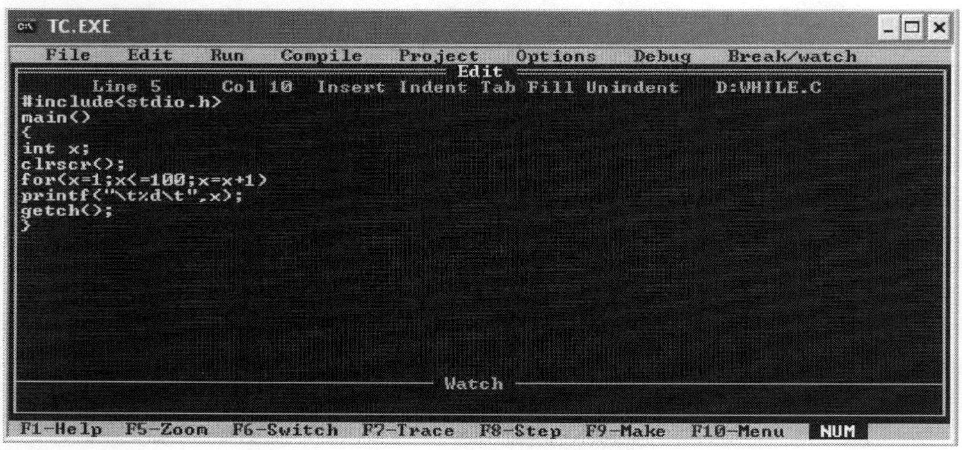

Figure 3.27 Usage of the function clrscr() in coding

The functions **getch()** and **clrscr()** are two of the console I/O functions, which would be used in all our programs. The function **getche()** would be very rarely used and the function **getchar()** would be used when we discuss character arrays.

Even though these functions are grouped under **conio.h**, they can be used without even including the header in the program, as it is purely optional for the programmers to do so.

3.4 FIRST PROGRAM IN C

After learning the building blocks and the essentials, we are now ready to write simple programs using the syntax of C. The following will be the very first and full-fledged program that we are going to discuss.

```
1.  /* program to add two integers and display the result */
2.  #include<stdio.h>
3.  main()
4.  {
5.  int x,y,z;
6.  clrscr();
7.  printf("Enter the values of x and y : ");
8.  scanf("%d%d",&x,&y);
9.  z=x+y;
10. printf("\n\nThe sum of %d and %d is %d",x,y,z);
11. getch();
12. }
```

We can see that the program has used some features, which we have not discussed at all. Let us consider and analyse the program line by line. The **first line** of the program has a brief description about the program and it starts with /* and ends with */ (a slash and an asterisk and an asterisk and a slash). The C compiler will simply ignore anything that it finds within these two pairs of symbols. This marks the **comment** line in the program and the programmers are free to use comments, whenever they feel like using. In fact, the above style of comment is **single line comment**. If the programmer wants to use a comment running over more than one line, then he/she has to follow the **multiline comment**, which starts with the same pair of symbols and ends with the same pair of symbols. Consider the following code which illustrates the way a multiline comment appears in the program code.

```
/* Name    :  Program to add two integers and display the result
Author     :  XYZ
Date       :  03-07-2008 */
```

We see that the information regarding the author of the program and the date on which the program is written is also provided along with the name of the program. This way we can provide a multiline comment in our programs. One thing which the programmers must keep in mind is that the comments can spread over any number of lines, but they cannot be nested. Consider for example, following comment.

```
/* Name   :  Program to add two integers /* and display the result*/*/
```

The above comment is invalid and the user would get syntax error, if such a comment is provided in the program. Figure 3.28 shows such an error. Another important point to be followed by the professional C programmers is that they must not use comments unnecessarily in programs and too much of their usage would make the program inelegant. The programmer has to take care that the number of comments in a program is not exceeding the actual number of lines of coding. This will surely spoil the look of the program and the program would definitely lose the professionalism.

If we analyse the **second line** of the program, we see that we have used #include<stdio.h>, which is called as a **preprocessor** in C language. Actually, the programmers include standard **header files** that are available in C library in order to use the built-in functions. For example, in our program we have used functions like printf() and scanf() in the lines 7, 8 and 10. These functions are already defined by the language designers and the user has to just use it and need not bother about the way they are designed.

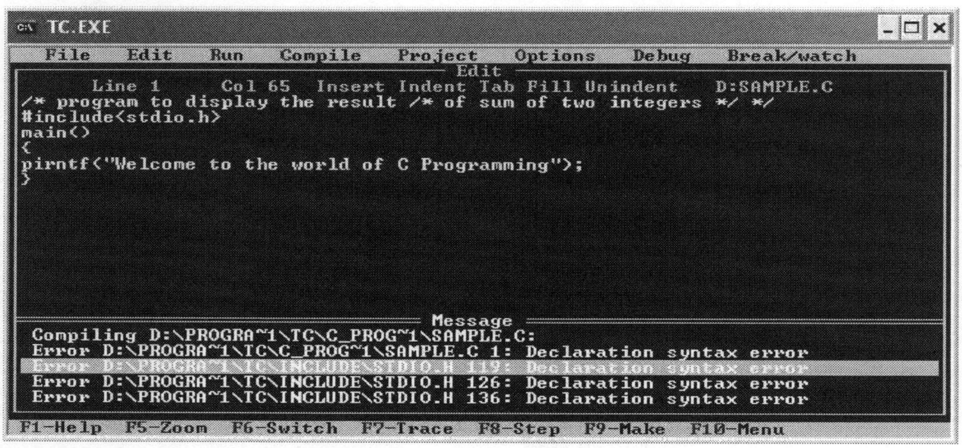

Figure 3.28 Error due to nesting of comments

The language designers have designed many such functions and they have grouped them into different categories. As many as 30 such categories are available in C language and each one is called as a header file and they have the extension .h (h stands for the word header). When the programmer wants to include a particular built-in function, he/she must be aware of the header corresponding to it and include the same in the program before the program's main() is coded. In the previous section, we had seen some of the console I/O functions, which are categorized under the **conio.h** header. But still we had stated that it is not necessary to include that header, since the same I/O functions could be defined in **stdio.h** as well. But some of the functions like **strcat(), sin(), cos(), etc.** are used only when their header files are included in the program. Otherwise they may not work properly. As for example, consider the following program in which we are trying to find out the square root of a given number using the built-in function **sqrt()**, which will take a value corresponding to **double** data type.

```
#include<stdio.h>
#include<math.h>
main()
{
double x=9.0;
clrscr();
printf("%lf",sqrt(x));
getch();
}
```

The output of the above program would be as shown in Figure 3.29.

Figure 3.29 Output for the program finding the square root (header included)

When the header file, namely **math.h**, is not included in the above program, the operation of finding the square root would not have been performed at all and hence we would get a different answer (in the case we are discussing we get 0.000000). This is shown in Figure 3.30. We must note another point here that there will not be any error message flashed during the compilation stage of the program and hence the programmer has to be very careful when using the built-in functions.

Thus, it is important to include the header files in programming and avoid getting answers like the one we had discussed and shown in Figure 3.30.

Figure 3.30 Output for the program finding out the square root (header not included)

The **third line** of the program indicates the beginning of the main section. One of the highlighting features of C language is that it is a compiled language containing many built-in functions. Users can also define their own functions and these functions are called as user-defined functions. One of the functions that would be used in all our programs is the function main() and we had discussed the importance of this function already in the Chapter Two (section 2.2.8). I would suggest the programmers to recall the corresponding section to know about the function main() before proceeding.

The **fourth line** of the program contains an opening brace ({) and it marks the beginning of the coding section of the program and also the beginning of a **block** of statements. When we were discussing the topics like conditional branching, loop control instructions, etc. we used the set of braces (opening and closing braces) and we denoted the lines of coding inside the set of braces as a block. Whenever we write a function too, we must include the lines of coding inside the block and any coding outside the block will not be considered as the part of the function and will be compiling to an error and hence execution of the program will get affected. Figure 3.31 shows this fact.

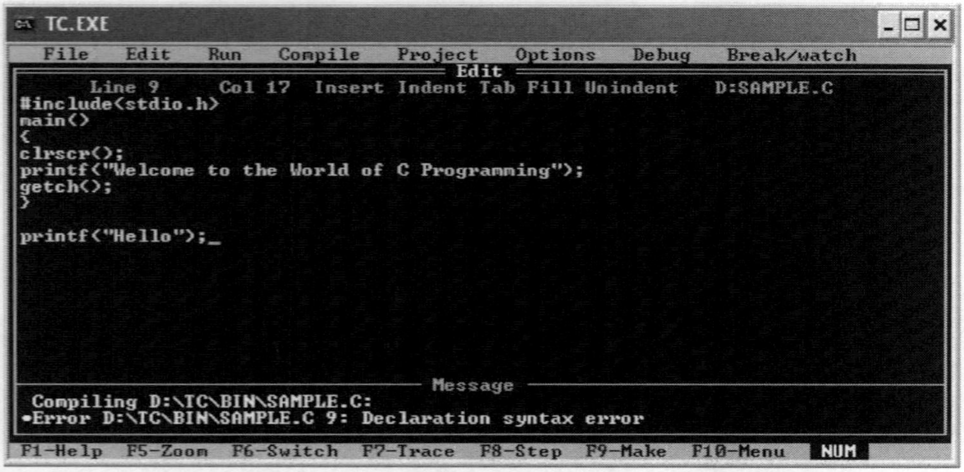

Figure 3.31 Coding outside the function main()—error generated

In the **fifth line** of the program, we are declaring the variables that are used in the program and the programmer must keep in mind the fact that the variables that are used in a program have to be declared in advance, regardless of the point of their usage in the program. The section where we declare the variables of the program is called as **Type Declaration Section**. We call the corresponding line(s) of coding that we use for declaring the variables as **Type Declaration Instruction(s)**.

The **sixth line** of the program has the function **clrscr()**, which is one of the console I/O functions and we had a detailed discussion on thm in this Chapter in the section 3.2.

The **seventh line** of the program has the function printf(), which interacts with the user and prompts him/her to enter the values of the variables x and y. The detailed discussion on the function printf() could be recalled from the section 2.2.7.2.

The **eighth line** of the program scans the input given by the user and we already knew from the section 2.2.7.2, how the function scanf() works and here in this program we see that we are getting input for two integers and hence we are using two %d format specifiers and two corresponding addresses of arguments.

The **ninth line** of the program has the coding for the **addition logic** and we see that it is an assignment expression having two sides, the Right Hand Side and the Left Hand Side. The values that are given by the user for the variables 'x' and 'y' are added on the RHS and the result is stored in the variable 'z' on the LHS. The variable on the LHS is always the target and the variables on the RHS are always the source.

The **tenth line** of the program displays the result of addition with the function printf(). We notice that the input values are also displayed along with the actual result, to make sure that the inputs given by the programmer are accepted and stored properly. This way, the programmer can have a check on any logical errors that he/she could commit.

The **eleventh line** of the program uses another console I/O function **getch()** whose details we have seen in this Chapter, in the section 3.2.

The last and **twelfth line** of the program has the closing brace, which indicates the end of the block and no more coding could be added to main after this. This point has been already discussed, while we were discussing the **fourth line** of the program.

We see that a simple program has these many details. We will discuss more programs of such kind and as we do so, we will learn a lot of new things and wonders in C.

3.5 SIMPLE PROGRAMS INVOLVING CALCULATIONS

As a beginner in programming, one would always like to learn simple concepts and implement them as programs. If one wants to be a professional programmer in C and have a better understanding of the functions such as printf() and scanf(), the data types, format specifiers and console functions, it is better to write more programs on basics and practise. As they use the above functions and data types, they would get more errors, both syntax and logical, and they would start applying their technical mind to solve the errors and slowly they would start mastering the language. In this section, we are going to discuss the basic programs in C, that would involve calculations and we would be using the basic formulae that we had learnt in our elementary and high schools. I have given the list of programs that I would like you to practice and I have written two of those programs here.

List of programs that will make you to become familiar with the syntax of C:

1. Finding the areas and perimeters (or circumferences) of the following two-dimensional figures:

   Square, circle, rectangle, triangle, parallelogram, rhombus

2. Finding the surface areas and volumes of the following three-dimensional figures:

   Cone, cylinder, sphere, cube, cuboid

I have written the programs corresponding to the area and perimeter of a square and volume and surface area of a sphere below.

```
/* Area and Perimeter of a Square */
#include<stdio.h>
main()
{
float side,area,perimeter;
clrscr();
printf("Enter the side of tbe square : ");
scanf("%f",&side);
area=side*side;
perimeter=4*side;
printf("Area = %f\tPerimeter = %f",area,perimeter);
getch();
}
```

The output for the above program is shown in Figure 3.32.

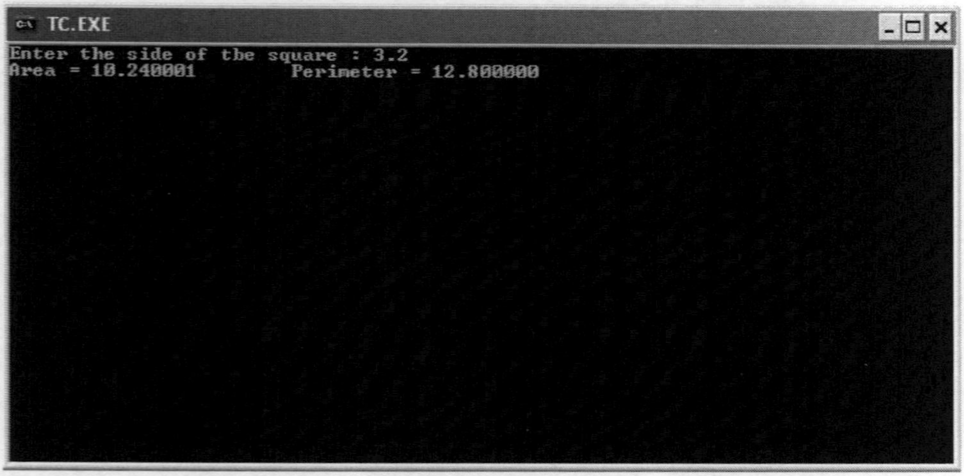

Figure 3.32 Program to find the area and perimeter of a square

The program for finding the surface area and the volume of a sphere is given below:

```
/* Surface Area and Volume of a Sphere */
#include<stdio.h>
main()
{
```

```
float surface_area,volume,radius;
clrscr();
printf("Enter the radius of tbe sphere : ");
scanf("%f",&radius);
surface_area=4*3.14*radius*radius;
volume=(4/3.0)*3.14*radius*radius*radius;
printf("Surface Area = %f\tVolume=%f",surface_area,volume);
getch();
}
```

The output for the above program is shown in the Figure 3.33.

Figure 3.33 Program to find the surface area and volume of a sphere

3.6 UNDERSTANDING AND DEBUGGING A "C" PROGRAM

Understanding how a C program works, and analysing the technical details involved is quite difficult initially. But as we do more and more programs and fix the errors, we may find the task much easier than ever. I had mentioned in the section 3.1.4 of this chapter that one has to maintain a good database of errors, which will surely facilitate the process of debugging the programs. We have given a few exercises too, at the end of each section of each chapter, only to test the technical maturity of the programmers. For example, consider the following program fragment.

```
float a=0.7;
if(a<0.7)
printf("Good");
else
printf("Bad");
```

In the above program, we are just using a simple conditional instruction and the conditional expression is also very simple. What would you as a user and a programmer anticipate as the output? A typical fresher or a beginner in programming shall surely say that the output is Bad and the reason, he/she would say is that the value of 'a' that is checked in the condition is equal to 0.7 and hence the conditional expression returns a false. But, what if I say the answer is not Bad but Good. Yes, of course the answer is Good and this obviously means that the conditional expression has returned a true much against the anticipation of the programmers. Let me justify the answer with a detailed explanation.

When we declare a variable of type float and store some value (in this case it is 0.7) in it, the value will be stored in the form of binary number. Remember that the value we had given for 'a' (0.7) is not binary and is a decimal. So, while storing it, the compiler follows the logic of converting that to its equivalent binary. (For knowing the conversion of a decimal fraction to a binary fraction, refer to section 1.9.3).

When we convert the number 0.7 to its decimal, we will get 0.1011 approximately. We have approximated the result to four bits as it will be convenient for our discussion and even otherwise the actual will be a lengthy sequence of binary digits. In our case the data type that we have used is a float and it will occupy 4 bytes (32 bits). Hence the conversion process will go on till all the 32 bits are filled up with either a 0 or a 1.

This conversion process happens internally when the variable 'a' is declared and initialized. When the value of 'a' is used in the conditional expression, its value is retrieved from the memory and it is converted back to its decimal value (since the user has given a decimal value during the initialization of 'a'). We know how to convert the value in binary to decimal. When we convert the above binary number (the approximated number 0.1011) to its decimal, we would get a value of 0.6875 and if we had considered all the 32 bits for the given decimal number, then upon converting the entire sequence of binary number, we would get 0.69999 and obviously it is less than 0.7 when we do the comparison in the expression. Thus the result returned by the expression would be true. This leads to the output Good.

The conclusion that one can make out of this discussion is that floating point values always create significant issues like this and hence it is advisable not to use them at least in the conditional expressions.

From a very small program fragment, we could get so many ideas about the compiler and the way it behaves. Thus debugging a program fragment needs a lot of practical expertise and patience. You will get a few questions at the end of this chapter which will really make you think and understand the way the C compiler works. More on debugging is discussed in detail in the Chapter 11.

SUMMARY

The idea behind this Chapter was to introduce the basic techniques in programming and using the editors for programming languages. By now, you would have gained a good amount of knowledge on the Turbo Editor and would appreciate its advantages when compared with the editors of other languages.

Installation of the C compiler in the hard drives of our computer systems was also discussed and the compiler is available at free of cost on the web. Also, that the installation process is easy and consumes less time and does not require much expertise.

Editing the source code and the freeness in using the editor is also understood. Saving the file and updating the modifications after saving the file is also discussed.

The C compilation model was also discussed and you must now be aware of the compilation process and would have understood the fact that the compiler alone does not perform all the tasks during the process of compilation and that the linker has also an equivalent role to play.

Execution, which is the final stage of any program validation, was discussed elaborately and the details like editing and user screens were clearly explained.

Following the basic things related to the creation of programs and source codes, we had a brief discussion on console I/O functions like clrscr(), getch(), getchar() and getche() and their significance in the program.

Then we had the very first full-fledged program, where we discussed some additional programming techniques like using a comment and regulations in using it. We had also discussed the header files and their importance in any C program. The importance and necessity of using the source codes within a block was also understood.

Then we had a couple of simple programs involving calculations that might help you in knowing the syntax of the language. Some exercises too have been given in this section for you to practice.

We have concluded the chapter with a hint on debugging and I have given a sample program and debugged the same. You would be surprised to know that C language has umpteen tricks and wonders like the one we had discussed and for you to master them, do a lot of practical work and apply your technical mind.

In all the sections of this chapter, I had used a lot of screen shots to make you understand the concepts under discussion in the clearest possible way. It also enables you to check your work and cross check the details that are included throughout the Chapter. I wish the readers of this chapter shall surely appreciate the technical efficiency of C language.

CRACK THE NUT

Identify the outputs for the following programs:

1. main() { int a=10, b=20;printf("%d\t%d\t%d\t%d",a=b,a>b,a<b,a==b); }
2. main() { printf(); }
3. main() { printf("%d\d%d\t%d\t%d",72,072,0x72,0x72); }
4. #include<stdio.h> main() { int i=10 ; #include<stdio.h> ; printf(«%d",i); }
5. main() { int i; printf("%d",i); }
6. main() { int a=32000+1536+10*1; printf("%d",a); }
7. main() { printf("Hello\n"); main(); }
8. main() { long a; a=65536+300; printf("%d",i); }
9. main() { int x; x=4%5+6%5; printf("%d",x); }
10. main() { int x=100; if(!!x) printf("%d",!x); else printf("%d",x); }
11. i = 2 + 3, 4 > 3, 1; printf("%d"i);
12. printf("%u",–1); what is the value printed out?
13. main() { int x,j,k; j=k=6;x=2; x=j*k; printf("%d", x); }
14. main() { unsigned int x = –1; int y; y = ~0; if(x == y) printf("same"); else printf("not same"); }
15. unsigned m=32; printf("%x",~m);

16. printf(5+"facsimile");

17. printf("%d,%d,%d", sizeof('3'), sizeof("3"), sizeof(3));

18. for(i=0; i<=10;i++,printf("%d",i));

19. main()

```
{
int y;
scanf("%d",&y); // input given is 2000
if( (y%4==0 && y%100 != 0) || y%100 == 0 )
    printf("%d is a leap year");
else
    printf("%d is not a leap year");
}
```

20. main()

```
{
int i=-1;
-i;
printf("i = %d, -i = %d \n",i,-i);
}
```

4

ARRAYS IN C

4.1 WHAT ARRAYS ARE AND WHY ARRAYS?

In principle, arrays in C are similar to those found in other programming languages. As we shall shortly see, arrays are declared and accessed slightly differently in C language when compared with others. Also, in C, arrays have a closer link with the pointers and we shall explore this later in the Chapter 6.

An array is defined as a structured data type that is capable of storing more than one value at a time, pertaining to a particular data type and these values are stored in contiguous memory locations. If you recall the Figure 2.4. from the Chapter Two, an array is listed in the secondary data types. The main advantage of using an array from the point of view of a programmer is that it can be referred with a single name.

In the programs that we had discussed in our earlier chapters, we had used variables of different data types and all of them could store only one value at a given time. But in reality, many applications require the processing of multiple data items that have common characteristics. In that case, the programmer cannot use the concept of a single variable. Even if the concept of a single variable is used by the programmer by declaring more number of variables in the program, he/she ultimately makes the program coding look chaotic. The programmer has to remember more variable names in such cases and obviously this is not possible in complicated applications. For example, if one is writing a program for storing the marks of a particular subject for a class of 50 students, if the concept of single variable is used, then the programmer has to declare and use 50 different variables and above all, he/she has to remember which variables represents whom in the class of 50 and this would be very tedious for the programmer. What if the programmer wants to remember and use just one common name for the entire class for that particular subject and store all the marks of that subject one after another? There comes the provision of using arrays in programming.

As said earlier, the programmer has the advantage of using and remembering just one name, called as the array name, and that would be shared by all the members of that particular array. Each data item of the array could be of any basic data type like integer, character, float, etc.

4.2 DECLARATION, USAGE AND CLASSIFICATION OF ARRAYS

While programming, the arrays, like other variables, shall be properly declared and accessed. This section would let you know how to declare an array and access the data items.

The following is the general syntax for declaring an array:

<Storage Class> <Data Type> <Array Name[Size of 1st Dimension] [Size of 2nd Dimension]...[Size of nth Dimension]>;

The idea of using the storage class is beyond the scope of this chapter and hence we will defer the discussion of the same here. Apart from that, we see that an array is declared just like an ordinary variable and the size of the array must be known to the C compiler well in advance and it must be given within square braces as given in the syntax. The following is how one may declare a one-dimensional integer array.

```
int a[100];
```

The above declaration tells the compiler that there is an integer array by name 'a' and it is capable of storing 100 elements (of type integer). Once we declare the array and execute the program, the compiler reserves memory space for the 100 elements (200 bytes in this case, since each integer occupies 2 bytes of memory). The way the above array would look like in the memory is shown in the Figure 4.1.

a[0]	a[1]	a[2]	--------------	a[99]
23	54	–98	------------	19

Figure 4.1 Memory picture of an array

Another important thing that the programmers must keep in mind while declaring the array is that they must always use a constant value within the square braces for the size of the array and they shall not use a variable there. The following declaration of an array is invalid and would generate an error while compilation.

```
int a[x];
```

The above declaration uses a variable 'x' and its value is not known to the compiler and hence would generate an error while compiling the program. The following declaration of an array is not valid, even one may argue that the value of 'x' is known to the compiler.

```
int x=100, a[x];
```

The error generated is shown in the Figure 4.2. How do we access the individual elements of the array? This is done by using the **subscript** or **index** of the array. The subscripts are normally unsigned integers. The C programmers shall know one fact before indexing the array or before learning the concept of indices that the indexing shall start from '0' (zero) and not from '1'. Initially, this would be difficult for the programmers who would have used FORTRAN or BASIC, but as they do more programs they would find it easy to use the idea.

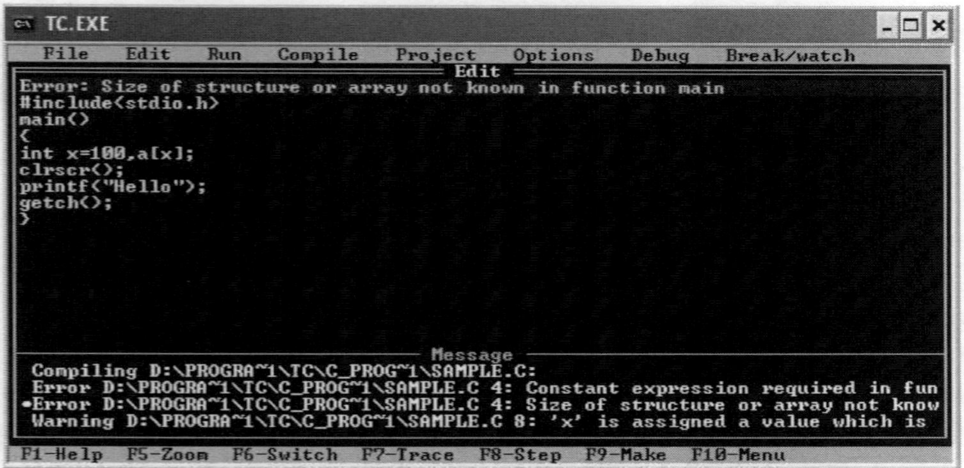

Figure 4.2 Error generated due to improper declaration of array

For example, if we want to access the **fifth** element of the above array, then the corresponding statement would be a[4]. We see that the value within the square braces is one less than the actual value we had specified and this is because the array index starts at '0' and not at '1'.

The programmers must process the array elements very carefully. Single operations involving the entire arrays are not permitted in C. For example, if 'a' and 'b' are two similar arrays (that is of same size and data type), the operations like comparing the elements of the arrays, finding the sum of the elements, etc. must be done on an element-by-element basis (that is the operations to be carried out on a piece-meal basis and not all together). The programmer could accomplish this using different loop structures that we had already learnt in the Chapter Two in detail.

Even though the arrays in C are most efficient, there is a slip. In C, there is no way to check whether the array subscript exceeds the array size. This means, there is no **bounds checking** in C. The programmers should take of this issue while programming and see to that the array subscript does not exceed the array size.

Arrays are classified based on two aspects, namely type of elements they store and their dimension. Based on the type of elements they store, they are classified as numerical arrays and character arrays. All those arrays that store elements of type integer, float, double and long double are called as numerical arrays. In our discussion of numerical arrays in the following section, we would consider the integer arrays only and whatever is being discussed equally holds good for the other numerical arrays. The other type of array is the character array, which is capable of storing alphanumeric characters and is otherwise termed as strings.

We had mentioned earlier in this section that if we would like to process the array elements, we must use loop control and in the loop control we would be obviously using a loop counter. The idea of loop counters had been already discussed in Chapter Two. The loop counters are normally incremented or decremented in steps of 'n' ('n' value depends on the nature of the program). When the loop counters are incremented or decremented in terms of '1', then we have an operator used for that purpose and the operator is called as the modification operator. This is a **unary operator** and hence takes only one operand. The two modification operators used in C are ++ (read as Plus Plus) and – – (read as Minus Minus). These operators could be used in two forms namely, pre-increment/pre-decrement (before the operand) and post-increment/post-decrement (after the operand). The programmers could use these operators only when they increment or decrement in steps of '1' and not in any other cases. Also, the programmers shall not be tempted to use an operator symbol '+++' or '– – –' for incrementing or decrementing a loop variable in steps of '2' and it is not logical to do so.

Let us see how these two operators work. For that consider the following program fragment.

```
int a=10;
printf("%d",a++);
```

What do you think the output of the program would be? If your answer is '11', it is wrong. Upon using the increment operator, the programmer should not come to the conclusion that it will simply increment and the output will be generated. The programmer shall see in which context the operator is used. Here in the above program, the operator is used in the function printf() in the post-increment form. Before the function printf() gets executed, the compiler breaks it into two statements as follows:

```
printf("%d",a);
a=a+1;
```

This means, the existing value of 'a' is used in the function printf() and following the execution of the function, the value of 'a' is incremented. So, the existing value of 'a' is used and then it is incremented. This is the real meaning of the **post-increment**. Had we used one more printf() and modified the above program fragment in the following manner, we would get the outputs '10' and '11'.

```
int a=10;
printf("%d",a++);
printf("%d",a);
```

Suppose that we have used the '++' operator in the pre-increment form as in the following program fragment.

```
int a=10;
printf("%d",++a);
```

What do you think about the way the compiler would treat the function printf() here and what would be the result? The function printf() would be treated as follows:

```
a=a+1;
printf("%d",a);
```

Thus the value of 'a' is first incremented and then the function printf() is executed. So, the existing value of 'a', which is 10, is incremented to '11' and passed to the function printf(), which in turn prints the same. So, the operation **pre-increment** means increment the existing value of the variable first and then use it.

Hence, wherever we have used (so far or henceforth) an increment or decrement operation in steps of '1', we could replace the respective expression with the modification operators. That is, if we had used, say, a=a+1 in a piece of coding, we may replace it with the expression 'a++' and if the expression that has been used is a=a-1, then it could be replaced with 'a– –'. (Readers are requested to check with this before they proceed further).

The same kind of explanation holds good for the decrement operator (– –). I would leave it to you for further discussion and conclusion. Let us proceed with the discussion of the arrays.

4.3 NUMERICAL ARRAYS

4.3.1 One-Dimensional Numerical Arrays

Consider the case of a one-dimensional numerical array, specifically an integer array. Let us learn how the integer array works through the programs. In all our programs, we will be using the memory picture of the arrays for a better understanding of their functions. Let us start with a program in which we read an array of 'n' integer elements and display the same on the output. The following is the program code for the process of reading and displaying 'n' integers.

```
/* Program to read and display an array of 'n' integers */
#include<stdio.h>
main()
{
/* Variables declaration */
int a[100],i,n;
clrscr();
/* Getting the value for number of elements in the array */
printf("\n\nEnter how many elements you want to enter : ");
scanf("%d",&n);
/* Reading the values for the individual elements of the array */
printf("\n\nEnter the elements of the array : ");
for(i=0;i<n;i++)
scanf("%d",&a[i]);
/* Displaying the elements read in Same Order */
printf("\n\nThe %d elements of the array are : ",n);
for(i=0;i<n;i++)
printf("%d\t",a[i]);
getch();
}
```

Figure 4.3 shows the output for the above program.

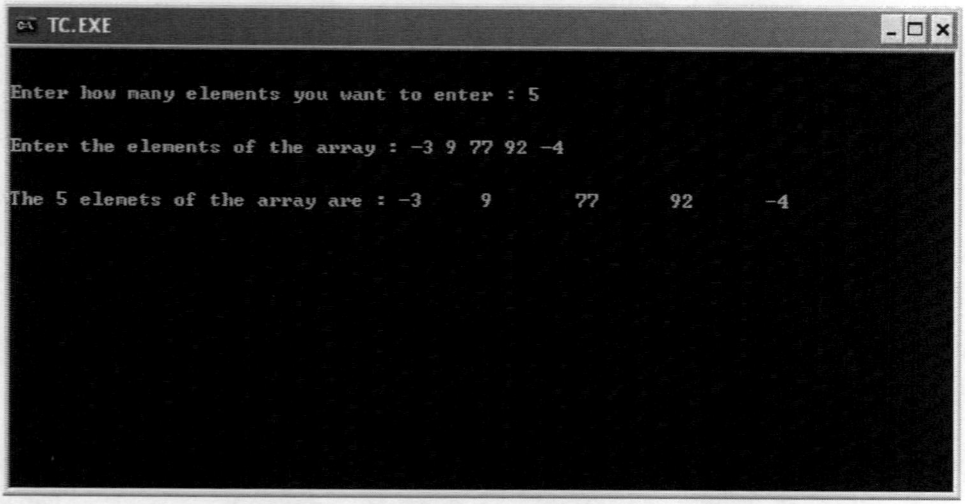

Figure 4.3 Reading and displaying an array of 'n' integers

Let me elaborate on how the program works using the memory picture of the array 'a'. The Figure 4.4 shows the memory picture of 'a', the loop counter 'i' and the variable 'n'.

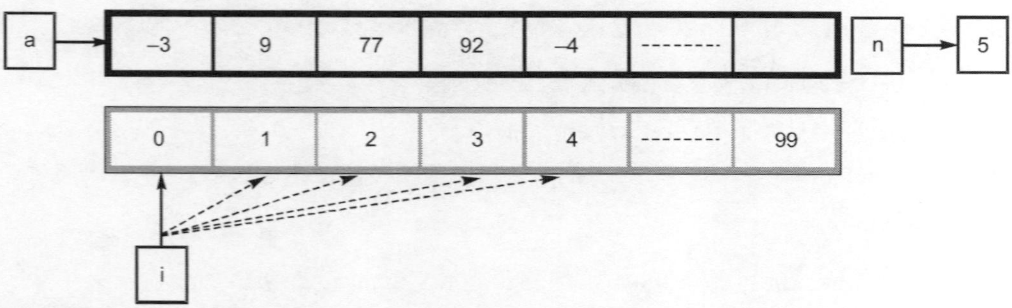

Figure 4.4 Memory picture of the array and the other elements

From the memory picture, we could see that the loop counter 'i' takes the initial value as '0' (the beginning of the array) and as we proceed incrementing the value of it inside the loop, it takes the values according to one less than the maximum limit 'n' (The values that are taken by 'i' during each cycle of the loop is shown as dotted lines). So, the value of the loop counter value would be from '0' to '4' (Even though the array size is 100 and is capable of storing 100 integers, we make use of only 5 of them. But memory will be reserved for all the 100 integers regardless of whether we use it or not.)

In the program, we are getting the value of 'n' first and this is considered to be the upper limit for the loop counter 'i'. Once the value of 'n' is obtained, we get the individual elements of the array using another loop and see that the function scanf() has been coded as scanf("%d",&a[i]) . This implies the real meaning of the function scanf(), which we had already seen elaborately in the section 2.2.7. The function scanf() here tries to access the address of the individual elements of the array based on the value of 'i' and drops in the values that the user gives in the corresponding addresses.

Once we get the values of the individual elements and store them in the memory, we are ready to print them by retrieving them from the memory. This is done in the next *for* loop which uses a similar loop structure as before, but uses the function printf() (instead of the function scanf()). See that when we use another loop, we reinitialize the value of 'i' since the last value of 'i' in the earlier loop would be '5'. If we want to print the elements of the array in the order of the inputs given, then we have to reinitialize the value of 'i'. You can see from the output shown in the Figure 4.3 that the elements are printed according to the order in which the user has given them as inputs. What would you do in case you want to print the elements in reverse order? That will be our next illustration.

In the above program code, do retain the coding for everything except the place where you print the elements. The following code could be replaced there for the existing code.

```
/* Displaying the elements read in Reverse Order */
printf("\n\nThe %d elements of the array are : " ,n);
for(i=i-1;i>=0;i--)
printf("%d\t",a[i]);
```

Here, we see that the initial value of 'i', the conditional expression inside the loop and the modification part of the loop have changed when compared to the earlier program. This is because, we wanted to print the elements that are read in reverse order. After the array elements have been read, the value of 'i' would be '5' and we would like to print the elements that are read from the last position, which is '4'. Hence we reinitialize the value of 'i' to the last but one position, that is '4' and we decrement it each time when it cycles through the loop and we stop the execution of the loop once it reaches a value below '0'. We see that the last valid value of 'i' would be '0' and if it becomes '–1', the execution stops. Figure 4.5 shows the output of the above logic.

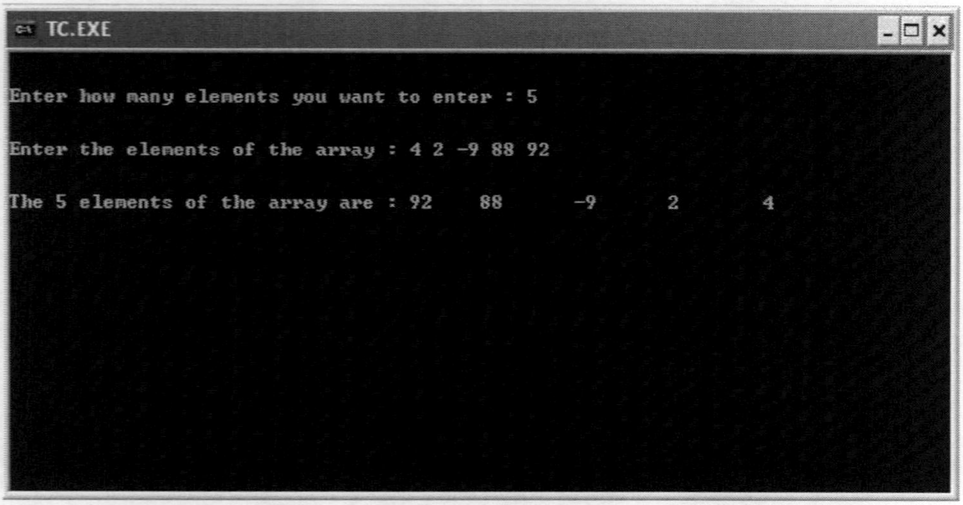

Figure 4.5 Reading and reverse displaying an array of 'n' integers

The memory picture in Figure 4.6 shows the way in which 'i' is reinitialized and the values are printed in reverse order. The dotted arrows show that the value of 'i' is decremented in each cycle and finally reaches '0'. Any further decrement would result in the value to be '–1' and invalidate the condition and terminate the execution of the loop.

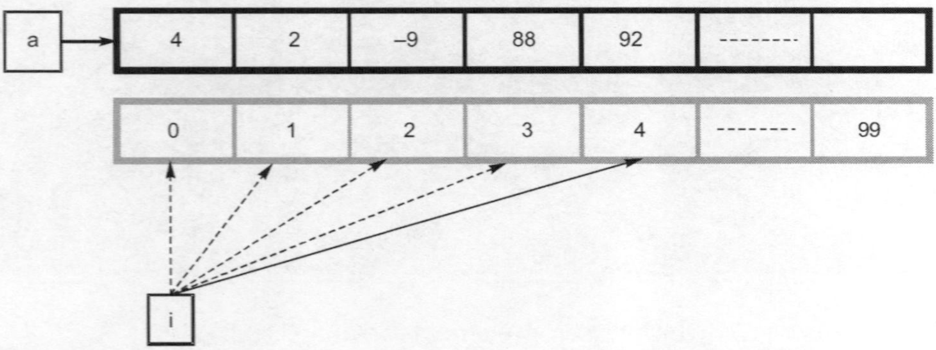

Figure 4.6 Memory picture of the array when the elements are reversed

In the above program coding, we could see that the array values are obtained from the keyboard using the function scanf(). One can initialize the array elements much in the same way as a variable is initialized during the time it is declared. The following is a program that initializes a one-dimensional array and displays the same.

```
#include<stdio.h>
main()
{
int a[5]={23,45,-7,453,89},i;
clrscr();
for(i=0;i<5;i++)
printf("%d\t",a[i]);
getch();
}
```

Figure 4.7 shows the output of the above program code and we could easily understand that array elements could also be initialized just like any other variable.

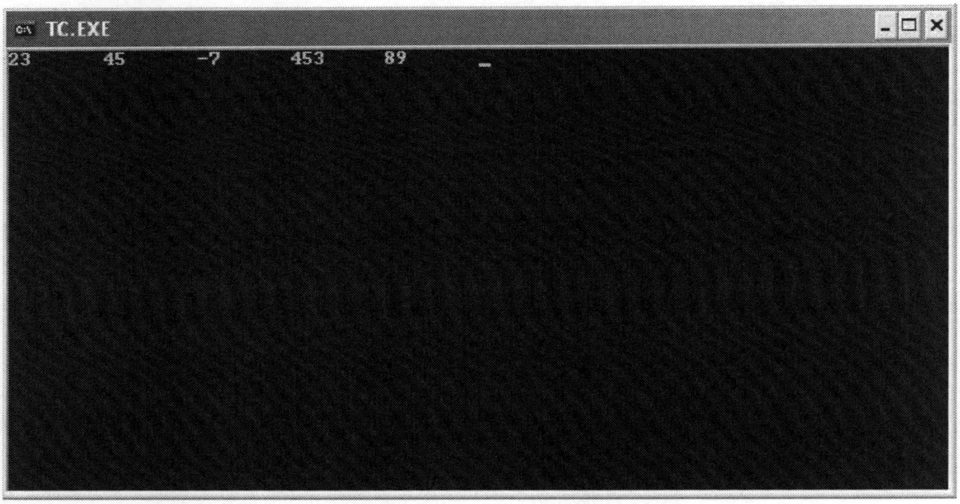

Figure 4.7 Program in which the array elements are initialized

4.3.2 Two-Dimensional Numerical Arrays

A two-dimensional array is declared in exactly the same way as a one-dimensional array, except that we need to supply an extra square braces as shown below, to imply the second dimension. A 2-D array can be thought of as an array of arrays.

```
int a[100][100];
```

In the above declaration we have the size of the first dimension to be 100 and that of the second dimension to be 100. In practice, one can imagine a two-dimensional array as a matrix that consists of rows and columns (any basic book on matrices and determinants would give you a fair idea on what matrices are). It is absolutely not necessary that both the dimensions shall be of same size. One can have a two-dimensional array declared as follows:

```
int a[20][15];
```

The above declaration means that along first dimension, we can store 20 elements and along the second, we can store 15 elements and all together, one store 300 elements (simply multiply the sizes to get the total number of elements that could be stored in the array).

Let us now analyse how the memory picture of a two-dimensional array would look like. For example, consider the following declaration of a two-dimensional array.

```
int a[2][3];
```

The above declaration of the two-dimensional array could store six elements. According to the general assumption of a 2-D array, there could be two rows and three columns in the above array. But when it comes to the storage of the elements in the memory, do you think they will be stored in rows and columns? If your answer is yes, then you are wrong. The individual elements would not be stored in the way that is shown in the Figure 4.8.

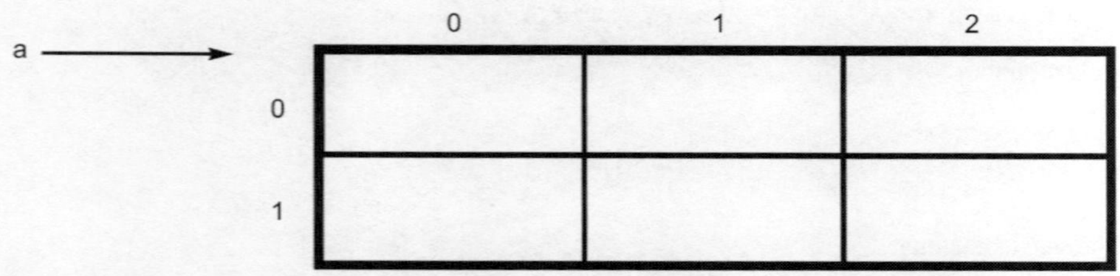

Figure 4.8 Memory picture of a 2-D array of size 2×3

Rather the array would be stored linearly and would be interpreted by the compiler as a combination of two one-dimensional arrays as shown in the Figure 4.9.

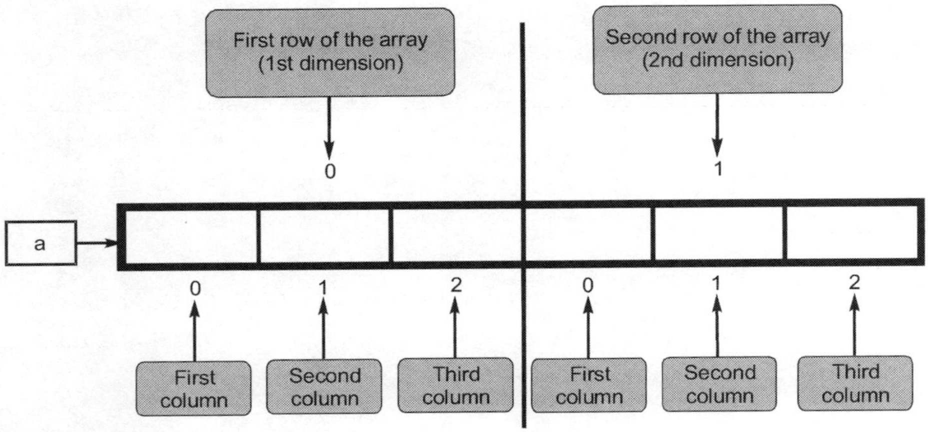

Figure 4.9 Arrangement of rows and columns of a 2-D array

It is easily understandable from Figure 4.8 that regardless of the dimension, the elements are stored in a linear fashion in the memory. The programmer would, for his/her easy understanding and for writing programs, consider the arrangement as a matrix. But in reality, it is not so. Let us write a few programs using the 2-D arrays.

As always, we will consider reading and displaying a 2-D array of size m × n, where 'm' is the size of the first dimension (number of rows in terms of matrices) and 'n' that of the second dimension (number of columns in terms of matrices). The following would be the program for the same.

```
/* Program to read and display a 2-D array of size m × n */
#include<stdio.h>
main()
{
/* Variables declaration */
int a[100][100],i,j,m,n;
clrscr();
/* Getting the size of the first and second dimension (number of rows and columns) */
printf("\n\nEnter the size of the first and second dimension : ");
scanf("%d%d",&m,&n);
/* Reading the values for the individual elements of the array */
printf("\n\nEnter the elements of the array :  ");
for(i=0;i<m;i++)
```

```
for(j=0;j<n;j++)
scanf("%d",&a[i][j]);
/* Displaying the elements read */
printf("\n\nThe elements of the array are\n");
for(i=0;i<m;i++)
{
for(j=0;j<n;j++)
printf("%d\t",a[i][j]);
printf("\n");
}
getch();
}
```

The output of the above program coding would be like the one shown in the Figure 4.10.

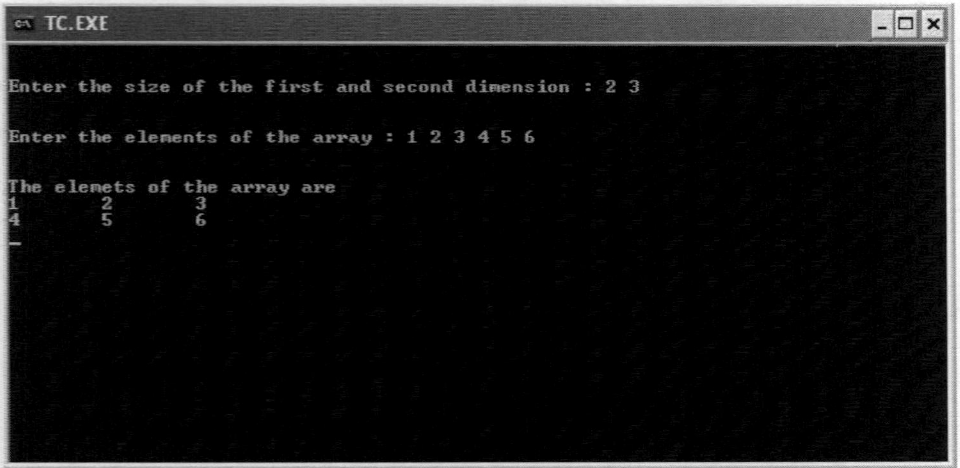

Figure 4.10 Reading and displaying the elements of an array of size 2 × 3 in matrix format

Let us closely analyse the coding written above. As said earlier, the array is declared with two dimensions (with two square braces following the array name). Apart from the array, we have declared four integer variables, two for holding the values of number of rows and columns (m and n) and two for accessing the rows and columns (i and j) while reading and displaying the values.

Initially we get the values of 'm' and 'n', which in turn would be the values corresponding to the number of rows and columns (values corresponding to the first and second dimension) in the array. Following that we get the array elements. See that we use two *for* loops, one after another (the concept of nesting *for* loops is implemented here), one for accessing the row elements and one for accessing the column elements. The outer loop has the access to the first dimension or the row and the inner loop has the access to the second dimension or the column. For each value of the outer loop, the inner loop will execute for three times, since the size of the first dimension is **two** and that of the second dimension is **three**. The outer loop executes **twice** (since the size of the first dimension is two) and hence there will be totally **six loop cycles** in all (two for the outer loop and for each cycle of the outer loop, the inner loop executes thrice).

As a general case, for any 2-D array of size m × n, where 'm' is the number of rows (size of the first dimension) and 'n' is the number of columns (size of the second dimension), there will be 'm' cycles in the outer loop and for each of the cycle, there will be 'n' cycles in the inner loop and hence, in total, there will be 'm × n' cycles and hence there will be 'm × n' elements in the array.

We could see from the above piece of coding that the function scanf() is coded as **scanf("%d",&a[i][j])**. This means, the base address of the entire array is passed to the function scanf(), and it happens to be the location of the first one-dimensional array. So, one can conclude that a two-dimensional array is nothing but a combination of many single-dimensional arrays. In the example given above, the array is of order 2 × 3 and so there are two one-dimensional arrays and each has three elements. You will know this idea more clearly when we discuss pointers and 2-D arrays in the Chapter 6. When we try to print the addresses of the individual elements of the above 2-D array, we would get the following as the output shown in Figure 4.11.

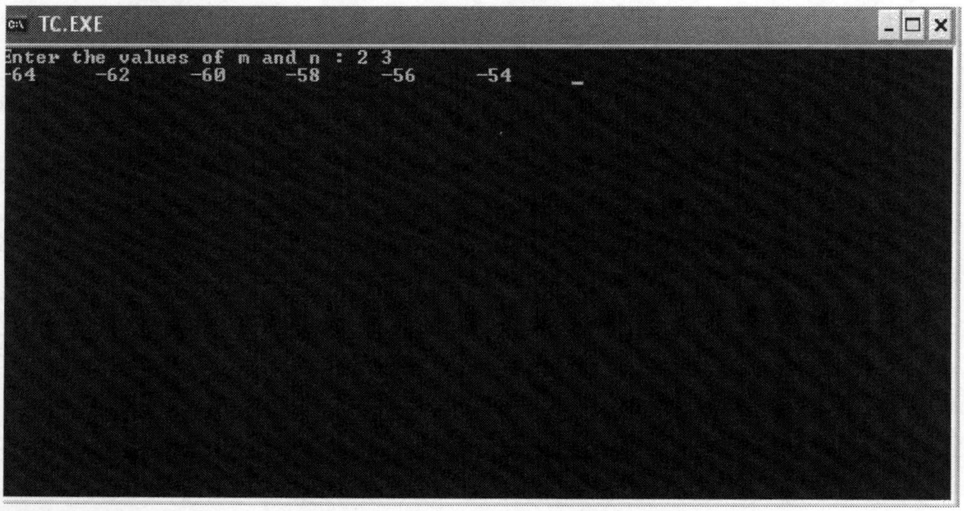

Figure 4.11 Reading and displaying the elements of an array of size 2 × 3

The following coding is responsible for generating such an output:

```
#include<stdio.h>
main()
{
int a[2][3],m,n,i,j;
clrscr();
printf("Enter the values of m and n : ");
scanf("%d%d",&m,&n);
for(i=0;i<m;i++)
for(j=0;j<n;j++)
printf("%d\t",&a[i][j]);
getch();
}
```

We could easily understand from the output that the base address of the entire array is –64 and it happens to be the base address of the first one-dimensional array and then follows the three elements in the first one-dimensional array. Then we have the address –58, which happens to be the base address of the second one-dimensional array. Likewise, if we have varied sizes for the first and second dimensions, the base addresses of the one-dimensional arrays would also vary accordingly. We also see that each element consumes two bytes of memory, since it is an integer array.

Similar to the one-dimensional arrays, the 2-D arrays could also be initialized when they are declared. This is done as shown below with the help of a program fragment.

```
#include<stdio.h>
main()
{
int a[3][3]={{1,2,3},{4,5,6},{7,8,9}},i,j;
clrscr();
for(i=0;i<3;i++)
{
for(j=0;j<3;j++)
printf("%d\t",a[i][j]);
printf("\n");
}
getch();
}
```

The output of the above program code would be the one shown in Figure 4.12. We see that the elements are printed in the form of a matrix with rows and columns.

Figure 4.12 Output of a 2-D array that is initialized

4.3.3 Multi-Dimensional Numerical Arrays

Multi-dimensional arrays are simply an extension of one- and two-dimensional arrays and whatever we had acquired while learning them is to be logically extended to have a more clear understanding of these arrays. A typical multi-dimensional array is a three-dimensional array declared in the following manner. A multi-dimensional array can be thought of as an array of arrays of arrays.

```
int a[2][3][3];
```

In the above declaration, we have the outer array (the first square bracket) has 2 elements, each of which is a two-dimensional array of 3 rows, each of which has 3 elements. Thus the total number of elements in the array would be $2 \times 3 \times 3 = 18$ elements. This means, while getting the input for the array, we would use three **for loops**. The outermost loop would have **two cycles** and for each cycle second loop would execute **thrice** and for each cycle of the second loop, the innermost (third loop) would execute **thrice**. Thus we would have totally 18 cycles in all and hence 18 elements. Further discussion of the multi-dimensional integer arrays would be beyond the scope of this book.

4.4 CHARACTER ARRAYS OR STRINGS

4.4.1 How Character Arrays Differ from Numerical Arrays?

So far we have been discussing a numerical array and we have taken an integer array as the case. As told earlier, the discussion holds good for any other numerical array, the only difference being the data type used. Character arrays are entirely different in C and also have a different treatment. The following is how one may declare a one-dimensional character array.

```
char a[100];
```

The above declaration tells the compiler that there is a character array by name 'a' and it is capable of storing 100 elements (of type character). Once we declare the array and execute the program, the compiler reserves memory space for the 100 elements (100 bytes in this case, since each character occupies 1 byte of memory). The way the above array would look like in the memory is shown in Figure 4.13.

a[0]	a[1]	a[2]	a[3]	a[4]	a[5]	---------	a[99]
'h'	'e'	'l'	'l'	'0'	'\0'	---------	

Figure 4.13 Character array elements

From the figure, we can easily understand the basic difference between the numerical array and the character array. The end of the character array is terminated with a '\0' (NULL Character), which is the bookmark for the compiler for recognizing the end of the array. In numerical array we know well in advance about the number of values that we would input and hence there is no need for keeping a separate terminating character for such arrays. But in character arrays, we would not be in a position to know the number of characters that we are going to feed in, since we cannot sit and count the input for this. Counting the number of characters for small inputs would be easy, but that would be impossible if the input is too lengthy and too bulky (as is the case with any text input). So, to let the compiler know that the array has terminated and to let the user feel comfortable when the array is read back, the character '\0' (NULL character) is appended with any character array. Null character is the one that differentiates a character array from a numerical array and it is a combination of two characters a backslash ('\') and a zero ('0'). The programmer shall supply the NULL character explicitly when the program is coded with a character array. The ASCII value of the NULL character is ZERO. One more thing the user must keep in mind is that, when a character array of size 'n' (where 'n' is a positive integer) is declared, the array could accommodate only n–1 characters due to the fact that one location is reserved for the NULL character. This is unlike the numerical array where one can store exact number of elements as declared.

4.4.2 One-Dimensional Character Arrays

Let us now start programming with the character arrays straight away to understand it more clearly and in a better way. To start with, we will read and display a string (as we did for an integer array).

The following is the program coding for the above task. Remember the fact that the accession of a character array is very much similar to that of a numerical array, except for the NULL character appendage at the end of the array after it has been read.

```
/* Program to read and display a string */
#include<stdio.h>
main()
{
/* Array and other variables used are declared */
char a[100];
int i;
clrscr();

/* The string is read */
printf("\nEnter the string : ");
i=0;
scanf("%c",&a[i]);
while(a[i]!='$')
{
i++;
scanf("%c",&a[i]);
}
a[i]='\0';

/* The string that is read is displayed */
printf("\n\nThe string you have entered is : ");
i=0;
while(a[i]!='\0')
{
printf("%c",a[i]);
i++;
}
getch();
}
```

Figure 4.14 shows the output for the above program coding.

Figure 4.14 Output for the program that reads and displays a string

We can see so many new things from the above program coding. We see that the array is declared as character array, whose size is 100 and it could accommodate only 99 characters given by the user of the program and one space would be used for the NULL character (this point has been discussed earlier).

We also see that the loop counter (array subscript) 'i' is declared as integer. The users of the program shall understand the fact that regardless of the type of array that is used in the program, the loop counter or the array subscript is always a positive integer.

We have used a *while* loop for reading the input string and displaying the read string. We had followed the general syntax of the loop and hence, we had initialized the value of 'i' to be 'zero'. Following this we read the first character of the input string using the function scanf() and see that we have used the format specifier '%c', since the input read is a character. Coming to the conditional expression in *while*, we have used a[i]!='$'. What does this mean? As the user proceeds giving the inputs one after another and at one point exhausted with the inputs, he/she must have a provision to make the compiler know that there are no more inputs. In integer array or any numerical array, we have achieved this by getting the number of inputs for the array ('n') in advance from the user. But in character array, as discussed earlier in this section, there is no way to know the exact number of inputs that would be given by the user. Thus, the user tends to use a character as criteria for stopping the process of giving inputs. The user must be careful in choosing this character and must ensure that it is not an integral part of the array. That is why I have used a special character ('$') instead of an alphabet.

Thus, the conditional expression will always return **true** whenever the character entered is not a $ and when it is a $, further inputs would not be stored (mind you the characters that are given after $ may be accepted while the user gives from the user screen, since the enter key is used as the key for passing the control to the compiler to execute the logic) in the array. Figure 4.15 (a modified version of the Figure 4.14) proves this fact. When the conditional expression returns **true**, the character that is entered from the keyboard gets stored in the array and inside the loop, we first increment the value of 'i', so that the array is readied for accepting the second character and this process continues as long as we do not give a $ from the keyboard. See that the function scanf() is repeated inside the loop, since the first scanf() is considered to be the initialization part of the loop (along with the statement i=0) and hence would be executed only once in its entire life cycle (recall the procedure for executing loops).

Figure 4.15 Excess input characters after '$' are ignored while display

Once the inputs exhaust, then the user has to store a NULL (\0) character to mark the end of the array. That is what is done at the end of the reading process in the above program coding. After getting the input, we have to display the string that is read. For that, we again use a *while* loop in which we reinitialize the array subscript (i), since it will be pointing to the location of the NULL character. This means we begin to display the string or the content of the array from the beginning and we go ahead by incrementing the loop counter. Figure 4.16 shows the memory picture for the above program.

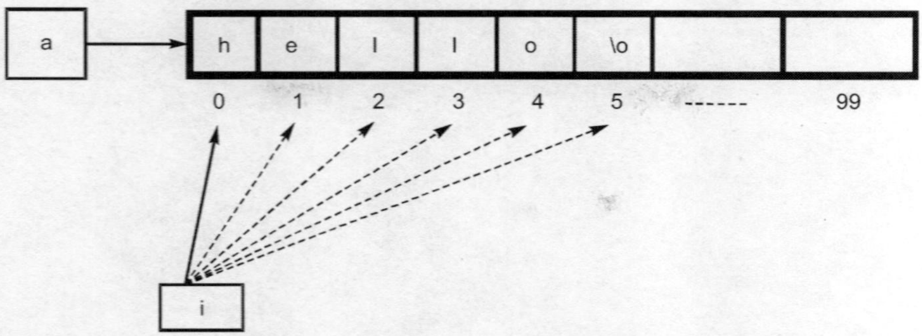

Figure 4.16 Memory picture for reading and displaying a string

We can see that the variable 'i' has been reinitialized to '0' (shown using dark arrow in Figure 4.15) and as we cycle it through the loop, it changes its value (in steps of '1' and shown using dotted arrows in Figure 4.15) and finally after we have given all the inputs, we store the NULL ('\0') character in the array and mark the end of the array. Once we read the array back, we check whether the array has terminated or not, using this NULL character only and that is what is being done in the *while* loop that is used for printing. Each time, the array is checked for the existence of a NULL character and if it is not, then the corresponding character will be printed in the output. Now, what would you do when the above is to be reversed? (like what has been already done with the numerical). The above program code will remain the same, except for the last section, where we print the string that is read. The following shall be the code for printing the reverse of the string that is read.

```
/* Reverse printing the string that is read */
printf("\n\nThe reversed string is : ");
i=i-1;
while(i>=0)
{
printf("%c",a[i]);
i--;
}
```

Figure 4.17 shows the output when the above coding is replaced for the section where the string that is read is displayed, in the program given.

Figure 4.17 Reading and reversing a string

Figure 4.18 is the memory picture for the above logic (see that in the memory picture we have used a different string for reversing, due to convenience and not the one shown in the Figure 4.17.)

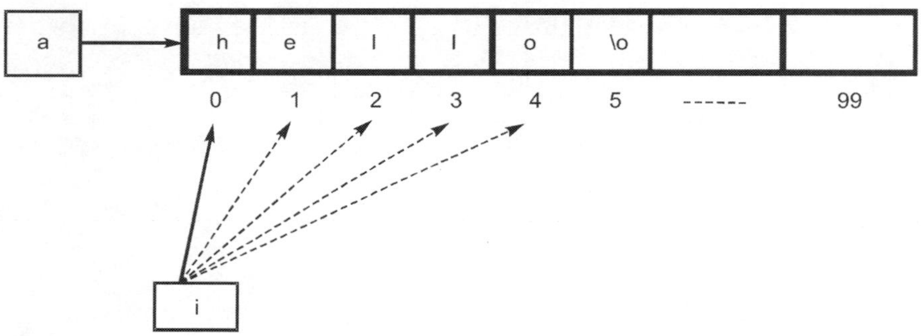

Figure 4.18 Reading and reversing a string

We could see from Figure 4.18 that after the string is read, the variable 'i', which would have been pointing to the NULL character at the position '5' of the above array, is brought to the position just before it, i.e., to '4', since our aim is to print the string that is read from the character 'o' and not from the NULL character (this is shown with a dark arrow) and as we go back, we decrement the value of 'i' and finally reach '0' and as we do so, we print the characters one after another (shown by the dotted arrows). The last character that would be printed is 'h' and after that the variable 'i' would take value '–1', thereby making the conditional expression in the *while* loop false and hence terminating the execution of the loop. We will get the reversed form of the word 'hello' in the output screen.

All such manipulations are possible with the strings and a few of them include finding the length of the string, converting a string in uppercase to lowercase, converting a string in lowercase to uppercase, finding the number of occurrences of a letter in the string and so on. You will be getting all these manipulations in the self-learning exercises section of this chapter.

The same programs could be done with *for* loop too and the following program shows how it can be done.

```
/* Program to read and display a string using for loop */
#include<stdio.h>
main()
{
/* Variables declaration */
char a[100],i;
clrscr();

/* Reading the characters of the array */
printf("\n\nEnter the string : ");
for(i=0,scanf("%c",&a[i]);a[i]!='$';i++,scanf("%c",&a[i]));
a[i]='\0';
/* Displaying the string */
printf("\n\nThe string you have entered is : ");
for(i=0;a[i]!='\0';i++)
printf("%c",a[i]);
getch();
}
```

From the above coding, we can see that the same task of reading and displaying the string could be achieved using *for* loop and we also notice that it is simpler than using the *while* loop.

Don't you think that for loop in the program above seems to be lengthy? Could it be shortened? Yes. C has provision to cut short the *for* loop that is used in the above program. Consider the following modified version of the same *for* loop.

```
for(i=0;(a[i]=getchar())!='$';i++); (a compressed for loop)
a[i]='\0';
```

We could see that a console I/O function, which we had discussed in the earlier Chapter, has been used here. The function getchar(), as you know, will reflect the character that is given from the input device (keyboard) on the screen, unlike the function getch(). It is more or less equivalent to using the function scanf() while the latter is used for getting a character input from the keyboard. Also, we see that a semicolon is supplied at the end of the loop (in both the versions of the *for* loop) and it is much against the basics of *for* loop. Why does a semicolon appear there? The answer is quite simple. If you can recall the general procedure of executing a loop, you can notice that there are four steps or components and you may remember that all the four steps or components are essential for the loop. Just go through the above loop where we have the initialization of the loop variable (i = 0), conditional-cum-assignment expression ((a[i]=getchar())!='$') and modification of the loop variable (i++). Obviously the component loop statement or loop body is missing. The compiler when compiling the program and executing it would expect the same to be present, as it goes in a logical order only. Also, not providing a proper loop body is a logical error and not a compiler error. So that would not be flashed as an error to you, but when you give inputs and execute the program, you would get an output that would not be anticipated by you. Figure 4.19 shows the output of such a program to justify this argument. See that the inputs are accepted without any problem.

Figure 4.19 O/P of the program in which the *for* loop has no; at the end

On providing the semicolon, we let the compiler know that there is a statement for the loop and it will be considered as a NULL statement by the compiler. A NULL statement is that which has no valid C syntax, except for a semicolon. We knew already that any statement is terminated with a semicolon in C programs and the presence of a semicolon itself is an indication that there is a statement. The compiler would treat the coding for the *for* loop given above as the one given below.

```
for(i=0;(a[i]=getchar())!='$';i++);
a[i]='\0';
```

Thus the semicolon given is not for the loop, but for the compiler and it is treated as a statement and hence one of the components of the loop. If one types the semicolon after the loop in a separate line, there won't be any problem, since the compiler itself would treat the semicolon to be available in a separate line. The *for* loop when used with the character arrays can be compressed in the above manner and could be used throughout. It is not possible to do such a compression with the integer arrays.

Another point the programmers who use the arrays should keep in mind is that, when they access the array from the program, the accession is done as a[i]. But the same expression could be written as i[a] too. This point is discussed in detail in the Chapter 6 which deals with pointers. This means when you try to access the 4th element in an array, you would type a[3] from your program and it will be equally fine if you type 3[a] from the coding.

Like the way we initialize an integer array, we can initialize a character array too, as shown below:

```
char a[4]={'x','y','z'};
```

Normally when we use a character array and scan the characters from the keyboard, we would be using "%c" as the format specifier. But when we want to print the string (which is a collection of characters) where all the characters are printed in one go, we could use a special format specifier "%s", which does not require a loop at all. For example, in the program, in which we read and displayed a string, we could have used a single line of coding given below for displaying the string, instead of displaying it character-by-character.

```
/* Displaying the string */
printf("\n\nThe string you have entered is : %s",a);
```

If you test the above coding after replacing this in the corresponding program, you may probably know that the result is the same as you got for the earlier version of the program. When you execute the program, you would get the output similar to the one shown in the Figure 4.19. But one should be cautious in not using the %s format specifier with the function scanf(), while getting the input due to technical reasons. Consider the following program coding as an example.

```
/* Program to read and display a text */
#include<stdio.h>
main()
```

```
{
char x[100];
clrscr();
printf("Enter the string : ");
scanf("%s",x);
printf("\nThe string you have entered is : %s",x);
getch();
}
```

In the above program, we get a string using the format specifier %s and printing the same. When we give an input text "world is beautiful", the output will have only the word "world" in it and the remaining part of the input will simply be ignored. The Figure 4.20 justifies this fact. This is because of the technicality involved with the format specifier %s. This format specifier will have its default delimiters as space or tab or enter. Thus the programmer when giving the input would give a space between the words (three words in our case and hence two spaces and one enter) and so the format specifier %s would take only one word of the three and stores it in the memory. When we execute the program, we get only the first word of the input and the remaining words are not reflected.

Figure 4.20 O/P of a program when %s is used with scanf()

After doing some good programs from the exercises that I have given at the end of this book, you would surely appreciate the way the *for* loop is useful and would like to do more such activities. We could justify the fact that the *for* loop is the most versatile and handy loop in C language.

4.4.3 Two-Dimensional Character Arrays

A two-dimensional character array is nothing but an array of strings and one can also term it as a multi-dimensional array. Consider the following example program that initializes 5 strings in a two-dimensional character array and prints them.

```
/* Two-dimensional character array or an array of strings */
#include<stdio.h>
main()
{
char strings[5][15] = {"Protein", "Amino Acid", "Nucleotide", "DNA", "RNA"};
int i;
clrscr();
for(i=0;i<5;i++)
printf("%s\n\n",strings[i]);
getch();
}
```

Figure 4.21 shows the output for the above program. From the output we could understand that the initialized strings are stored in more than one dimension and printed using *for* loop.

Figure 4.21 Strings that are stored in a 2-D character array

Figure 4.22 shows the memory representation of how the above strings are stored. One can easily justify the concept behind a 2-D character array from the given memory picture.

	0	1	2	3	4	5	6	7	8	9	10	11	12	13	14
0	P	r	o	t	e	i	n	\0							
1	A	m	i	n	o		A	c	i	d	\0				
2	N	u	c	l	e	o	t	i	d	e	\0				
3	D	N	A	\0											
4	R	N	A	\0											

Figure 4.22 Memory picture of the 2-D character array

In the above program the variable "strings" is an array of strings. It is represented by the size [5] [15]. This representation is shown in the Figure 4.22. The size of the second dimension is given as 15 and it specifies the longest one-dimensional array (including the terminating character \0) and each of the strings that are stored are accessed by the first dimension and with the help of the variable i. One can explore more into a 2-D character array and could easily understand the working of the same. This is possible had they understood two things: one, the way the 2-D integer array works and two, the way a 1-D character array works. These two things have been discussed in length and breadth in this chapter.

SUMMARY

We have come to the concluding part of this Chapter and now it is time to summarize what we have discussed. We started the discussion with the basic understanding of arrays and their usage in C. Their importance was highlighted and the way they are declared and used in programs was also discussed in detail.

The classification of the arrays based on their dimension and on the type of elements they store was explained. Based on the dimension, we can classify them as single- and multi-dimensional arrays and based on the type of elements, we classify them as numerical arrays and alphanumerical arrays. We went further in studying the details of declaring an array and how elements in an array could be accessed and used.

In this chapter, we could learn the concepts behind the one-dimensional and two-dimensional integer and character arrays. We have discussed the topics with examples and these examples were supported by proper screen shots and explanations. We had used what is called a **memory picture** in learning and accessing the individual elements in the array. We learnt how to initialize the one- and two-dimensional arrays, instead of getting values for them during the execution of the program.

While we started with the character arrays, we first understood the basic differences between the numerical array and a character array. We also learnt the importance of using a NULL character and how to use it. We discussed a few illustrations in both one- and two-dimensional integer and character arrays and more work is given in the **Self-Learning Exercises** at the end of this book.

We could also see how the *for* loop could be compressed when it is used with the character arrays and prove the fact that it is the most versatile loop in C language. We had learnt how a NULL statement could be used in programming, especially when we would like to supply all the components of a loop.

We had also discussed the two operators ++ and – –, which are unary and could be used in postfix and prefix forms with the operands. We also came to understand that these operators are very much handy when we use them in loops.

Thus, after studying this Chapter, you will be able to grasp and understand the way the one- and two-dimensional integer and character arrays work and even carry out with better programming with them. If you had thoroughly understood the concepts of these arrays, especially the character arrays, you will fall in love with them. I would suggest you try to work out all the programs that are given at the end of this book to make yourself a master in handling the arrays.

CRACK THE NUT

Identify the output for the following programs:

1. main()

```
{
int i,a[5]={10,20,30,40,50};
for(i=0;i<5;i++)
printf("%d",i)
}
```

2. main()

```
{
int i,a[5];
0[a]=1;
1[a]=2;
2[a]=3;
a[3]=4;
a[4]=5;
for(i=0;i<5;i++)
printf("%d",a[i]);
}
```

3. int a[10]; what will be the sizeof(a)?

4. main()

```
{
int a[3][9]={ {1,1,1}, {2,2,2}, {3,3,3} };
int i,j;
for(i=0;i<2;++i,printf("\n"))
for(j=0;j<2;++j)
printf("%d",a[i][j]);
}
```

5. main()

```
{
Char str[]="ABCD";
str[1]='\0';
printf("%s\t%s",str,str+2);
}
```

6. main()

```
{
char a[4]={'x','y','z'};
```

```
printf("%s",a);
}
```

7. main() { int x=20,y=35; x = y++ + x++; y = ++y + ++x; printf("%d %d\n",x,y); }

8. main() { int a=2, b=3; printf("%d", a+++b); }

9. int a=–3, b=2, c=0, m; m=++a && ++b || ++c; printf("%d,%d,%d,%d", a,b,c,m);

10. int a=2; printf("%d,%d", ++a, ++a);

11. main()

```
{
char p[ ]="string";
char t;
int i,j;
for(i=0,j=strlen(p);i++)
{
t=p[i];
p[i]=p[j-i];
p[j-i]=t;
}
printf("%s",p);
}
```

12. main()

```
{
int rows=3,colums=4;
int a[rows][colums]={1,2,3,4,5,6,7,8,9,10,11,12};
i=j=k=99;
for(i=0;i<rows;i++)
for(j=0;j<colums;j++)
if(a[k][j]<k) k=a[i][j];
printf("%d\n",k);
}
```

13. main()

```
{
char s1[]="Ramco";
char s2[]="Systems";
s1=s2;
printf("%s",s1);
}
```

14. main()

```
{
int i, a[100];
for (i = 0; i < 100; ++i)
a[i] = (scanf, ("%d", i));
}
```

15. main()

```
{
int a[100], n;
for(n=0;n<15;n++)
scanf("%d",a[n]);
for(n=0;n<15;n++)
    printf("%d",a[n]);
    }
```

5

STRUCTURES AND UNIONS

5.1 STRUCTURES

A structure is a data structure in which the individual elements can differ in type. A single structure may contain simple type data like integer, character and float elements. It also includes pointers, arrays and other structures. The individual elements of the structure are called as its members. A structure is a heterogeneous data type whereas the array is a homogeneous one. The main use of structure is to lump together collections of disparate variable types, so they can conveniently be treated as a unit.

The concept of structures is so important that when we learn C++, we would study what is called a class for which we require all the ideas behind a structure. Classes in C++ inherit many properties of the structure in C.

5.1.1 Why Structures?—Are they Advantageous Over Arrays?

We had been discussing arrays in the earlier chapter and we knew how significant they are in programming with elements of similar data type. Suppose we want to store elements of different data types in a single array, is it possible to do so? The answer is no. One can store elements of similar type in an array and clubbing different types is not possible. What if we write a program to store the bio-data of a person? The bio-data would contain attributes like name, age, sex, address, salary (if the person is employed) and similar ones. We could see that each attribute is of different type and if we use the attributes in our program by declaring them as usual, then we would be able to write a program but the program would be technically weak, even though it works fine. For example, the following would be the coding for storing the bio-data of a person.

```
/* Program to store the bio-data of a person */
#include<stdio.h>
main()
{
/* Variables declaration */
char name[100],sex,add[100];
int age,i;
float sal;
clrscr();
/*  Getting the name of the person */
printf("\nEnter the name :  ");
for(i=0;(name[i]=getchar())!='$';i++);
name[i]='\0';
/* Getting the age of the person */
printf("\n\nEnter the age of the person :  ");
scanf("%d",&age);
fflush(stdin);
/* Getting the gender of the person */
printf("\n\nEnter the sex :  ");
scanf("%c",&sex);
fflush(stdin);
/* Getting the salary of the person */
printf("\n\nEnter the salary :  ");
scanf("%f",&sal);
fflush(stdin);
/* Getting the address of the person */
printf("\n\nEnter the address of the person :  ");
for(i=0;(add[i]=getchar())!='$';i++);
add[i]='\0';
/* Printing the bio-data of the person */
```

```
printf("\n\nName : %s",name);
printf("\n\nAge : %d",age);
printf("\n\nSex : %c",sex);
printf("\n\nSalary : %f",sal);
printf("\n\nAddress : %s",add);
getch();
}
```

In the program, we see that the variables pertaining to the bio-data are declared accordingly and used. But will they be grouped together in a particular memory chunk? No, they won't. Instead they would be scattered here and there and the overheads involved in the execution of the program would be high. Also, if we would like to store bio-data of more than one person, we have to declare as many set of variables as the number of persons for whom we are storing the bio-data. Thus, we must have some alternative where all the variables related to the bio-data are grouped together and stored in a single variable, so that they are stored one after another and not scattered in the memory. Also, if we can do like this, we would be able to reduce the overheads of the compiler and the execution time for the program would be comparatively less. C language provides a special aspect to deal with this idea. They are nothing but structures. It is a user-defined data type. Once we define a structure, we can declare for it and use it very much like a variable declared for built-in types. In the forthcoming sections, we would see more structures. Thus structures are far more advantageous over the arrays.

5.1.2 The Concept of Structures

The concept of the structures is simple and in fact easy to understand. The structure is defined in a program using the keyword **struct**. Any structure would have its name or tag and following that, we would declare the members of the structure. The members of the structure are nothing but the individual attributes that we would have used, had the structure not been defined. The following is the general syntax of defining a structure.

```
struct <name of the structure>          ——> Structure name/tag
{
<data type> member-1;  \
<data type> member-2;   }               ——> Structure members
_____________________   |
<data type> member-n;  /
   }<variable(s)>;
```

If we want to declare the structure for the above program, then the following code would be the corresponding coding.

```
/* Program to store the bio-data of a person using structure */
#include<stdio.h>
/* Structure definition */
struct bio
{
char name[100],sex,add[100];
int age;
float sal;
};
main()
{
/* Structure variable declaration */
struct bio A;
int i;
clrscr();
/* Getting the bio-data of the person using the structure variable */
printf("\nEnter the name : ");
for(i=0;(A.name[i]=getchar())!= '$';i++);
A.name[i]='\0';
printf("\n\nEnter the age of the person : ");
scanf("%d",&A.age);
fflush(stdin);
printf("\n\nEnter the sex : ");
scanf("%c",&A.sex);
fflush(stdin);
printf("\n\nEnter the salary : ");
scanf("%f",&A.sal);
fflush(stdin);
printf("\n\nEnter the address of the person : ");
for(i=0;(A.add[i]=getchar())!='$';i++);
A.add[i]='\0';
```

```
/* Printing the bio-data of the person using structure variable */
printf("\n\nName : %s",A.name);
printf("\n\nAge : %d",A.age);
printf("\n\nSex : %c",A.sex);
printf("\n\nSalary : %f",A.sal);
printf("\n\nAddress : %s",A.add);
getch();
}
```

We see from the above program that a structure by name **bio** is defined in the beginning before the function main() starts. It has the variables (members) pertaining to the bio-data of a person and we see that all are grouped in one container and that container is nothing but the structure. The name of the structure (bio) is a new data type for which we can declare variables and access the members of the structure. Since the structure is defined outside the function main(), we can say that the structure is global in scope and can be used by any other function that may be defined by the user in the same program. We would see more about global scope in the chapter related to Functions.

The variable for the structure is declared inside the function main(). It could be declared along with the syntax too, in which case it would also be having a global scope as the structure. Here, the variables that are declared inside the function main(), are local to main() and would not be visible outside it.

Once we declare the structure variable, we are ready to access the members of the structure.

After the declaration, we get the values of name, age, sex, salary and address of a person (A) and finally print them. If you closely observe the program, you may notice that while accessing a member of the structure, we have used a **dot operator** (.) and this is one of the operators used in structures. Thus for reading in the name and other details of the person A, we have used the dot operator. See that the dot operator is used for both reading and displaying the details of the person.

We can declare as many variables for the structure **bio** as we wish. If we need more variables corresponding to the structure defined, then we would go in for an array of structure. Figure 5.1 shows the memory map for the structure defined and used in the above program. We could see from the figure that the members of the structure **bio** are distributed continuously in the memory, unlike the earlier version of the same program, where the same variables would have been distributed randomly. This is the primary advantage of using structures in our programs.

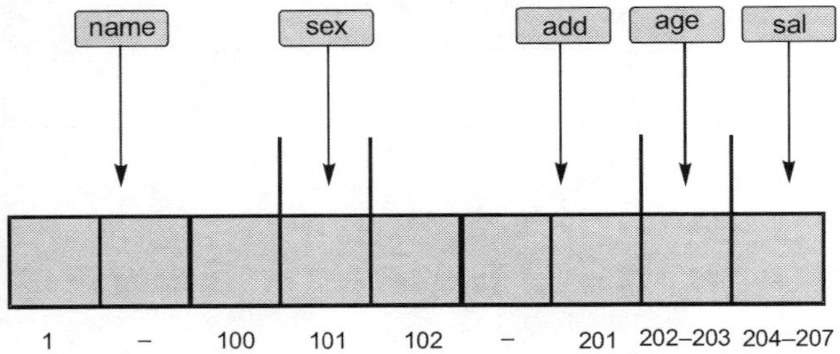

Figure 5.1 Memory map of the structure bio

5.1.3 Structure Operators

As we had seen in the earlier section, dot operator is one of the operators used in structures. Apart from the dot operator, we have one more structure operator denoted by the symbol –>, called arrow operator. This operator is used when we use pointer for a structure and hence will be discussed in the Chapter Six and we will defer the discussion of this operator here. The dot operator is the fundamental operator needed for accessing any member of the structure from the functions including main(). This concept has laid the foundation for the object-oriented programming and the concept of structure was extended to class later and the accessibility of class members was made possible by using the dot operator.

5.1.4 Forms of Structure

Even though there is general syntax available for a structure (as given in the section 5.1.2), there are three forms of a structure as discussed below.

Structure without tag (anonymous structure) The following is an example of a structure without a proper tag and is called technically as an anonymous structure.

```
struct
        {
        char name[100];
        int a,b,c;
   } A, B;
```

Normally an anonymous is not preferably used by the professional C programmers. One shall be aware of the fact that a declaration like the one above is possible in C. The accessibility rule for an anonymous structure is the same as for any other structure.

Structure with tag The following is an example for a structure that has a proper tag.

```
struct bio
        {
        char name[100];
        int a,b,c,;
        } A, B;
```

We have already discussed about such a structure in the section 5.1.2 with an example program.

Structure as a new data type Consider the following definition of a structure.

```
struct book
        {
        char bname[100], aname[100], pname[100];
        float price;
        int no_of_pages;
        };
struct book A;
typedef struct book BOOK;

BOOK B;
```

In the above definition BOOK is a user-defined structure type. It is just an alias for the structure book and one can declare variables for the structure either using the tag book or using the tag BOOK. This is a provision available in C language and this is achieved by using the keyword typedef. Even for built-in types, one can use this typedef and create alias. For example, the following program is perfectly valid.

```
#include<stdio.h>
main()
{
typedef int I;
I a;
clrscr();
scanf("%d",&a);
printf("%d",a);
getch();
}
```

Here we first create an alias I for the type int and then use the alias to declare variables for the type I.

5.1.5 Initializing Structure Variables

Like any ordinary variable, a structure variable could be initialized and used. The following program code illustrates how it is done.

```
/* Program to initialize a structure and store the bio-data */
#include<stdio.h>
/* Structure definition */
struct bio
{
char name[100], sex, add[100];
int age;
float sal;
};
main()
{
/* Structure variable declaration */
struct bio A,B={"johnson",'m',"coimbatore",25,23000};
int i;
float sal;
};
main()
{
/* Structure variable declaration */
struct bio A,B={"johnson",'m', "coimbatore",25,23000};
int i;
clrscr();
/* Getting the bio-data of the person using the structure variable */
printf("\nEnter the name : ");
for(i=0;(A.name[i]=getchar())!='$';i++);
A.name[i]='\0';
printf("\n\nEnter the age of the person : ");
```

```
scanf("%d",&A.age);
fflush(stdin);
printf("\n\nEnter the sex : ");
scanf("%c",&A.sex);
fflush(stdin);
printf("\n\nEnter the salary : ");
scanf("%f",&A.sal);
fflush(stdin);
printf("\n\nEnter the address of the person : ");
for(i=0;(A.add[i]=getchar())!='$';i++);
A.add[i]='\0';
/* Printing the bio-data of A using structure variable */
printf("\n\nName : %s",A.name);
printf("\n\nAge : %d",A.age);
printf("\n\nSex : %c",A.sex);
printf("\n\nSalary : %f",A.sal);
printf("\n\nAddress : %s",A.add);
/* Printing the bio-data of B using structure variable */
printf("\n\nName : %s",B.name);
printf("\n\nAge : %d",B.age);
printf("\n\nSex : %c",B.sex);
printf("\n\nSalary : %f",B.sal);
printf("\n\nAddress : %s",B.add);
getch();
}
```

We see that the structure bio is defined and two variables are declared for it namely A and B. The attributes for A are obtained from the keyboard and displayed, whereas the attributes for B are initialized during the program coding time itself. This is the way to initialize a structure. See that we have followed the same order as we have defined the members of the structure, when we initialize the attributes for B. It is compulsory that we have to follow the same order. Even in the memory map in Figure 5.1, you can see that the order of the variables is very much the same as they are declared inside the structure.

What if we want to initialize one structure variable and would like to assign the same values to another one, i.e., copy one structure value to another? When we want to achieve this for built-in types, we use the assignment operator (=). Will it be applicable for the structures, since they are user-defined types? You would just like to know whether the following code is accepted in the above program.

```
struct bio A,B={"johnson",'m',"coimbatore",25,23000};
A=B;
```

It is acceptable to use such a code. This is another aspect, which we would say as the highlighting feature of any object oriented programming languages, namely operator overloading, is implicitly available in the structures of C language. But the concept is not fully supported in C. Again I would justify how it is not fully utilized in C. Consider the following fragment of a program, for example.

```
/* Program to justify that operator overloading is not supported fully in C structures */
#include<stdio.h>
struct one
    {
    int a;
    }A;
struct two
    {
    int a;
    }B;
struct three
    {
    int a;
    }C;
main()
{
A.a=4, B.a=5; /* initialize the variables A and B */
clrscr();
C=A+B; /* perform the addition of the variables A and B */
printf("%d",C.a);
getch();
}
```

In the above coding, A, B and C are the variables pertaining to the structures one, two and three respectively. See that we are initializing the variables A and B with the values 4 and 5 respectively. In the next line of coding, we add them and try to store the result in C. This kind of addition is performed having a presumption that the values will be added exactly like those of built-in types. But this will flash an error; since the operator '+' is designed only to add the values pertaining to built in types and direct usage of it on the structure variables is not allowed. But when the same expression is used in C++, it would be correct, since the operator '+' would be considered as overloaded (provided overloading function is written). The above line of coding C=A+B shall be modified in the following way to make the program work fine.

```
C.a=A.a+B.a;
```

The error flashed by the above program is shown in the Figure 5.2.

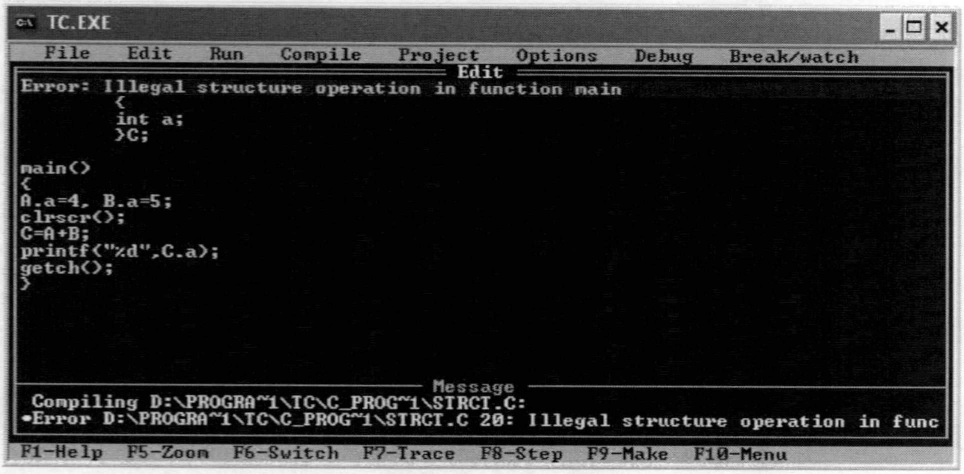

Figure 5.2 Error due to illegal usage of '+'

Once you make the correction that is mentioned above in the error-prone line, the program will work fine and get you the output.

5.1.6 Structures within Structures

One can declare a structure within another structure and thereby nest structures to any level. The following program code illustrates how it is done.

```
/* Structure within another structure */
#include<stdio.h>
struct bio1 /* PARENT STRUCTURE */
```

```
    {
    struct bio2 /* CHILD STRUCTURE */
        {
        int age;
        }A;
    float sal;
    }B;
main()
{
B.A.age=22;
B.sal=25000;
clrscr();
printf("\n\nAge : %d",B.A.age);
printf("\n\nSalary : %f", B.sal);
getch();
}
```

We could see that the structure bio2 is declared within the structure bio1 and we have the corresponding variables A and B declared for them respectively. When we try to access the member of the structure bio2, then we have to use the variable A, but since that structure is within the scope of bio1, the variable corresponding to that structure must be primarily used. That is why the protocol is B.A.age. The member **sal** is in the parent structure and hence it could be directly accessed using its own variable. This way, we could nest the structures to any level and provide the correct variable to access the individual members of the structure.

5.1.7 Mutually Referential Structures

Two or more structures that are defined in each other are called as mutually referential structures. For example, if we have two structures A and B, if we define a variable for the structure B in the definition of the structure A and if we define a variable for the structure A in the definition of structure B, then we would say that they are mutually defined. This concept was later used in C++ for the classes and is given the name **friend classes**. Here we can term one structure as a friend of another. This kind of using a structure is more prevalent when structures are used with pointers. The following program fragment shows how it is done.

```
/* Mutually Referential Structures */
#include<stdio.h>
struct B; /* Forward Declaration */
struct A
    {
    int a;
/* Declaring a pointer for the structure B in the structure A */
    struct B *bptr;
    };
struct B
    {
    int b;
/* Declaring a pointer for the structure A in the structure B */
    struct A *aptr;
    };
```

We shall discuss more on this when we deal with linked lists later in the Chapter 9. We could see from the above fragment that a pointer for structure B is declared in the definition of structure A. So, the compiler should know that there would be a structure by name B later in the program. This is the reason why we have used what is called as **forward declaration** and by doing so, we let the compiler know that the definition of structure B would follow later in the program and even before that, there is a pointer or a related variable that would be declared in another structure. In the absence of such a forward declaration, the compiler would be confused about the status of the variable *bptr that is declared in the structure A and hence would flash an error while compiling the program. The real usage of such structures would be in linked lists which we will discuss later in this book.

5.1.8 Self-referential Structures

A self-referential structure contains a pointer member that points to a structure of the same structure type. Consider the following definition of a structure.

```
struct node {
  int data;
  struct node *nextPtr;
};
```

The coding above denotes that a pointer variable of type node is created within its own definition and hence we say that the structure refers to itself. This is something like a recursive function where the function calls itself from within its own code. This type of structure is useful when we write programs for implementing linked lists and binary trees, which we will discuss in the Chapter 9.

5.1.9 Structures with Bit Fields

A bit field is an additional declaration in a structure that can be used to specify the number of bits. Its length is set off from the declarator for the field name by a colon. A bit field is interpreted as an integral type. This means a bit field is used in expressions in exactly the same way as a variable of the same base type would be used, regardless of how many bits are in the bit field. If the bit fields are <=1 byte, then it takes one byte. Similarly if bit fields are >1 and <=2, then it takes two bytes. If it is >2, it will lead to error.

Consider the following program.

```
/* Bit Fields */
#include<stdio.h>
struct with_bits
	{
	unsigned first : 5;
	unsigned second : 9;
	};
main()
{
union bits
	{
	struct with_bits b;
	int i;
	}a;
	a.i=0;
	a.b.first=9;
clrscr();
printf("%d,%u",a.i,a.b.first);
getch();
}
```

We see that the bit fields are treated as unsigned integers. When we execute the above program, the output would be 9, 9. The identifier with_bits is a structure with two members, **first** and **second**. The member **first** is an unsigned integer with 5 bits and **second** is also an unsigned integer with 9 bits. Both the numbers can be stored in a 16-bit entity now rather than two separate integers. Combined with unions, such structures can be used with great effectiveness, as in the above example. We will not delve into the details of bit fields as it is beyond the scope of this book.

5.1.10 Array of Structures

One can form an array of structures exactly in the same way an array of integers or float is formed. The array thus formed will have the individual structures as its elements. Similar to declaring structure variables, an array of structures can be declared in two ways, either following the structure definition or in the function where we would like to use it. Consider the following example.

```
/* Array of Structures - Processing Employee Records */
#include<stdio.h>
struct date
    {
    int day, month, year;
    };
struct person
    {
    char name[20];
    struct date bday;
    float salary;
    };
main()
{
struct person emprec[10];
int i,n;
clrscr();
printf("\n\nEnter how many records you want to store : ");
scanf("%d",&n);
printf("\n\nPress any key to continue....");
getch();
```

```
clrscr();
for(i=0;i<n;i++)
{
printf("\n\nYou are now going to enter the details of the %d employee\n",i+1)
printf("\n\nPress any key to continue....");
getch();
printf("\n\nEnter the name of the person : ");
scanf("%s",emprec[i].name);
printf("\n\nEnter the date of birth : ");
scanf("%d",&emprec[i].bday.day);
printf("\n\nEnter the month of birth : ");
scanf("%d",&emprec[i].bday.month);
printf("\n\nEnter the year of birth : ");
scanf("%d",&emprec[i].bday.year);
printf("\n\nEnter the salary of the person : ");
scanf("%f",&emprec[i].salary);
printf("\n\nPress any key to continue....");
getch();
clrscr();
}
for(i=0;i<n;i++)
{
printf("\n\nYou are now going to view the details of the %d employee\n",i+1);
printf("\n\nPress any key to continue....");
getch();
printf("\n\nName of the person : %s",emprec[i].name);
printf("\n\nDate of birth : %d\\%d\\%d",emprec[i].bday.day,
emprec[i].bday.month,emprec[i].bday.year);
printf("\n\nSalary of the person : 0.2%f",emprec[i].salary);
printf("\n\nPress any key to continue....");
getch();
clrscr();
}
```

Figure 5.3 Getting the input for the number of employee records

In the above program, we define two structures namely **person** and **date**. The variable for the structure **date** is declared within the structure **person** (recall the concept of the structure within another structure). Thus we have the definition of structure **date** followed by that of **person**. Then we declare an array variable **emprec[10]** for the structure **person**, so that it can take the details of the employees according to the value of 'n' that we give from the keyboard (number of records of employees). We get the name of the person directly using the variable declared for the structure **person** and amidst getting the values for the other members, we get the date of birth (date, month and year) of the person through another structure **date**. The program has been written interactively so that the user knows what happens at each point. The user could easily understand the program since the program uses *for* loop for getting the inputs and displaying the records. Figures 5.3–5.8 show the various stages of the execution of the above program. The program is executed for getting the details of two employees and hence would have two records stored in the given structure variable **emprec[10]**.

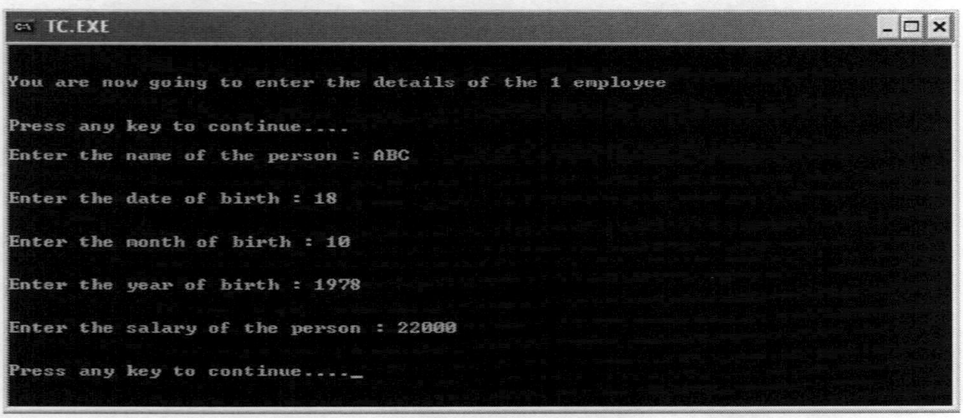

Figure 5.4 Getting the input for first employee record

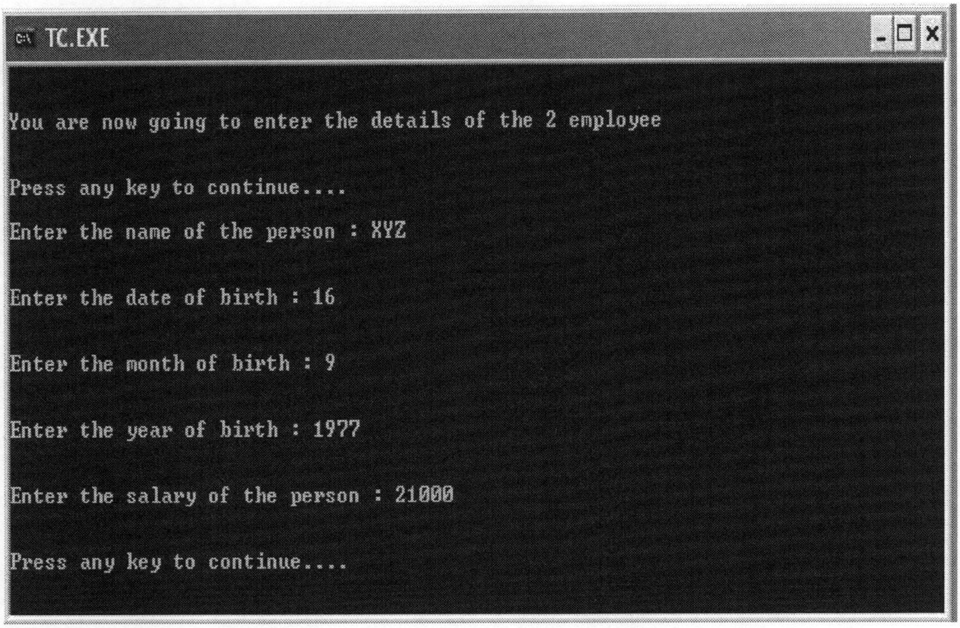

Figure 5.5 Getting the input for second employee record

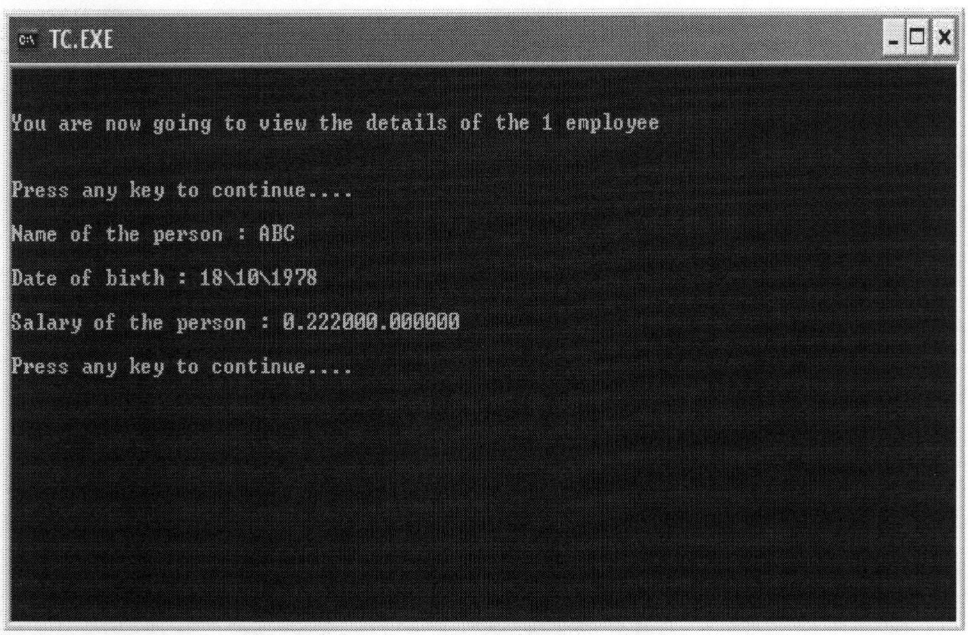

Figure 5.6 Displaying the first employee record

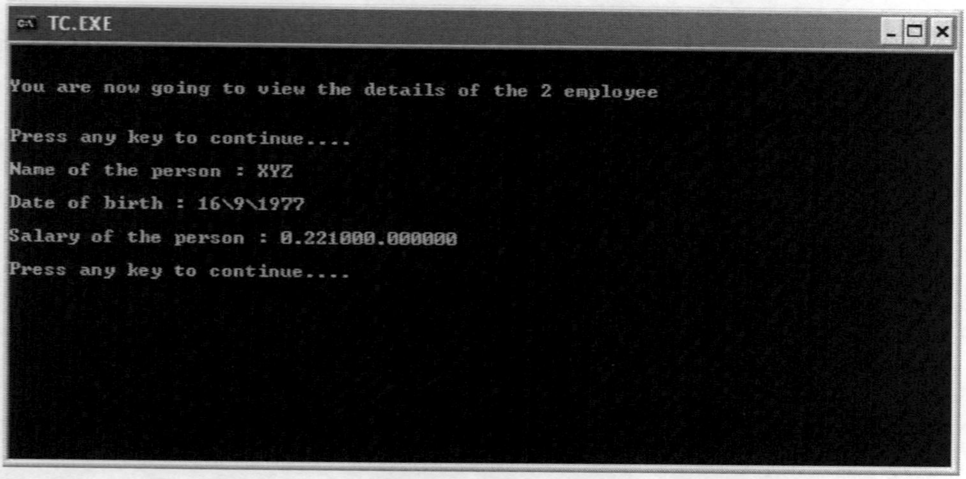

Figure 5.7 Displaying the second employee record

Figure 5.8 Winding up the program

5.2 UNIONS

Unions are similar to structures and they are declared using the keyword **union**. This means, a union would also contain members of different types. A union declaration specifies a set of variable values and optionally a tag naming the union. As said earlier, the same set of rules govern the declaration and usage of both the structure and union members. Like structures, unions could also have bit fields. A union type declaration is also a template like that of a structure and memory would not be reserved until the variable is declared for it and used.

The primary and the only difference between a structure type and union type is the memory space used by their members. When we declare a structure and declare a variable for it, the memory space used by that variable would be equal to the total of size of the individual members of that structure template. Consider the following declarations of a structure and a union.

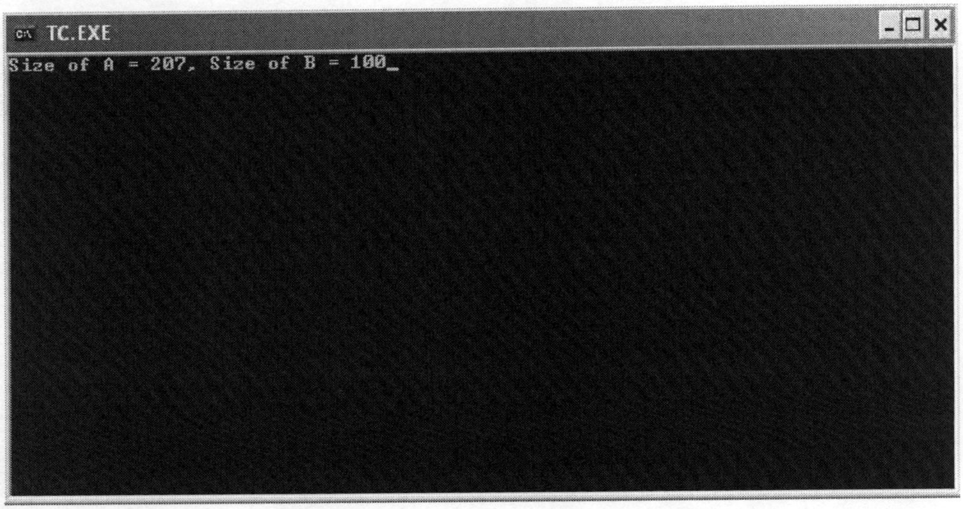

Figure 5.9 Size of the structure variable A and union variable B

```
/* Structures and Unions - Primary Difference */
#include<stdio.h>
struct bio_1
    {
    char name[100],sex,add[100];
    int age;
    float sal;
    }A;
union bio_2
    {
    char name[50],sex,add[100];
    int age;
    float sal;
    }B;
```

```
main()
{
clrscr();
printf("Size of A = %d, Size of B = %d", sizeof(A), sizeof(B));
getch();
}
```

The output for the above program would be 207 and 100 as shown in Figure 5.9. Note that we have used a new operator called as sizeof, which will find the size of the variable pertaining to a data type and returns that value. Thus it is obvious that the storage associated with a union variable is the storage required for the largest member of the union. When a smaller member is stored, the union variable can contain unused memory space. All the members are stored in the same memory space and start at the same address. The stored value is overwritten each time a value is assigned to a different member of the union.

Consider the following program for getting the information regarding the storage of members of the union type.

```
/* Union - Storage of members */
#include<stdio.h>
main()
{
union test
    {
    int i;
    char ch[2];
    };
    union test t;
    t.i=64;
    clrscr();
    printf("\n\ntest.i = %d",t.i);
    printf("\n\ntest.ch[0]=%d",t.ch[0]);
    printf("\n\ntest.ch[1]=%d",t.ch[1]);
    getch();
}
```

This data type would occupy only 2 bytes in the memory, whereas the same would occupy 4 bytes with the **struct** definition. Here in this program, we are initializing the value of 'i' for the structure variable't' using a statement t.i and the value is 64. The same memory locations used for t.i are also being used by t.ch[0] and t.ch[1]. This means the memory locations used by t.i can also be accessed using t.ch[0] and t.ch[1]. Figure 5.10 shows the output for the above program when the value of t.i is 64.

Figure 5.10 Output of the program when the value of t.i is 64

We know that 64 is an integer and hence requires 2 bytes of memory. Figure 5.11 is the memory representation of the number 64 (see that the number is converted to its binary equivalent and stored in memory).

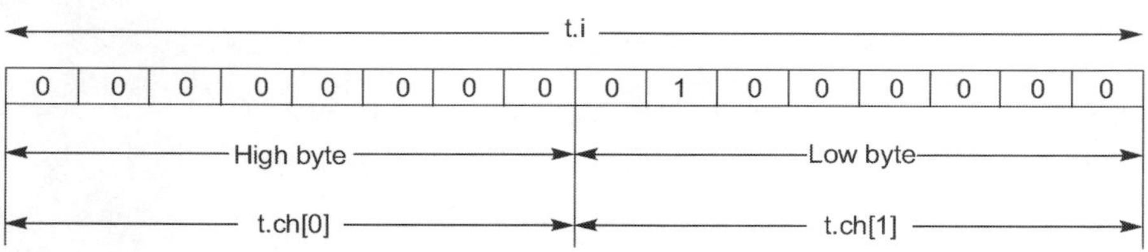

Figure 5.11 Memory representation of 64

The member 'i' is an integer and hence is represented in two bytes in Figure 5.11, namely a low byte and a high byte. We can see that the low byte is occupied before the high byte. Thus, the output in Figure 5.10 has the value of t.i as 64, t.ch[0] as 0 and t.ch[1] as 64. Considering the value of t.i to be 256 in the program above, we will have the memory picture as in Figure 5.12. In this case, we have the output for t.i as 256, for t.ch[0] as 0 and ch[1] as 1 as shown in the Figure 5.13.

You may try out with other values for t.i and check the values of t.ch[0] and t.ch[1]. This obviously means that the union type allows its members to share the memory space. In unions the variables will have only one member at any time and hence any reference to the wrong type of data will produce meaningless results. Unions are used when all the members need not exist in the memory at the same time. All the union members are stored in the same memory space, with the same starting address, but not at the same time.

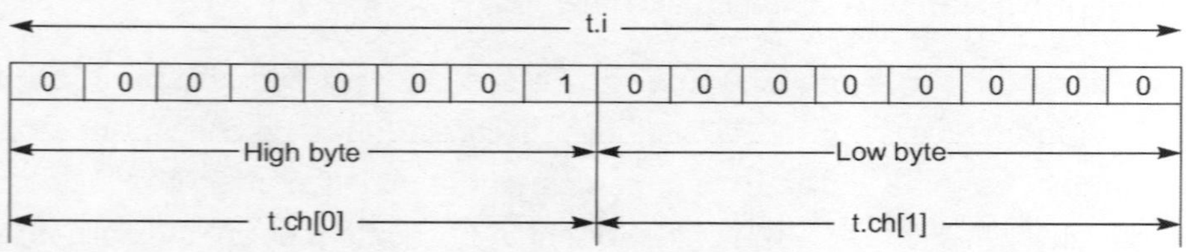

Figure 5.12 Memory representation of 256

Figure 5.13 Output of the program when the value of t.i is 256

SUMMARY

This chapter has dealt in detail with two of the user-defined data types namely structures and unions. Even though they are similar in many aspects, technically they are different and we saw that memory occupied by the union variables is comparatively lesser than that occupied by the structure variables. Both of them are synonymous in almost all ways. We had seen this important significant difference in detail and I am sure that the figures and the screenshots would have been of great help to you in learning this technical difference very clearly.

When we started off with the introduction to structures, we discussed the importance and advantage of structures over arrays. We had also discussed how the concept of structures was useful in extending the same to classes later in the programming language C++. The discussion of structures also dealt with a few of the object oriented concepts that are supported by it and its limitations have been explained in a lucid manner.

We then discussed the different forms of structure. We also learned how to initialize a structure variable and what are the technical details involved in it. We learnt the concept of using and accessing a structure within another structure. We then discussed different types of structures like mutually referential structures and self-referential structures. We also understood that the ideas of mutually referential structures are extended in C++ as friend classes. We then discussed the usage of bit fields in structures and how useful it is while programming. We concluded the discussion of structures with a program and explanation of how to declare an array of structures. We processed the employee records and had a few figures explaining how the array of structure is accessed. The learners of this book shall try out in getting the memory picture of an array of structure and you may use the Figure 5.1 in getting the same.

We concluded the discussion with a very brief note on the unions, and the idea of memory sharing by the members of the union was explained using proper pictures and surely it could have provided the readers with a very clear understanding of how advantageous a union is in most complicated programming applications, where the user intends to use a structure.

The readers have to keep in mind that this chapter has given an overall view of using structures and unions and I have tried to give most of the technical issues related to them. The concept of structure is further extended in the Chapters 6 and 7, where I had discussed pointers and functions. The structure is a very big ocean in C. In fact, C ends in structures, where C++ begins. When one starts learning the idea of class in C++, he/she must have the background knowledge of structure and this is where the migration begins. So, I would advise the readers to do more programs in structures (they may try out all those at the end of this book) and get practical exposure to them. Once they have a thorough knowledge on these, I am sure they would not have any difficulty while migrating from C to C++.

CRACK THE NUT

1. What is wrong in the following structure declarations?

```
struct ss
{
int i;
char s[10];
};
main()
{
________
}
```

Identify the output for the following programs:

2. struct s1 { float f2; };

```
struct s2 { char c; };
main()
{
struct s1 v1,v2;
struct s2 v3,v4;
clrscr();
printf("%d,%d,%d,%d",sizeof(v1),sizeof(v2),sizeof(v3),sizeof(v4));
getch();
}
```

3. struct { int a,b; } aa={3,4,5};

```
main()
    {
    printf("%d",sizeof(aa));
    }
```

4. struct aa

```
    {
        int i:2;
        int j:4;
        int k:2;
    };
    union
        {
        struct aa a;
        unsigned char c;
        }bb;
    main()
    {
    bb.c=199;
    clrscr();
    printf("%d,%d,%d",bb.a.i,bb.a.j,bb.a.k);
    getch();
    }
```

5. union c

```
    {
    int i;
    char c;
    }cc.c=' ';
    main()
```

```
    {
clrscr();
printf("%d",cc.i);
getch();
    }
```

6. main()

```
{
    struct student
    {
    char name[30];
    struct date dob;
}stud;
    struct date
      {
      int day,month,year;
      };
    scanf("%s%d%d%d", stud.rollno, &student.dob.day, &student.dob.month,
    &student.dob.year); }
```

7. main()

```
    {
    struct date;
    struct student
    {
    char name[30];
    struct date dob;
    }stud;
    struct date
    {
      int day,month,year;
    };
```

```
scanf("%s%d%d%d", stud.rollno, &student.dob.day, &student.dob.month,
&student.dob.year);
}
```

8. main()

```
{
struct employee
{
unsigned id: 8;
unsigned sex:1;
unsigned age:7;
};
struct employee emp1={203,1,23};
clrscr();
printf("%d\t%d\t%d",emp1.id,emp1.sex,emp1.age);
getch();
}
```

9. void main()

```
{
struct bitfield
{
signed int a:3;
unsigned int b:13;
unsigned int c:1;
};
struct bitfield bit1={2,14,1};
clrscr();
printf("%d",sizeof(bit1));
getch();
}
```

10. main()

```
{
struct bitfield
{
unsigned a:3;
char b;
unsigned c:5;
int d;
}bit;
clrscr();
printf("%d",sizeof(bit));
getch();
}
```

11. main()

```
{
struct world
{
int a;
char b;
struct india
{
char c;
float d;
}p;
};
struct world st ={1,'A','i',1.8};
clrscr();
printf(«%d\t%c\t%c\t%f»,st.a,st.b,st.p.c,st.p.d);
getch();
}
```

12. main()

```
{
struct india
{
char c;
float d;
};
struct world
{
int a[3];
char b;
struct india orissa;
};
struct world st ={{1,2,3},'P','q',1.4};
clrscr();
printf(«%d\t%c\t%c\t%f»,st.a[1],st.b,st.orissa.c,st.orissa.d);
getch();
}
```

6

POINTERS

6.1 INTRODUCTION

If you ask a professional C programmer as to which concept he/she found very difficult to understand when they learned C, the answer will be very simple and obvious. Yes, the answer would be **pointers**. The significance of pointers in C is the flexibility it offers in programming. It is C's clever use of pointers that has made it an excellent and an efficient language. As a professional programmer, you can play with the pointers long enough until you finally mess it.

The basic problem with the C programmers in understanding the pointers would be in not programming with them, but in understanding the concept and the terminology involved in it. So, the point lies in understanding the terminology very clearly.

Simply stated, a pointer is nothing but an address. Instead of being a variable, it is a pointer to a variable stored somewhere in the address space of the program. For the C programmers to understand a pointer more clearly, they must understand the organization of memory in computers. We discuss the organization of the memory in the following paragraph.

Memory is organized as a sequence of byte-sized (1 byte = 8 bits) locations. These bytes are numbered beginning with a zero. The number associated with a byte is known as its address or memory location. A pointer, therefore, is an entity that contains a memory address. Thus, a pointer is a number which specifies a location in memory.

Thus pointer variables are defined as variables that contain the memory addresses of data or executable code. Like any other variable, the pointer variable must also be defined, so that it will work only with data of a given type. Just as an integer variable can hold only integer values, character variable can hold only character values and so forth, a pointer variable can point only to a specific type of data.

6.1.1 Why Pointers?

Although a lot of programming can be done without using the pointers (as we have done in the earlier Chapters), their usage would enhance the capability of the program to manipulate data very easily and efficiently. In C, we have a concept called Dynamic Memory Allocation where the application and usage of pointers becomes indispensable. For example, consider that we are reading marks of a set of students and processing them. For the process, we have to declare an array as the one given below:

```
float marks[100];
```

But, the obvious limitation of this declaration is that the maximum number of students is limited to 100. If during run time, we need one more student's marks to be entered, it may not be possible, since the size of the array is fixed. On the other hand, using the dynamic memory allocation, we can design the program such that the limit for the maximum number of students is restricted only by the amount of available memory and the number could be decided by the user during the execution of the program.

By using the pointer effectively, variables can be created and defined when the program is being executed. Memory can be obtained from the system, if available, and allocated for the creation of a new variable or a data structure. This is called dynamic memory allocation. Dynamic variables so created can only be accessed by the use of pointers.

Pointers offer tremendous flexibility in the creation of dynamic variables, accessing and manipulating the contents of memory locations, and releasing the memory occupied by the dynamic variables which are no longer needed. Some of the uses of pointers in C are:

- Accessing the array elements
- Passing arguments to functions by reference
- Passing arrays and strings to functions
- Creating data structures such as linked lists, trees, graphs and so on
- Obtaining memory from the system dynamically

6.1.2 Pointer Operators—Storing Addresses using Pointers

In pointers and related discussions, we will be using two operators designated by & (ampersand) and * (asterisk). These two symbols are already familiar to you. But their meaning is different in pointers and it is not the same as that discussed in the earlier chapters. The symbol '&' was used to imply the bitwise logical AND operator and '*' was used as multiplication operator. Here in pointers, they have entirely a different name and operation.

The symbol '&' is called as **address of** operator. We have already used this operator in the syntax of the function scanf() and I had explained clearly about this operator's function in the Chapter 2, when we were discussing I/O instructions. In the function scanf(), it is provided to fetch the addresses of the arguments. That is why we call this operator as 'address of' operator. Consider the following piece of program. What do you think will be the output of the printf() statement in this?

```
int a=10;
printf("%d",&a);
```

So far, we have been using '&' only in the function scanf() and in the above piece of coding, we have used this in the function printf(). Will it flash any error? No. The above coding is perfectly valid and would give an output. It would print the address of the variable 'a'. As we have used '&a' as the argument to the function printf(), it is printing the address, since we knew that '&' will fetch the address of the corresponding variable that is used after it.

Consider the following declaration in a program.

```
int a=10;
```

What exactly happens when this line of program is executed by the C compiler? Three things happen at the compiler level. They are:

1. Allocate space in memory to hold an integer value
2. Associate the name 'a' with that location
3. Store a value 10 in that location

The following memory picture would imply the above three steps very clearly.

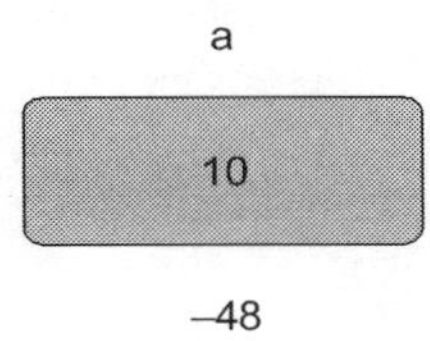

Figure 6.1 Memory map of the variable 'a'

We can see from Figure 6.1 that the variable 'a' is allotted with a memory space (in terms of bytes) and given an initial value (as given in the piece of coding). When no value is given for 'a', it would assume any garbage value within the range for an integer. We further see that the address of the variable 'a' is –48 and it is the one that is selected by the C compiler and not by the programmer or the user. When you declare 'a' and use it, the compiler may select a different location for 'a'. The point worth noting here is that the address of 'a' in memory is a number. As discussed in the earlier paragraph, this address can be printed using the function printf().

The symbol '*' is called as '**value at address**' and it will fetch the 'value' from the corresponding address as instructed. It is otherwise called as 'indirection' operator. Consider the following piece of coding.

```
/* Program for demonstrating the function of '*' */
#include<stdio.h>
main()
{
int a=10;
clrscr();
printf("\n\nValue of a = %d",a);
printf("\n\nAddress of a = %d",&a);
printf("\n\nValue of a = %d",*(&a));
getch();
}
```

When we execute the above program, we can get the output as shown in the Figure 6.2.

Figure 6.2 Output for the program—'&' and '*' in action

Thus, we see that printing the value of *(&a) is the same as printing the value of 'a'. We conclude that the operators '*' and '&' complement each other.

We had defined earlier a pointer as a variable that stores the address of another variable. So, let us see how they are declared and used in programs. Is it possible to store the address of a variable in another? Yes. We can do it by using the assignment operator (=). The following program just does that.

```
/* Program to demonstrate the function of '&' and '*' */
#include<stdio.h>
main()
{
int a=10,*b;
b=&a;
clrscr();
printf("\n\nValue of a = %d",a);
printf("\n\nValue of a = %d",*b);
printf("\n\nValue of a = %d",*(&a));
printf("\n\nValue of b = %d",b);
```

```
printf("\n\nAddress of a = %d",&a);
printf("\n\nAddress of a = %d",b);
printf("\n\nAddress of b = %d",&b);
getch();
}
```

Figure 6.3 shows the output of the above program.

Figure 6.3 '&' and '*' in action

The following memory picture in Figure 6.4 would provide more idea on how the above output is generated.

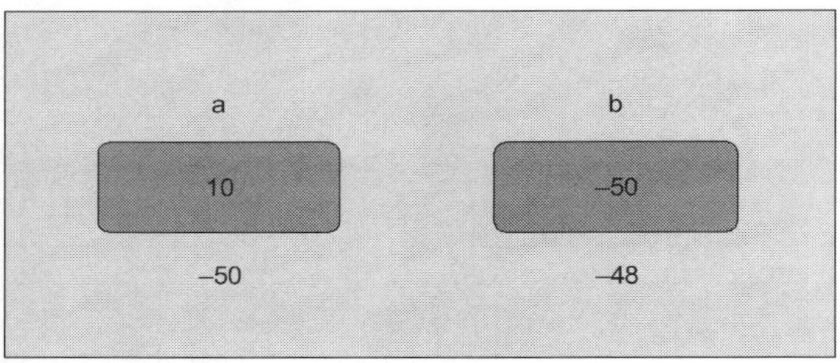

Figure 6.4 Memory map of 'a' and 'b' of the above program

We can understand from Figure 6.4 that a's value is 10 and b's value is a's address. According to the convention in C, no variable could be used without declaring it. If you closely observe the program, we have defined both 'a' and 'b'. But, 'a' is an ordinary variable and 'b' is a pointer variable and it is declared as **int *b**. This declaration tells the compiler that 'b' would be storing the address of an integer value and in other words, 'b' would point to an integer. Can you justify the usage of the operator '*' in the declaration of 'b'?

For understanding this, let us recall what actually is meant by '*'. It is 'value at address' operator. Thus the declaration **int *b** would mean, the value at the address contained in 'b' is an **integer**. Go through the program above and correlate the same with the Figures 6.3 and 6.4 and you will easily get to know the concept of pointers.

To put in a nutshell, we would say that pointers are variables which contain addresses and the addresses are always whole numbers. Hence pointers would always contain whole numbers. For example, if we have a declaration **float *b** in a program, it does not mean that 'b' is going to contain a float value and all that it means is 'b' is going to contain the address of a float variable or it is going to be a pointer for a float variable.

The idea of pointers could be extended to any level and let us see with an example how it is extended to the next level. Consider the following program.

```
/* Pointer to Pointer */
#include<stdio.h>
main()
{
int a=10,*b,**c;
b=&a;
c=&b;
clrscr();
printf("Value of a = %d",a);
printf("\n\nValue of a = %d",*b);
printf("\n\nValue of a = %d",*(&a));
printf("\n\nValue of a = %d",**c);
printf("\n\nValue of b = %d",b);
printf("\n\nValue of c=%d",c);
printf("\n\nAddress of a = %d",&a);
printf("\n\nAddress of a = %d",b);
printf("\n\nAddress of a = %d",*c);
printf("\n\nAddress of b = %d",&b);
```

```
printf("\n\nAddress of b = %d",c);
printf("\n\nAddress of c = %d",&c);
getch();
}
```

Figure 6.5 Output for the program illustrating pointer to pointer

In the program, we see that three variables are declared and used, namely 'a', 'b' and 'c', out of which 'a' is an ordinary variable and 'b' and 'c' are pointer variables. 'b' is a pointer to 'a' (from the statement b=&a), since it stores the address of 'a', and 'c' is a pointer to 'b' (from the statement c=&b), since it stores the address of 'b'. Since, 'c' points to not just an ordinary variable, but to a pointer, 'c' is called a pointer to pointer. We can go on extending this idea to any level and the most comprehensive version is a pointer to pointer. The output of the above program is shown in Figure 6.5. The memory map in the Figure 6.6 will help you better understand the way the variables and their addresses are stored in the memory.

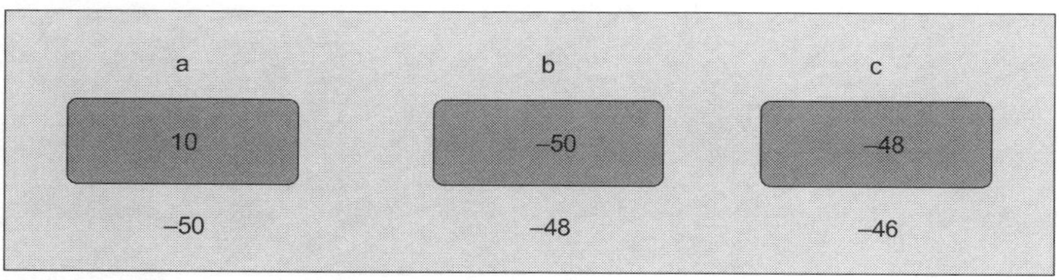

Figure 6.6 Memory Map showing the status of variables a, b and c

6.2 BASIC PROGRAMMING USING POINTERS

Now that we know how a pointer variable stores the addresses and how they could be accessed, we are ready to do some basic programs using the pointer notations and operators. Let us begin with a very simple example of adding two integers using pointers and display the result. The following is the coding.

```
/* Adding two integers and displaying the result */
#include<stdio.h>
main()
{
int a,b,c,*pa,*pb,*pc;
pa=&a;
pb=&b;
pc=&c;
clrscr();
printf("\n\tEnter the values of a and b : ");
scanf("%d%d",pa,pb);
*pc=*pa+*pb;
printf("\n\tThe sum of %d and %d is %d",*pa,*pb,*pc);
getch();
}
```

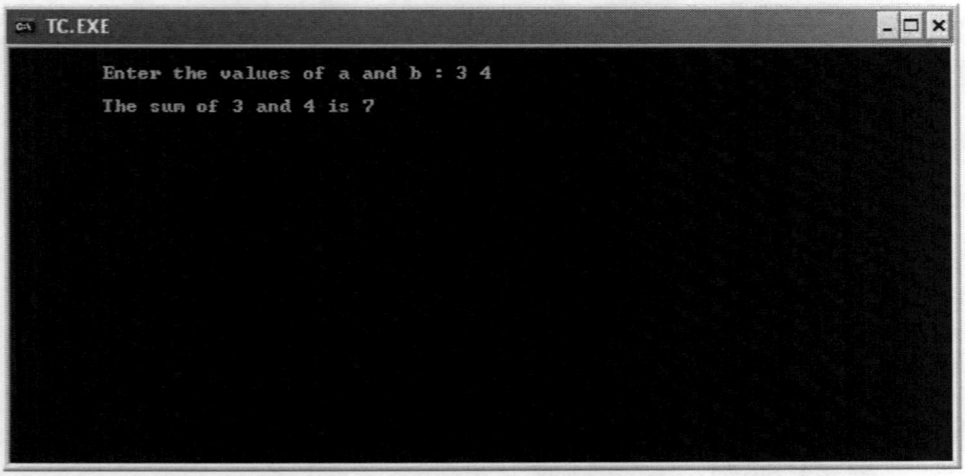

Figure 6.7 Output of adding two integers

The output of the above coding is shown in Figure 6.7. From the program, we can understand that a, b and c are the integer variables and *pa, *pb and *pc are pointers and they point to a, b and c respectively. It is a convention while using pointers that the letter 'p' is used as prefix and it denotes that the variable is a pointer and it is followed by the name of the variable to which the pointer is actually pointing. Thus, 'pa' points to 'a', 'pb' points to 'b' and 'pc' points to 'c'. Figure 6.8 shows the memory map of the variables.

To make yourself familiar with pointers you may try using pointers in some of the basic programs which I have given as self-learning exercises. This practice is essential since as we go further in learning pointers, we will be using these kind of jargon in a much advanced manner and as a programmer one shall not feel the difficulty in either writing programs or understanding them.

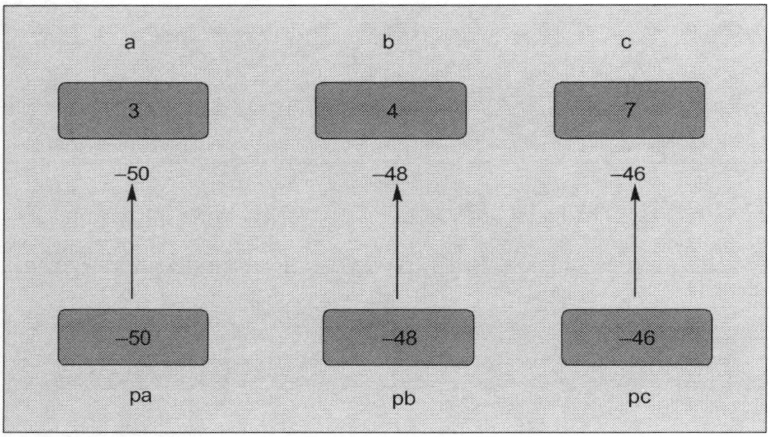

Figure 6.8 Memory picture for the program adding two integers

Let us do one more program here to make the concept clearer. Consider the program in which we find the temperature in Celsius for the given input temperature in Fahrenheit. The following would be the program coding for the same.

```
/* Program to convert temperature in Fahrenheit to Celsius */
#include<stdio.h>
main()
{
float cel,fa,*pcel=&cel,*pfa=&fa;
clrscr();
printf("\n\tEnter the temperature in Fahrenheit : ");
scanf("%f",pfa);
```

```
*pcel=(*pfa-32)*5/9;
printf("\n\tThe temperature in Celsius is %0.1f for %0.1f in Fahrenheit",*pcel, *pfa);
getch();
}
```

Figure 6.9 shows the output of the above program. We could see from the program that the output is formatted corresponding to one place after the decimal point. This is possible while we print the output using the corresponding format specifier (%f in this case) and this is possible with only float and its associated data types like double and long double. Also from the program, we notice that when the pointers are declared, they are assigned with the addresses of the respective variables. This is unlike the earlier program, where we declare the pointers in one step and assign the addresses in the subsequent step. This type of initializing the pointers when declaring them is allowed in C. This is very much similar to initialization of a variable during its declaration, which is possible while writing programs in C.

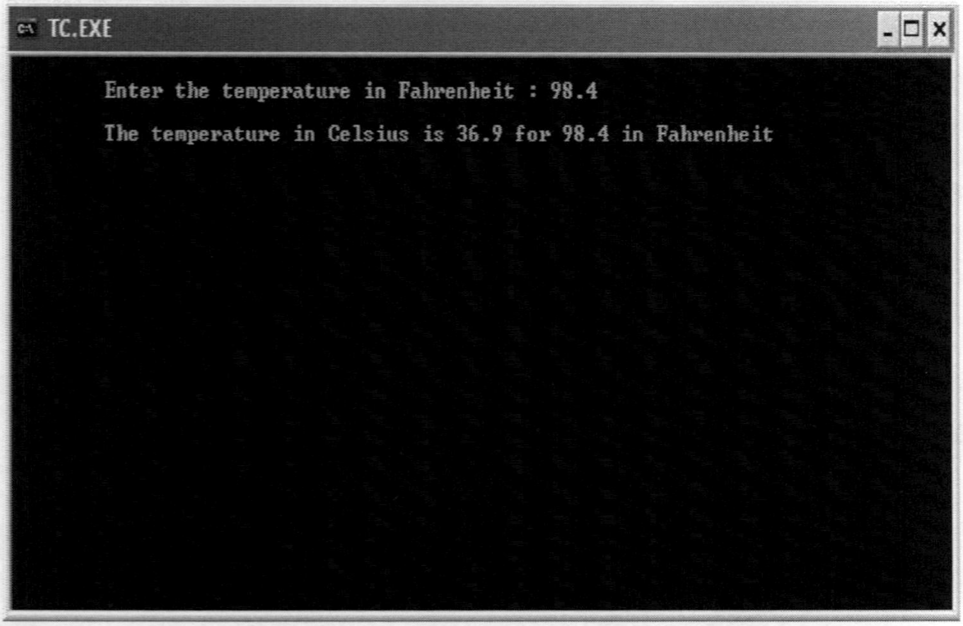

Figure 6.9 Output for the temperature conversion using pointers

Figure 6.10 shows the memory picture of the variables declared and their corresponding pointers in the above program.

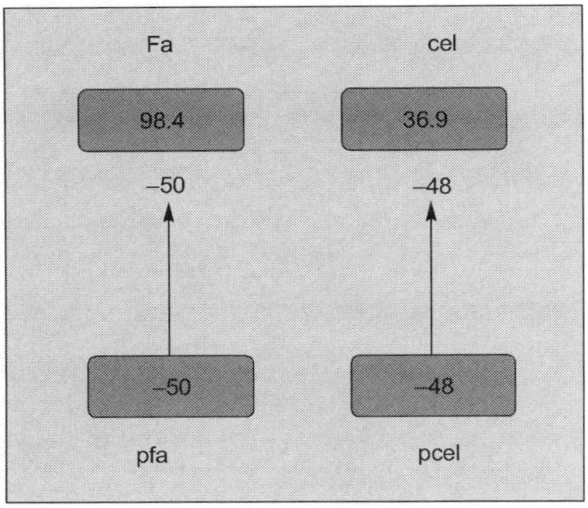

Figure 6.10 Memory picture of variables in temperature conversion program

6.3 POINTER ARITHMETIC

Pointer arithmetic is important for a C programmer to understand how a pointer works with an array, since pointers are most powerful when they are combined with arrays. Moreover, pointers and arrays are like couples and they are inseparable. For knowing about the arithmetic involved with the pointers, consider the following program.

```
/* Pointer Arithmetic */
#include<stdio.h>
main()
{
int a,*pa=&a;
float b,*pb=&b;
char c,*pc=&c;
double d,*pd=&d;
clrscr();
printf("\n\nBefore incrementation");
printf("\n*********************");
printf("\n\nAddress of a = %u",pa);
printf("\n\nAddress of b = %u",pb);
printf("\n\nAddress of c = %u",pc);
```

```
printf("\n\nAddress of d = %u",pd);
pa++;
pb++;
pc++;
pd++;
printf("\n\nAfter incrementation");
printf("\n*******************");
printf("\n\nAddress of a = %u",pa);
printf("\n\nAddress of b = %u",pb);
printf("\n\nAddress of c = %u",pc);
printf("\n\nAddress of d = %u",pd);
getch();
}
```

What do you guess about the output of the above program, assuming that the addresses of the variables a, b, c, and d are 65468, 65470, 65475 and 65478 respectively? See that we have used the format specifier %u for printing the addresses, since we would sometimes like to have the addresses in terms of positive integers.

We have used two similar sets of function printf() in the program. For the first set of the function printf(), we would get the output where the corresponding addresses of the variables would be printed. For the next set of the function printf(), we would get the output as 65470, 65474, 65476 and 65486 respectively for a, b, c and d. This is because of the arithmetic involved with the pointers and due to the four increment statements (pa++, pb++, pc++ and pd++) following the first set of function printf().

Since 'pa' is a pointer to an integer and an integer occupies two bytes in memory, when we increment 'pa', it would increment and point to the next integer that would be available two bytes after the present position, i.e., the variable 'a' which is pointed to by 'pa' is presently occupying the bytes 65468 and 65469 and when we increment the pointer by one, it would increment by two bytes and hence point to the next location of the integer, which is 65470. Similar justifications could be given for the other three variables b, c and d. Thus, when 'b' occupies the addresses 65470, 65471, 65472, 65473 and it is pointed to by 'pb' and when 'pb' is incremented by one, it would point to the next location, which is four bytes after the present location and hence 65474. Since a character variable occupies only one byte, when we increment it we get the result as 65476. A double takes 8 bytes and hence the value is incremented from 65478 to 65486 (variable 'd' occupies the bytes 65478, 65479, 65480, 65481, 65482, 65483. 65484, and 65485).

Figure 6.11 shows the output for the above program code. The main point that the programmers are expected to understand and assimilate from the above program and its output is that when we increment/decrement a pointer, it would increment/decrement according to the data type it is pointing to. Upon incrementing it by '1', when it is pointing to an integer, it would increment by 2 bytes, when pointing to a float, it would increment by 4 bytes and so on. This point is to be remembered from here, as we will be using pointers with the numerical and character arrays and there we would apply this concept.

Figure 6.11 Output of the pointer arithmetic program

Apart from this, a C programmer shall also know what operations are possible with pointers. The following list would be helpful to you when you use pointers in your programs. Keep in mind the points, since any misuse of the pointers will ultimately result in malfunctioning of the program and in the worst case may lead to the program or system crash. Figure 6.12 shows the memory picture of the above program.

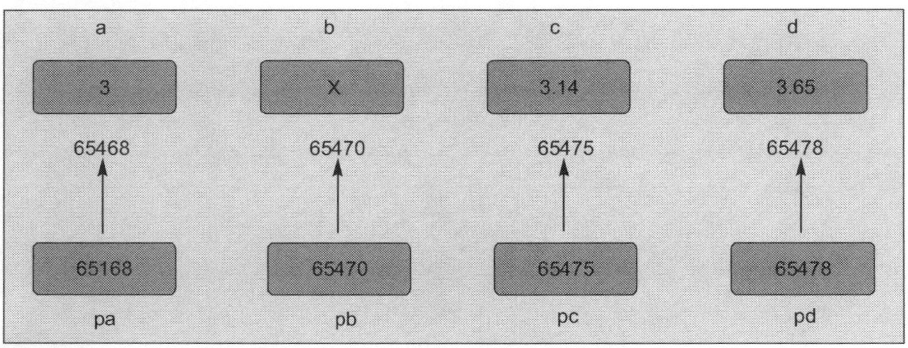

Figure 6.12 Memory picture of the pointer arithmetic program

Do's and Don'ts with the pointers

1. Do not add, subtract, multiply or divide two pointers (as it is equivalent to performing arithmetic operations between two memory addresses which will not yield anything fruitful). Consider pa, pb, and pc which are pointing to integers a, b, and c respectively. We are not allowed to carry out the following arithmetic operations with these pointers.

   pc=pa+pb, pc=pa-pb, pc=pa*pb and pc=pa/pb

2. You can add a constant with a pointer or subtract a constant from a pointer. But you cannot multiply a constant with a pointer or divide a pointer by a constant. Consider for example, 'n' is a constant. We can perform the following operations between the pointer 'pa' and the constant 'n'.

   pa=pa+n and pa=pa-n.

   We cannot perform the following operations between 'pa' and 'n'.

   pa=pa*n and pa=pa/n

6.4 POINTERS AND ARRAYS

For a C programmer to master the concept of pointers, he/she must be efficient in handling the pointers not just with ordinary variables, but with the arrays. This is because, arrays are the structured data types in C language and they implicitly use pointers to access the individual elements of the array. If you recall the discussion on arrays, we would have used what is called as a subscript. An array subscript is essential for a programmer to identify the individual elements of the array and normally alphabets like i, j, k are used for designating them. When we want to access the i'th element of an array 'a', then we say a[i] from our program coding and give various values for the variable 'i' by allowing it to go through the loop. But when we say a[i] from the program, it is implicitly converted to pointer notation by the C compiler and it never uses the notation that is used and understood by the programmer. Let us see how this is done. The expression would be converted to pointer notation as shown below.

```
a[i] = *(a+i)
```

When the value of 'i' is '0' the above expression would be like the one below.

```
a[0] = *(a+0) = *a
```

Thus we see that the expression is reduced to *a. When we refer the array name anywhere, in a program we will get the base address (starting address) of the array . This is a very important fact to be remembered by the programmer whenever he/she uses any type of array. Thus, in the above expression, we get the base address of the array 'a' and assume it to be the '1000th' byte of memory, then we would get the result of *(a+0) as *(1000+0) which is equal to *1000 and you know what the operator '*' does in the pointers. Yes, you are right. It will fetch the value at the address '1000'. Thus, on increasing the value of 'i', we would get the corresponding values at the subsequent addresses. If the base address that is quoted above is that of an integer array 'a', then upon giving a value of '1' for 'i' we will get the expression as *(1000+1), which is equal to 1002. The reason behind this will be easily understood from the earlier section, where we had been discussing the pointer arithmetic. Since the array 'a' is an integer array and its base address is '1000' and when we say a[1], it is converted to an equivalent expression *(a+1) and when we have the name of the array, we will get the base address of the array and hence the above expression gets further reduced to *(1000+1) and since an integer takes 2 bytes in memory, the result of the above expression is 1002 and not 1001.

We have mentioned earlier in this section that the expression a[i] would be converted internally by the compiler into *(a+i). Now the question is, can you express a[i] as i[a]? Yes. Obviously, you can. Let me justify how this could be done.

Consider the expression a[i], which could be written according to the compiler as *(a+i) and *(a+i) is the same as *(i+a), due to the commutative property of addition. If we can express a[i] as *(a+i), then why not *(i+a) could be expressed as i[a]? As we had discussed earlier, the compiler will not bother about whether it is a[i] or i[a], since it converts either of these expressions to their pointer equivalents. I would advise you that, you check the programs that you have done in the Chapter 4 by rewriting the expression a[i] as i[a] to understand this concept more clearly. It is just a simple mathematics, which is summarized below.

```
a[i]=*(a+i)=*(i+a)=i[a].
```

I would suggest you go through the earlier sections till this point once again so that you make yourself to get prepared for learning pointers and arrays.

6.4.1 Pointers and Numerical Arrays

Pointers, as mentioned earlier, work more efficiently with arrays and invariably pointers and arrays are interlinked. In the previous section, we discussed how the C compiler converts the notation a[i] to the pointer notation internally and whether the user likes or not, the conversion happens internally and the C compiler does not bother about the array subscript that is being used by the programmer. Let us see how pointers work with numerical array with the help of a few programs. We will consider the integer array for our discussion, which will hold good for any type of numerical array. Consider the following program code, where we read and display 'n' integers using pointer to the integer array declared in the program.

```
/* Read and display an array of 'n' integers using pointers */
#include<stdio.h>
main()
{
/* Array and a pointer and 'n' are declared */
int a[100],*pa,n;
clrscr();
/* Getting the value of 'n' */
printf("\n\nHow many elements you want to enter : ");
scanf("%d",&n);
/* Validating the value of 'n' */
if(n<0 || n>100)
{
printf("\n\nInvalid value entered for n");
getch();
return;
}
/* Getting the values of the individual elements of the array */
printf("\n\nEnter the elements one after another : ");
for(pa=a;pa<(a+n);pa++)
scanf("%d",pa);
/* Displaying the values read one after another */
printf("\n\nThe values entered are :  ");
for(pa=a;pa<(a+n);pa++)
printf("%d\t",*pa);
getch();
}
```

We see that the program is more or less the same as we had written earlier without using pointers. The variable declaration part has the array declared and a variable 'pa', which is used as a pointer to the array. We also have a variable 'n' which is used to get and store the number of elements in the array. Once we get the value for 'n', we check whether it exceeds the limit of the array or takes an invalid value. If so, we flash an error message and return. This has to be done explicitly in any program using any kind of array since, in C there is no check to see if the array subscript exceeds the array size or not. If the 'n' value is acceptable, we proceed getting the elements of the array. This is done using *for* loop.

We can see that in the loop, we initialize the pointer with the base (starting) address of the array. This is achieved with the statement pa=a. We know that when we refer the name of the array at any point of the program, we would get its base address. This fact is used in writing such an expression in the loop. Since the variable 'pa' is a pointer, it stores the base address of the array 'a'. Following the initialization, we check the condition using the expression pa<(a+n). This expression, again, appears quite strange. When we use the array name, we would get the base address and we see that we add the value of 'n' with it in the above condition. We are allowed to do like this because we have learnt earlier in pointer arithmetic that a constant value could be added to a pointer. Thus the right hand side of the above expression becomes a constant and the value of 'pa' changes in each cycle of the loop. Assuming the base address of the array to be 1000 and the value of 'n' to be '5', the initialized value of 'pa' would be 1000 and the RHS of the expression would be 1000+5=1010. This is because, when we add five integer values with the base address of the array, it is equivalent to adding 10 bytes to it, since an integer occupies two bytes in memory. On incrementing the value of 'pa' in steps of '1', it would point to the next memory location that an integer would occupy. Thus upon incrementing 'pa' for the first time, it would take the value 1002, since that will be the location of the next integer in the array 'a'.

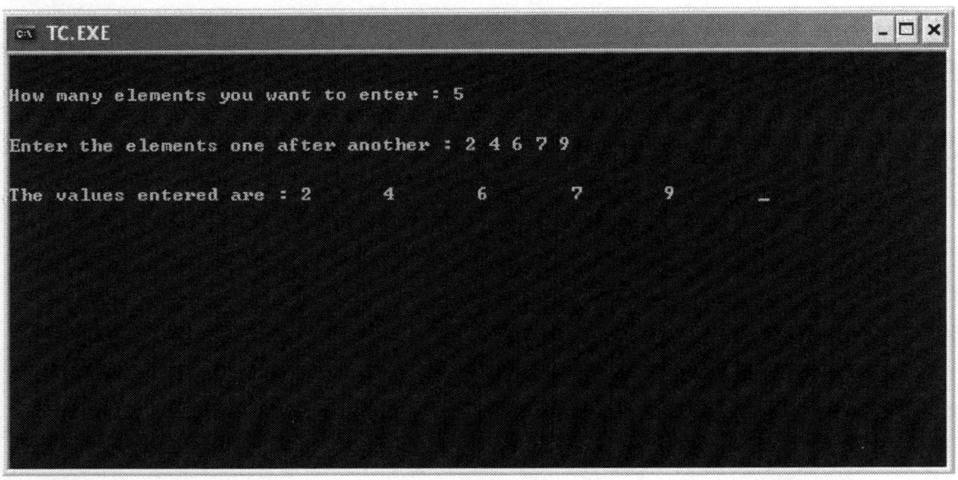

Figure 6.13 Output of the program that reads and displays 'n' integers

As we get the n'th value in the array, the value of 'pa' would increment from 1008 to 1010, thereby making the expression to return a false and hence that will cause the loop to terminate. Once we get all the required inputs, we would like to display them and hence use another *for* loop to achieve this. Since at the end of the reading process, the pointer would be at 1010, we have to bring it back to the initial position or the base address of the array and hence we reinitialize it by writing pa=a. Then we proceed with the process of printing the array elements one by one, in the order they have been read. This is very much done like the earlier loop. Except for the function scanf(), we are using the function printf() in that part of the loop. Also, see that we use the escape sequence '\t' in order to leave one tab spacing in between the elements when they get printed. This will increase the clarity of the output of the program. Figure 6.13 is a sample output of the above program.

Figure 6.14 shows an output of the same program when we give a value for 'n' which is not acceptable and hence an error flashed. Figure 6.15 shows the memory picture of the array 'a' and the pointer 'pa' used in the above program. The memory picture shall make the things clearer and better.

Figure 6.14 Output of the program when 'n' value is invalid

How will you modify the above program in case you are reversing the array? You would find the task easier, if you have clearly understood the way the pointer is accessing the array elements using their memory addresses. The only place where you would be modifying the above program would be the section where you display the array elements. Instead of displaying them in the same manner as you have read, you would reverse-display them. This is achieved by the following loop (that you would have replaced in the appropriate position of the above program).

```
printf("\n\nThe values in reverse order are : ");
for(pa=pa-1;pa>=a;pa--)
printf("%d\t",*pa);
```

We see that the initial value of 'pa' is 'pa-1'. This is because, once we have read the array, 'pa' would be at 1010 and we have to bring it one position less than that it is actually pointing to. We know that we can subtract a constant from the pointer and hence this kind of initialization is allowed. We see the condition is written as pa>=a and the modification part is written as pa– –. As we decrement the value of 'pa' in steps of '1', it keeps pointing to the preceding locations and at one point of time would reach the base address. That is the reason why we have used the expression pa>=a, since we know very well that mentioning the name of the array would give its base address. Figure 6.16 would help you in understanding things in a better way.

In Figure 6.16, we have given the memory map for the above program that is drawn in the interest of the learners of this book and you try to figure out the ideas from the memory map, on your own.

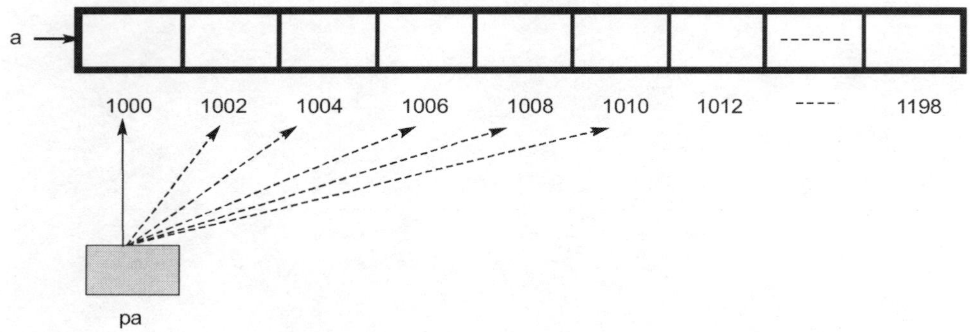

Figure 6.15 Memory picture—program to read 'n' integers

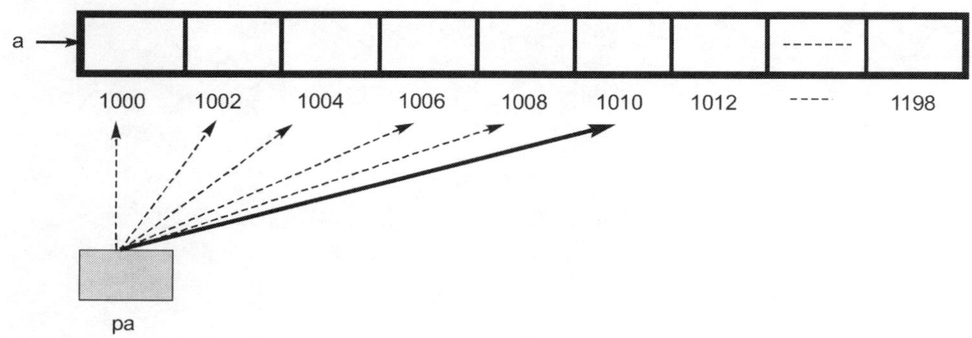

Figure 6.16 Memory picture—program to read and reverse 'n' integers

I am sure that you would have understood how the pointers work with the arrays. I would suggest that you work with all the programs that are given in the self-learning exercise in the Chapter 4, which will make you better understand the pointers and you will ultimately master them.

6.4.2 Pointers and Character Arrays

Pointers work with character arrays in the same way as with a numerical array. Since we knew the way the characters are stored in an array and how the character array is terminated, we will straight away take a program to illustrate the way the pointer is used with the character array. Consider the program where we read and reverse a string or a text. The following is the program code for the same.

```
/* Program to read and reverse a string using pointer */
#include<stdio.h>
main()
{
char a[100], *pa;
clrscr();
printf("\n\nEnter the string : ");
for(pa=a;(*pa=getchar())!= '$';pa++);
*pa='\0';
printf("\n\nThe string you have entered is : %s",a);
printf("\n\nThe reversed string is : ");
for(pa=pa-1;pa>=a;pa--)
putchar(*pa);
getch();
}
```

We can see that the pointer variable 'pa' is used in exactly the same way as it was used earlier with the numerical array. Figure 6.17 shows the output of the above program and Figure 6.18 shows the memory picture of the array and pointer variables that are used in the program. The memory picture will give you a clearer explanation of the program and hence the way the pointer works with the character array.

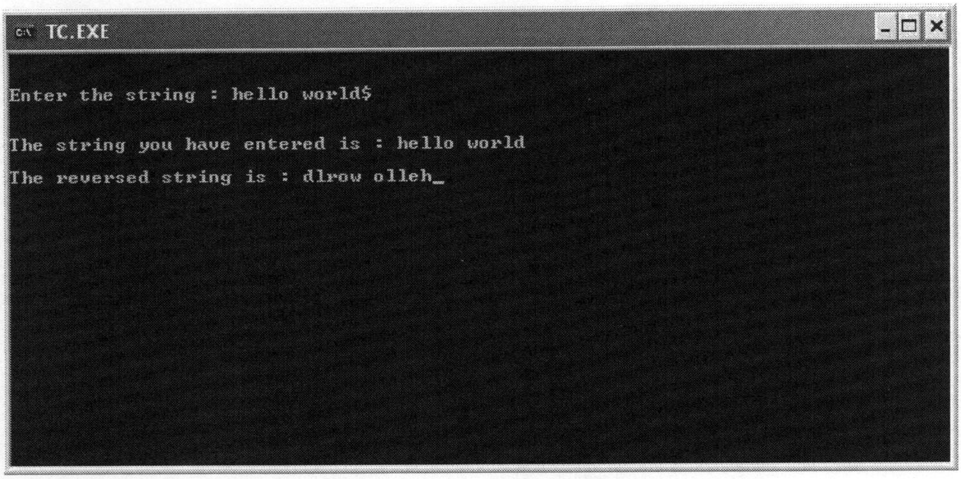

Figure 6.17 Output of the program—reversing a string

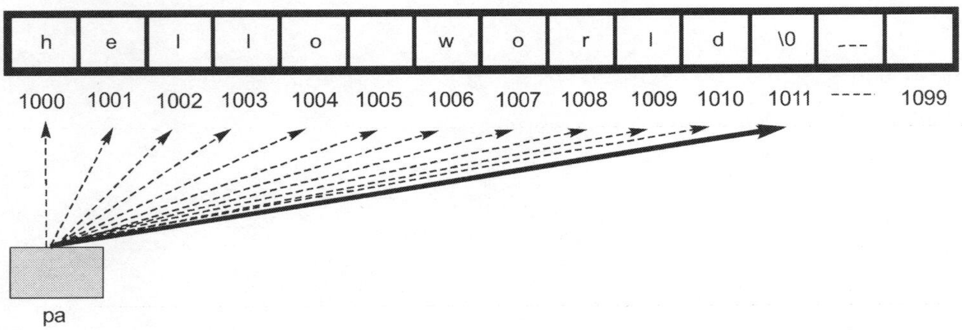

Figure 6.18 Memory picture—reversing a string

We have assumed that the array 'a' has the base address 1000 and hence the end address would be 1099 (since the array is of size 100). We get the input text for the array using *for* loop and once we finish getting all the input characters, we terminate the array using a NULL character, which is essential for any character array. Since in the above program we reverse the string that is read, we reinitialize the pointer with the value pa-1 in the second *for* loop. This is because, once we read the array elements one after another, the pointer would be having the address 1011, when we store the NULL character. When we want to reverse the array, we must store the address 1010 in the pointer and hence we deduct '1' from the present value of 'pa' and hence we use the expression pa-1. I am sure that this explanation would help you in understanding the way the pointers work with character array. See also that we have used the function putchar() (works in exactly the opposite manner to the function getchar()), which will display one character at a time and the advantage of using this is that we need not use a format specifier for the character type values we are displaying (unlike the function printf() which essentially needs format specifier for any type of values used in it).

Let us do another program, in which we will convert the lowercase alphabets of the given array to their equivalent uppercase and display. The following is the program code for the same.

```
/* Program to convert the lowercase alphabets to uppercase using pointer */
#include<stdio.h>
main()
{
char a[100], *pa;
clrscr();
printf("\n\nEnter the string :  ");
for(pa=a;(*pa=getchar())!='$';pa++);
*pa='\0';
printf("\n\nThe string you have entered is : %s",a);
printf("\n\nThe uppercase string is :  ");
/* Logic for Conversion */
for(pa=a;*pa;pa++)
if(*pa>=97 && *pa<=122)
putchar(*pa-32);
else
putchar(*pa);
getch();
}
```

Please observe the segment of the program where we convert the lowercase string to uppercase. There we have used a condition, if(*pa>=97 && *pa<=122), which means the input when given as mixed (both uppercase and lowercase), only that part of the input which has lowercase character(s) has to be converted with this logic, otherwise the character has to be displayed as such. Figure 6.19 shows the output of the program.

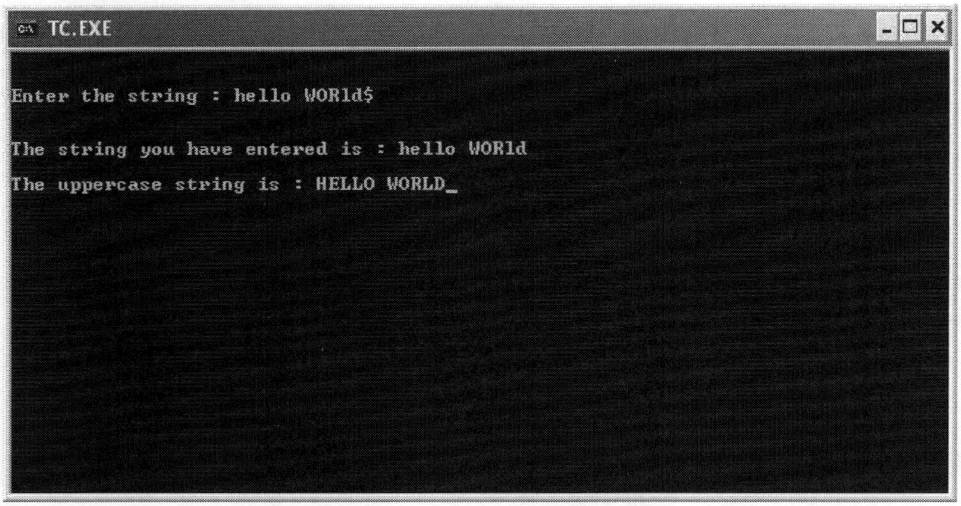

Figure 6.19 Output of the program—lowercase to uppercase

You may also observe the logic used for the conversion process, wherein we subtract a value 32 from the existing alphabet. This is because, the standard difference between the ASCII values of uppercase and lowercase alphabets is 32. Thus, subtracting 32 from an alphabet will yield its corresponding uppercase equivalent. Similarly when we want to convert an uppercase alphabet to the equivalent lowercase, we have to add 32 to the existing alphabet.

I hope the above programs would clearly imply to you, how efficiently we can work with the arrays clubbed with pointers. Try to rewrite all the programs corresponding to the character array that are given as exercise to you at the end of this book using pointers and it will surely make you to use pointers more with arrays more efficiently.

6.5 POINTERS AND STRUCTURES

We have dealt with the concept of structures already in Chapter 5. This section will let you know how to declare a pointer to a structure and access the same. Let us declare a structure called book where we store the details of a book such as book name, author name, publisher's name, year of publication, number of pages and price. Of these attributes, the first three are character arrays, fourth and fifth attributes are integers and the last one is a float type. Since a structure allows different types of data to be clubbed together, the above declaration of the book structure is acceptable. The following is the complete program which reads details of a book and displays the same using pointer to the declared structure.

```
/* Declaring and using a pointer to a structure */
#include<stdio.h>
/* Inclusion of Floating Point Emulator */
void LinkFloat(void)
    {
    float a=0,*b=&a;
    a=*b;
    }
/* Declaration of the book structure */
struct BOOK
    {
    char bname[100],aname[100],pname[100];
    int nop,yop;
    float price;
    };
main()
{
int i;
/* Declaration of variable and pointers of type of BOOK */
struct BOOK A,*PA;
/* Assigning the addresses of A to PA */
PA=&A;
clrscr();
/* Getting the details of the BOOK A using the pointer PA */
printf("\nYou are now ready to enter the details of A");
getch();
printf("\n\nEnter the name of the book : ");
for(i=0;(PA->bname[i]=getchar())!='$';i++);
PA->bname[i]='\0';
fflush(stdin);
printf("\n\nEnter the name of the author : ");
for(i=0;(PA->aname[i]=getchar())!='$';i++);
```

```
PA->aname[i]='\0';
fflush(stdin);
printf("\n\nEnter the name of the publisher : ");
for(i=0;(PA->pname[i]=getchar())!='$';i++);
PA->pname[i]='\0';
fflush(stdin);
printf("\n\nEnter the number of pages in the book : ");
scanf("%d",&PA->nop);
printf("\n\nEnter the year of publication : ");
scanf("%d",&PA->yop);
printf("\n\nEnter the price of the book : ");
scanf("%f",&PA->price);
/* Displaying the details of the BOOK A */
printf("\n\nBook Name : %s", PA->bname);
printf("\n\nAuthor Name : %s", PA->aname);
printf("\n\nPublisher Name : %s", PA->pname);
printf("\n\nNumber of Pages : %d",PA->nop);
printf("\n\nYear of Publication : %d",PA->yop);
printf("\n\nPrice : %0.2f",PA->price);
getch();
}
```

We see from the program that many new things have been included while the coding is written. The first and most important of them is the inclusion of a segment called **Floating Point Emulator**. This is a simple fragment of coding, which will enable us to use floating point data type in the structure flexibly. When we do not use this, we may get an error saying **floating point formats not linked**, especially when we use the floating point data types in a structure. The inclusion of the above coding would enable the compiler to link the floating point formats and values properly.

The next important aspect of the program is the usage of a pointer to the structure declared, unlike the normal usage of the structure variable to access its members. Once we declare a pointer to the structure and assign the base address of the structure to it, further accession of the members of the structure could be easily done using it.

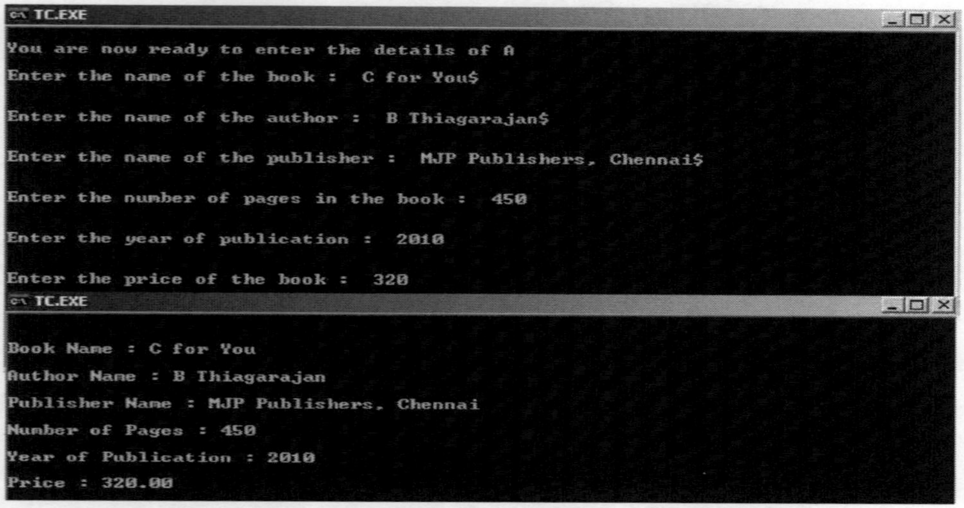

Figure 6.20 Using pointer to a structure

Closely observe the program. You will get the next significant aspect. We have used a new operator (–>) called as arrow operator. This is again an operator in C and could be used for the unions too. Figure 6.20 shows the output of the above program. The memory picture of the structure and the members of the structure are shown in Figure 6.21.

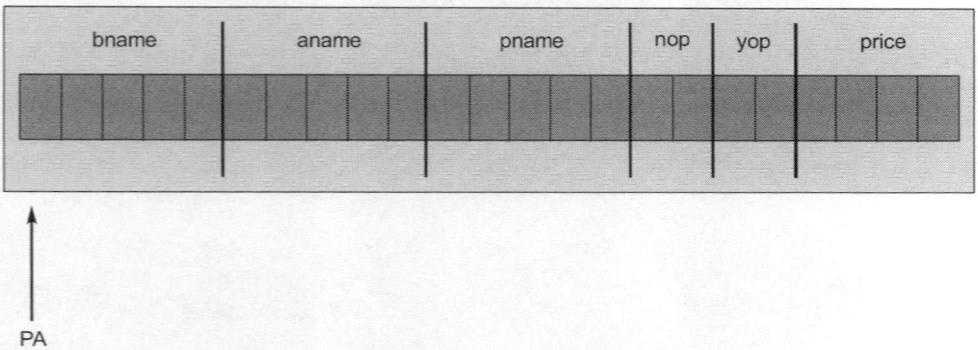

Figure 6.21 Memory picture of the structure BOOK

See from the memory picture that the base address of the structure BOOK is the same as that of the first member of the structure. In our case the first member of the structure is the character array **bname** and the base address of **bname** would be the base address of the entire structure.

From the discussion and the memory picture with a sample program, the concept of using pointers with structures would have been clear to you. You may try pointers with array of structures and also unions, which would be very much easier, since the concept of array of structures and unions are already known to you. Read this section of the chapter, which will facilitate you in understanding the concept of linked lists.

SUMMARY

We started off this chapter with an introduction to pointers. Why a programming language shall use pointers was discussed and it was also highlighted that C language efficiently uses pointers. We then followed with the explanation of the operators used in pointers and how these operators are used in storing and retrieving the values and data.

We then did a few basic programs using pointers and used the memory pictures and screenshots of the outputs to understand the way the pointers function in C. We then learnt a very important aspect in pointers, namely, the Pointer Arithmetic and in that, we discussed how a pointer would increment or decrement and how this concept is useful in arrays, where pointers are used much.

The discussion of pointer arithmetic followed that of the usage of pointers with numerical and character arrays. We did a few programs on these two topics and as usual we had the screenshots of the output and memory pictures for more clarity in understanding the concept.

We concluded the chapter with a very important discussion on pointers and structures, which was the ultimate concept that would be used in implementing the linked lists.

CRACK THE NUT

Identify the output for the following programs?

1. main()

```
{
int a, *b, **c, ***d, ****e;
a=10;
b=&a;
```

```
c=&b;
d=&c;
e=&d;
printf("a=%d,b=%u,c=%u,d=%u,e=%u",a,b,c,d,e);
printf("%d,%d,%d",a, a+ *b, **c + ***d + ****e);
}
```

2. main()

```
{
char c, *pc;
int a;
long b;
float d;
c='X';
a=20;
b=67888;
d=3.145;
pc=&c;
printf("\n\nc=%c, pc=%u",*pc,pc);
pc=&a;
printf(("\n\na=%d,pc=%u",*pc,pc);
cc=&b;
printf(("\n\na=%ld,pc=%u",*pc,pc);
cc=&d;
printf(("\n\na=%f,pc=%u",*pc,pc); }
```

3. main()

```
{
char *c;
int *i;
long *l;
```

```
    float *f;
    double *d;
    printf("\n\nc=%u,c+1=%u",c,c+1);
    printf("\n\ni=%u,i+1=%u",i,i+1);
    printf("\n\nl=%u,l+1=%u",l,l+1);
    printf("\n\nf=%u,f+1=%u",f,f+1);
    printf("\n\nd=%u,d+1=%u",d,d+1);
    }
```

4. main()

```
    {
    int i=10,j=20,diff;
    diff=&j-&i;
    printf("\n\nAddress of i=%u\tAddress of j=%u",&i,&j);
    printf("\n\nDifference between addresses of i and j is : %d",diff);
    }
```

5. main()

```
    {
    int *i,*j;
    j=i*2;
    printf("\n\nj=%u",j);
    }
```

6. main()

```
    {
    int i=10;
    printf("\n\nValue of i=%d\tAddress of i=%u",i,&i);
    &i=7200;
    printf("\n\nNew value of i=%d\tNew address of i=%u",i,&i);
    }
```

7. main()

```
{
int a[]={10,20,30,40,50};
int j;
for(j=0;j<5;j++)
{
printf("%d",*a);
a++;
}
}
```

8. main()

```
{
float a[]={13.24,1.5,1.6,5.4,3.5};
float *j,*k;
j=a;
k=a+4;
j*=2;
k/=2;
printf("%f\t%f",*j,*k);
}
```

9. main()

```
{
int a[25];
a[0]=100;
a[24]=200;
printf("\n%d\t%d",*a,*(a+24),*(a+0));
}
```

10. main()
```
{
int a[]={10,20,30,40,50};
int i,*k;
k=&a[4]-4;
for(i=0;i<=4;i++)
{
printf("%d\t",*k);
k++;
}
}
```
11. main()
```
{
int a[]={2,4,6,8,10};
int i;
for(i=0;i<=4;i++)
{
*(a+i)=a[i]+i[a];
printf("%d\t",*(i+a));
}
}
```
12. main()
```
{
int a[]={2,4,6,8,10};
int *ptr;
for(ptr=&a[0];ptr<=&a[4];ptr++)
printf("%d\t",*ptr);
}
```

13. main()

```
{
int a[]={2,4,6,8,10};
int *ptr,i;
for(ptr=a+4,i=0;i<=4;i++)
printf("%d\t",ptr[-i]);
}
```

14. main()

```
{
int a[]={2,4,6,8,10};
int *pa[]={a,a+2,a+1,a+4,a+3};
int **ptr;
ptr=pa;
**++ptr;
printf("\n\n%d\t%d\t%d",**ptr,ptr-pa,*ptr-a);
}
```

15. main()

```
{
char s[]="C Language";
int i;
for(i=0;s[i];i++)
printf("\n%c\t%c\t%c\t%c",s[i],*(s+i),i[s],*(i+s));
}
```

16. main()

```
{
char s[]="C Language";
char *ps;
ps=&s[6]-6;
while(*ps)
```

```
printf("%c",*ps++);
}
```

17. main()

```
{
char s[]="Malayalam";
char *ps;
ps=s+8;
while(ps>=s)
{
printf("%c",*ps);
ps--;
}
}
```

18. main()

```
{
char a[]="Able was i ere i saw elba";
char *pa,*pb,*pc;
clrscr();
pb=a;
pc=a+strlen(a)-1;
pa=pc;
while(pb!=pc)
{
printf("%c",*pb);
pb++;
printf("%c",*pa);
pa--;
}
}
```

19. main()

```
{
int a[12];
printf("%d",sizeof(a));
}
```

20. main()

```
{
char a[]="C Language";
int i;
printf("%c\n",*a);
a++;
printf("%c",*a);
}
```

7

FUNCTIONS

7.1 WHY WE NEED FUNCTIONS?

The efficiency of any language is measured from the way it does the task by splitting it into different modules. **Divide and Rule** is the fundamental fact behind any successful programming language. C has the power of dividing the given task into different modules and accomplishing the same. The gravity of the programmers using the C language could be known from two of its most striking features. The first one is of course, the pointers and the second one is the functions.

In C, a function is a self-contained program segment that carries out some specific, well-defined task. We have invariably used many functions so far such as printf(), scanf(), getchar(), clrscr(), getch(), etc. Above all we have been using the function main() in all our programs and as said earlier, it is the essential one for any program to compile and execute; in simple words, it is the heart throb of programs written in C.

C language is a compiled language and it contains many functions—all of them are built-in, which means they come along with the language and the language designers have given to the users and the programmers of the language. They are categorized under different heads according to the context in which they are used. We have discussed earlier about header files and their usage in programming. Many of the functions are categorized as header files and there are as many as 30 header files in the language. All these headers were made available for the users when the language was compiled. But as the language was used across the programming community and started growing, these built-in functions were not sufficient and hence the language was modified in such a way that one can define his/her own function, what we call as user-defined function. This aspect has made C language most popular among the programming group.

Functions are thus considered an essential part of the programs. The user is at his/her liberty to create as many functions as he/she wants in a program. One more beauty is that a function that is created is not just restricted to the program where it is declared and defined, but could be used across different programs, just like the headers.

This aspect gave C one more face, namely, the idea of **code reusability**. The code was defined just once and used any number of times at any point of the particular program or across different programs. This feature was inherited later by C++. Thus C++ became popular not just because of its object oriented nature but also because of the procedure oriented features inherited from C. This idea of code reusability gave much power to the language and the users had a feeling that they were given freedom in using the language syntax.

7.2 TYPES OF FUNCTIONS

C provides us with two types of functions, as mentioned in the earlier section, namely, built-in/ library functions and user-defined functions. The header files contain the details of the library functions. We, in this chapter, are going to discuss the way the user-defined functions are declared, defined and used. As said earlier, the key advantage of user-defined functions is the code reusability.

7.3 FUNCTION HEADER AND FUNCTION PROTOTYPING

Any user-defined function must be properly declared and defined, before used, just like the variables. We call the declaration of functions as function prototyping, which is the essential part of any C program (even though some versions of C compilers do not expect a function to be prototyped). Prototyping is not an issue in the case of built-in functions, since their declarations are available in the standard header files and we would anyway include the corresponding headers, using the #include macro once we want to use a particular built-in function.

The C function has two parts, namely, the function header and body of the function. The function header typically has the following style.

<return type> <function name>([<type-1> <argument[s]>...<type-n> <argument[s]>])

The above style of using the function is the classical notation and initially when Dennis Ritchie and Brian Kernighan defined functions, the header was quite different and the following was their style and called as K&R notation and most of the modern compilers still accept this style too.

<return type> <function name>([argument[s]])

[<type-1> <argument[s]>...;

<type-n> <argument[s]>];

From both the styles, we can conclude that the usage of arguments inside the functions is purely optional and if they are used, we have to clearly specify their types. Also, we would be following the classical notation of declaring the function headers in our discussion.

A function basically accepts zero, one or more inputs or values, processes them and returns (or does not return anything at all) the result. Based on this aspect we can say a function:

- Accepts one or more inputs, but does not return anything
- Accepts one or more inputs, processes them and returns one or more results (remember an ordinary function cannot return more than one result, where a function that uses a pointer could do so)
- Does not accept anything, but may return some result
- Neither accepts nor returns anything

When we want a user-defined function to work, we would call them from some other function of the same program or a different program. In either case, the user-defined function is said to be **called** by the other function. Thus we can introduce two terms here, namely **caller** and **callee** (similar to an employer and an employee). A function which calls another function is called the caller and the function that is being called is the callee. Remember the fact that a function can be called any number of times, depending upon its requirement. Also remember that, in C language we can just call one function from another, but we cannot define the logic of one function in another (this is why C is technically not a block structured language). One can make use of any function in C language only by calling it. The compiler will recognize the call and compile the programs to success, provided the corresponding header is included in the case of built-in functions and the function is properly defined in the case of user-defined functions.

Now, I will put a question to you, for which the answer will be simple and I would give the answer in the later part of this section (Better do not read the later part of this section immediately after getting the question. I would advise you to give a serious thinking before you look for the answer). I had just told you that any function could be used in C only by calling it. Thus we make use of the functions printf(), scanf() and all such functions by calling them from within the function main(). Thus the function main() is the **caller** for these functions and they are the **callee** functions. But who is the caller for the function main()? Think seriously the answer for this question.

The reason why I had raised this question is that, I had told you any function in C could be used only by calling it and the function main() is no exception for this rule.

The answer for the above question is, the **Operating System**. Yes, the operating system under which the C compiler works is the boss for the function main() and hence it will be the caller and the function main() is the callee. If we use the C complier under the DOS platform, then DOS is the caller for the function main(). The caller would change according to the platform that we use for programming with this language. I had used DOS as the platform for all the programs in the book and hence the caller for all of them is invariably the DOS.

The functions in C could be categorized also according to the way they are called. You can call them with the values as the arguments, you call this as **call by value** and you can call them with the addresses as the arguments, where you call this aspect as **call by reference**. The latter method is more efficient and advantageous in many ways when compared with the former. The significant advantage is that when we use call by reference, we can implicitly return more than one value from the function to the calling function, whereas, when we use call by value method, we can return either nothing from the called function or a maximum of one value from it. There are many other technical advantages of using call by reference method, which you would understand as we do programming later in this chapter.

I am sure you are comfortable with the function header and prototyping concepts discussed so far. This would be the base for you to program using functions, from basic ones to the complicated ones.

7.4 WRITING SIMPLE FUNCTIONS

Let us write simple functions in C and validate them. To begin with, let us write a function that adds two integers and returns the result to the calling function. The coding for the same is as follows. We validate the function using a C program.

```
/* Program to add two integers using functions */
#include<stdio.h>
main()
{
/* Declaration of variables */
int a,b,c;
/* Function prototyping */
int add(int,int);
```

```
clrscr();
/* Getting the values of a and b */
printf("\n\nEnter the values of a and b : ");
scanf("%d%d",&a,&b);
/* Calling the function add() */
c=add(a,b);
/* Printing the result of addition */
printf("\n\nThe sun of %d and %d is %d",a,b,c);
getch();
}
/* Definition of the function add() */
int add(int x, int y)
{
int z;
/* Adding the values that are passed from the calling function */
z=x+y;
/* Returning the result of addition to the calling function */
return z;
}
```

In the above program, we have used a very simple user-defined function, by name add() and it is used for adding two integers and returns the result. Here the calling function is main() and the called function is add(). In other words, the function main() is the caller and the function add() is the callee. We see that the function add() is properly prototyped, just like the variables used in the program and we see that the prototyping follows the declaration of variables. The prototyping could also be done outside the function main() and in doing so, we would use it globally, again, just like a variable. The prototyping has the details like the return type, function name and the type of values the function accepts and this is obviously essential for any function that is prototyped. When a function does not accept or return anything, we use the type **void** to imply that. A typical function could be prototyped like the one below.

```
void add(int,int);
```

The above prototype means the function does not return anything, but accepts two integers. Another typical prototype could be like the one below.

```
void display(void);
```

Here, the function neither accepts nor returns anything. The keyword **void** could be optionally used within the parentheses following the function name.

The prototyping of functions itself is optional in C language, as its importance was not realized initially. Later when the concept of functions was inherited by C++, prototyping was made essential in all the C++ programs, thereby removing the anomalies created by the C functions.

Following the prototyping, we get the values of a and b, as usual. Following this, comes the function call **c=add(a, b)**. Earlier to the call, we have let the compiler know that there would be a function defined later in the program with the name add() and it would accept two integer type values and return one integer type value. Now, we call the function add() by passing the two values (a and b) that we have obtained using the function scanf(). This is exactly the call by value method of using the functions. These variables that are in the calling function (main()) are called as the actual arguments. When these values are passed on to the function add(), there must be two variables exactly similar to a and b in that function, which must be compatible to receive and store these values. See that we are using two more variables x and y in the function add() and these variables are called as formal ones or dummy arguments. They are just copies of a and b and their lifetime will be as long as the control is executing in the function add() and once the control goes back to the function main() or the calling function, these formal arguments vanish and that is why we call them as dummy arguments.

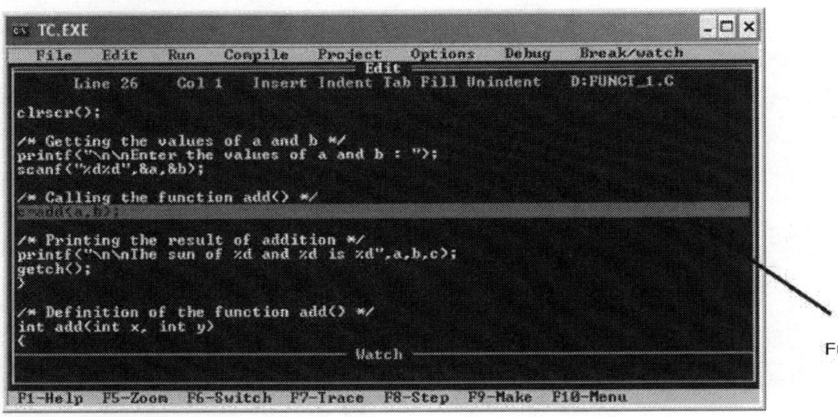

Figure 7.1 Function add() getting called during the trace of the program

When the function add() is called, the execution in the function main() is temporarily suspended and the control gets into the function add() and till it encounters the *return* statement in it, it remains there. The *return* statement explicitly takes the control back to the calling function and the execution in the calling function will be resumed from the place where the control last left. For that matter, in no C function, the control will remain in the called function. It must always come back to the calling function and terminate the program execution from there. I would suggest you check the above program code using the keystroke F7 (tracing the program execution). Figures 7.1 and 7.2 show the trace of the program execution just when the control goes from the calling function (main()) to the called function (add()). Figure 7.3 shows the trace of the program execution when the control resumes action in the calling function after coming back from the called function.

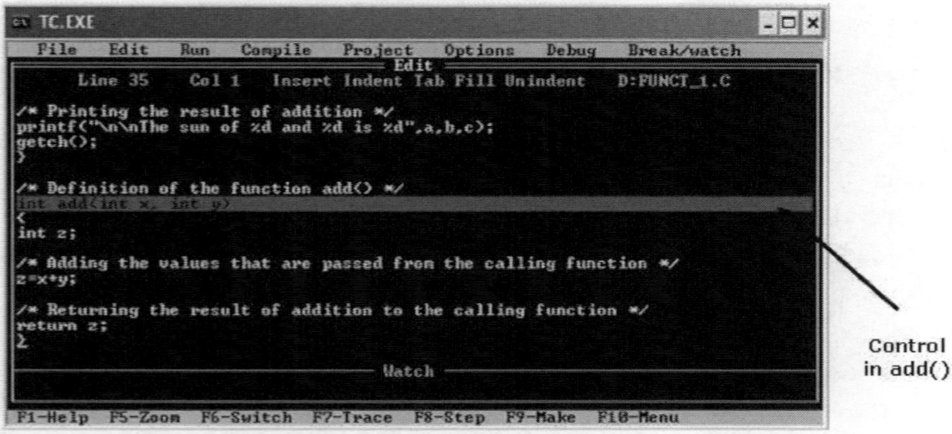

Figure 7.2 Control getting into the function add() from main()

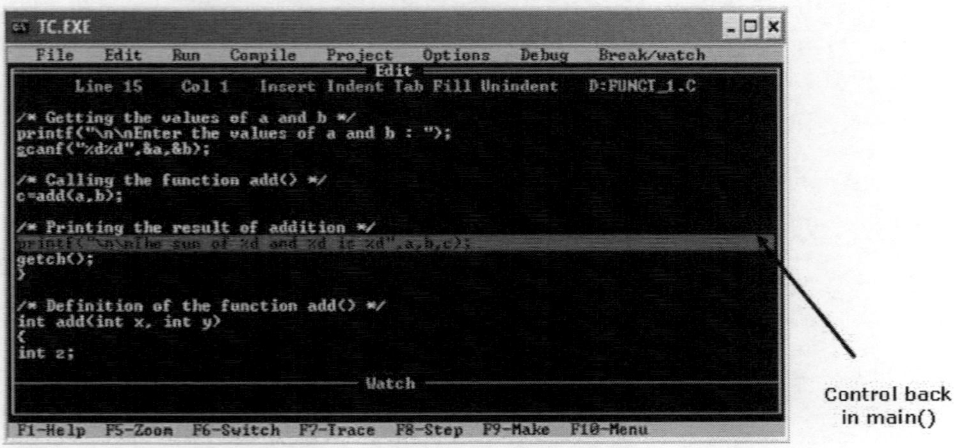

Figure 7.3 Control back in the function main()

I have given the memory picture of the above program in the Figure 7.4 for a better understanding of how the control flows throughout the program.

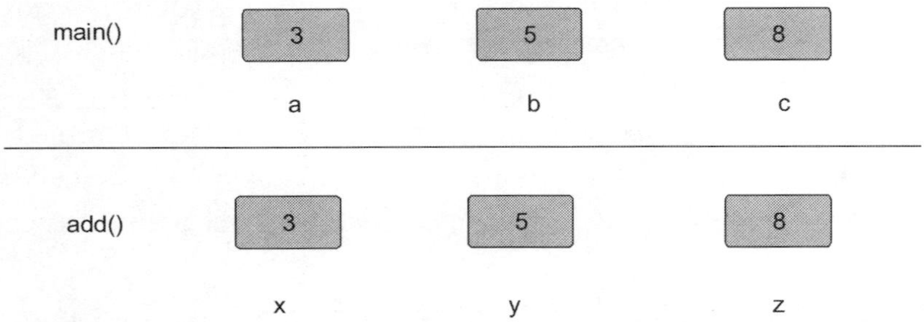

Figure 7.4 Memory picture—adding two integers

From Figure 7.4, we can observe that the values of a and b in the function main() are copied on to the variables x and y respectively in the function add(). The variables a and b are the actual parameters or variables and the variables x and y are formal or dummy parameters. The values of x and y are added and the result that is stored in 'z' is returned to the calling function (main()) using the *return* statement and it is collected in the variable 'c' in main() and then displayed. We see that there is no relation between the variables a and b and x and y, except that x and y are dummies for a and b respectively. There will be memory allocated for the dummy variables when the control is inside the function add() and their memory will be automatically released once the control gets back to the calling function. Another important fact is that the function add() will also be allocated with a memory space, since it is just like another variable. The memory allocated for the function will remain as such as long as the program is in the execution stage. Also, know that it is absolutely not necessary to name the variables in the function add() as x and y and we could have named them also as a and b, since there is no relation between the variables declared in the calling function and called function and hence there would not be any name collision.

In the function add() we could have avoided the declaration of the variable 'z' and we could have written the same coding as follows.

```
int add(int x, int y)
{
return (x+y);
}
```

This is very simple compared to the earlier version of the function add(). Figure 7.5 shows a sample execution of the above program.

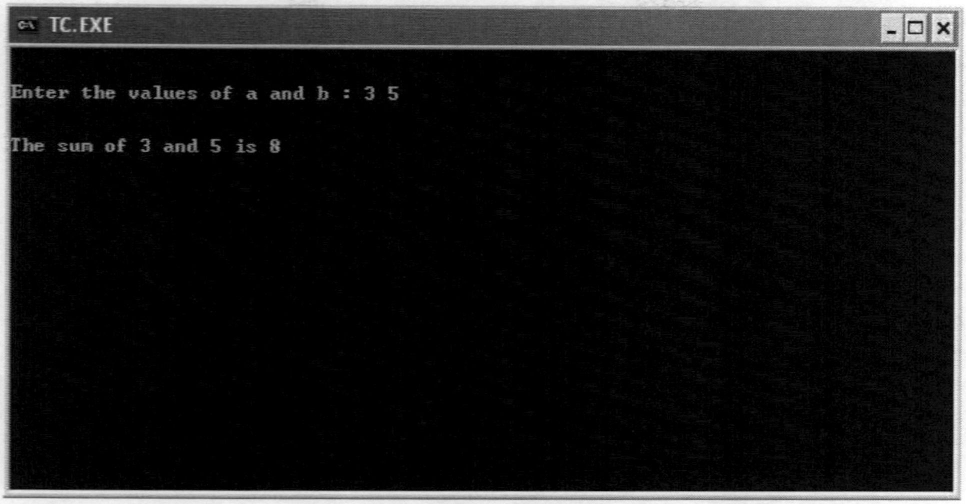

Figure 7.5 Sample run—adding two integers

Assume that we are doing a similar program to add two float values. The following coding for the same.

```
/* Program to add two floats using functions */
#include<stdio.h>
main()
{
/* Declaration of variables */
flaot a,b,c;
/* Function prototyping */
float add(float,float);
clrscr();
/* Getting the values of a and b */
printf("\n\nEnter the values of a and b :  ");
scanf("%f%f",&a,&b);
/* Calling the function add() */
c=add(a,b);
/* Printing the result of addition */
```

```
printf("\n\nThe sum of %f and %f is %f",a,b,c);
getch();
}
/* Definition of the function add() */
float add(float x, float y)
{
return (x+y);
}
```

Suppose that we do not use the prototyping for the function add() and we do not specify the return type of the function add(). Guess what would happen and what would be the result of adding two float values. The compiler will compile the program to success, but the result will be the one that is not anticipated at all by the programmers and some compilers truncate the result to an integer and return, thereby making the functions default return type as integer. The concluding fact is that all the functions must be properly prototyped to get correct results.

Now that we have a basic understanding of how to use functions, let us proceed in understanding how to use conditions and loops in the functions and how to return the values once the process is done in the function. Consider that we are writing a program to check the odd or even nature of a number. The following is the program code. Observe the program closely to get to know the new ideas that we have used in it.

```
/* Program to check the odd or even nature of a number */
#include<stdio.h>
main()
{
int a,flag;
int check(int);
clrscr();
printf("\n\nEnter a number to check : ");
scanf("%d",&a);
/* Function check() is called */
flag=check(a);
```

```
if(flag)
printf("\n\nThe number %d is even",a);
else
printf("\n\nThe number %d is odd",a);
getch();
}
/* Definition of function check() */
int check(int x)
{
return(!(x%2)?1:0);
}
```

We could observe that two variables 'a' and 'flag' are used in the function main(). We get the value of 'a' using the function scanf() and pass the same to the function check() and there we process the value of 'a' that is copied in 'x' and return the result, which is a Boolean true or false. Thus we have used the variable 'flag' in the function main() as a Boolean variable (since there is no Boolean data type in C, we have declared this as an integer and it is left to the programmer to make sure that this variable stores only a 1 or 0).

In the function check(), we are using the ternary expression inside the *return* statement and this is perfectly allowed. You may also observe a very important aspect both in the function main() and check(). In the function main() we have used a conditional expression if (flag). What is the meaning of saying like this? This has an implicit meaning and called technically as implicit checking of the conditions. When we write the code as if(<variable>), it means that if(<variable> !=0). Instead of using the inequality operator and a zero explicitly, C allows the users to make their presence understood through the implicit checking as given above. We have another kind of implicit checking, which is given as if(!<variable>). When we write a code like this, it means if(<variable>==0). This again reduces the usage of equality operator and the zero explicitly. Note that implicit checking in conditions is possible only when we find the equality or inequality of a variable with zero and no other values.

In the function check(), we have used the second form of implicit checking and return either a '1' or a '0' based on the input number. Figure 7.6 shows the sample run of the above program.

Figure 7.6 Sample run—odd or even nature of a number

Figure 7.7 shows the memory picture for the above program.

Let us see one more program using loop and at the end of this section, you must be in a position to answer question 112 and you will be ready to work with arrays and functions in the forthcoming sections. Consider that we are finding whether a given number is prime or composite using *for* loop. The following would be a program coding for the same.

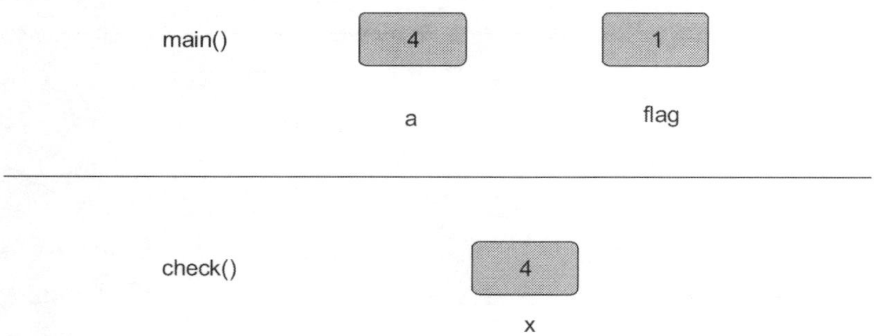

Figure 7.7 Memory picture—odd or even nature of a number

```
/* Program to find whether the number is prime or composite */
#include<stdio.h>
main()
{
```

```
int n,flag;
int prime_check(int);
clrscr();
printf("\n\nEnter the number to check : ");
scanf("%d",&n);
if(n==1)
{
printf("\n\n%d is neither prime nor composite",n);
getch();
return;
}
flag=prime_check(n);
if(flag)
printf("\n\n%d is a prime number",n);
else
printf("\n\n%d is a composite number",n);
getch();
}
int prime_check(int x)
{
int flag=1,i;
for(i=2;i<=x/2;i++)
if(x%i==0)
{
flag=0;
break;
}
return flag;
}
```

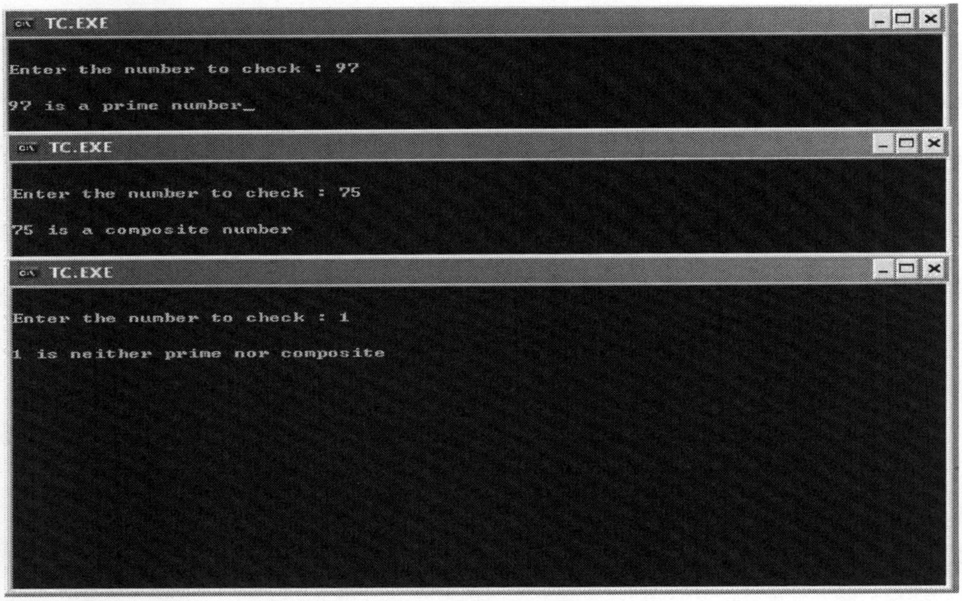

Figure 7.8 Sample run—prime or composite nature of a number

We get the value to check in the variable 'n' and immediately check it for unity. If it is '1' then we flash the corresponding message. If not, we call the function prime_check() and pass the value of 'n' to the dummy variable 'x' and allow it inside a loop and accordingly we set the value of the 'flag' variable (remember that we had used 'flag' variable in our earlier program also and it is used as a Boolean variable and will store a '1' or '0'). Finally we return the value of 'flag' and based on it, we print the result as to whether the given number is prime or composite. Figure 7.8 is a sample run of the above program (it shows the three different executions—prime, composite and neither prime not composite).

Figure 7.9 shows the memory picture for the above program for a prime number.

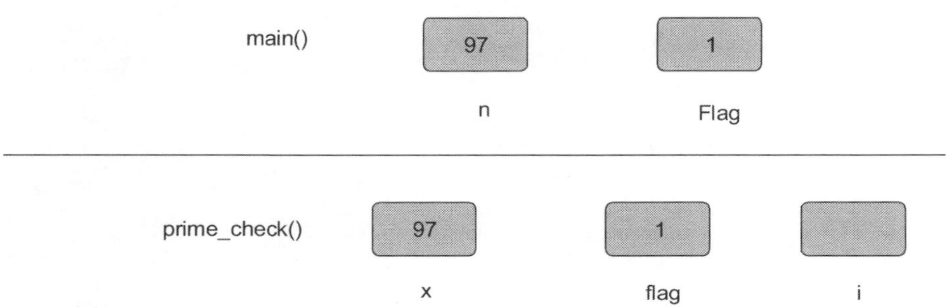

Figure 7.9 Memory picture—prime or composite nature of a number

I am sure that you would definitely be able to complete question 112 without any difficulty. It will just mould you up in declaring the function prototypes and define the functions. Later when we discuss the functions and arrays and functions, arrays and pointers, you will find that a child's play. Let us proceed with the discussion on functions and numerical arrays.

7.5 FUNCTIONS AND NUMERICAL ARRAYS

We can write functions by passing not just variables as we have seen in the earlier section. We can also pass arrays also as arguments and in this section we will discuss how functions work with single- and two-dimensional numerical arrays.

7.5.1 Functions and Single-Dimensional Numerical Arrays

Let us consider the program in which we read and display an array of 'n' integers. The following would be the program code for the program.

```
/* Functions and Numerical Arrays - Read and Display an array of 'n' integers */
#include<stdio.h>
main()
{
int a[100],n;
void read_arr(int[],int);
void print_arr(int[],int);
clrscr();
printf("\n\nEnter how many elements you want to enter :  ");
scanf("%d",&n);
printf("\n\nEnter the elements :  ");
read_arr(a,n);
printf("\n\nThe elements you have entered are :  ");
print_arr(a,n);
getch();
}
/* Reading the elements of the array */
void read_arr(int x[], int n)
{
int i;
```

```
for(i=0;i<n;i++)
scanf("%d",&x[i]);
}
/* Printing the elements of the array */
void print_arr(int x[], int n)
{
int i;
for(i=0;i<n;i++)
printf("%d\t",x[i]);
}
```

We see from the program that we are using two functions, unlike the earlier programs. One function read_arr() is used for reading in the elements of the array and the other function print_arr() is used for printing the elements that are read. We also see that when we send the array to the functions, we send it by referring its name (just like the other variables) and once we refer the name of the array name in the program, we know that its base address would be accessed (from the discussion of pointers and arrays). Thus it is immaterial to pass the size of the array to the functions, since the address is passed. The functions read_arr() and print_arr() hence could understand the size of the array very easily using the addresses passed. Also note the prototypes used for the functions read_arr() and print_arr(). They contain an array variable (denoted by square braces) and an integer variable. Figure 7.10 shows the sample run for the above program.

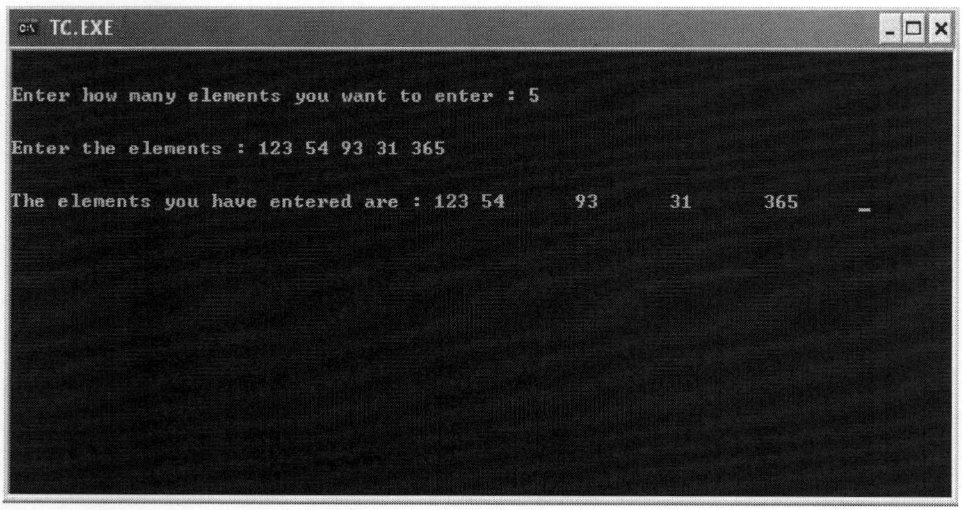

Figure 7.10 Sample run—read and display an array of 'n' integers

The memory picture is shown in the Figure 7.11.

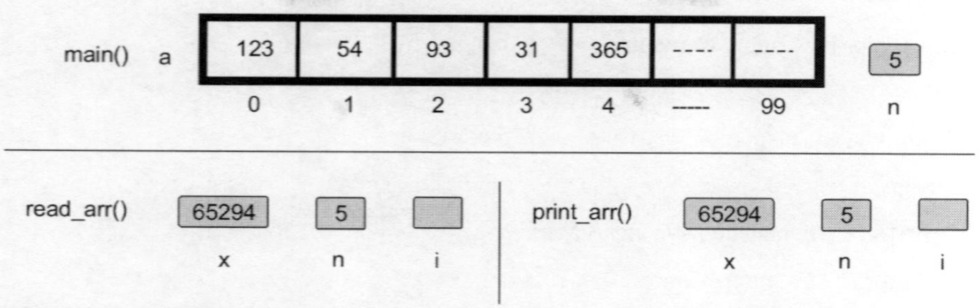

Figure 7.11 Memory picture—read and display an array of 'n' integers

We can see from the memory picture that when we call the functions read_arr() and print_arr(), we pass the address of the array 'a' and the value of 'n' and we get the input for the array and display the same using the respective functions. Note that when we pass the address of the array from the function main(), the dummy variable 'x' receives that and from there on it could access the individual locations of the array 'a'. This is possible for the variable 'x' since it has the base address of the array 'a'. Thus, even though it is not a pointer, it acts as an implicit pointer to the array 'a'. In the memory picture above, 65294 happens to be the base address of the array 'a' and you can see that the variable 'x' has that address when it receives 'a' from the function main(), when the functions read_arr() and print_arr() are called. You may check this by adding the following printf() statement in your program in any one of the two user-defined functions read_arr() and print_arr().

```
printf("%u",x);
```

The above printf() statement would result in printing the value in 'x', which is nothing but the base address of the array 'a'.

Let us see another program for a better understanding of how numerical arrays work with functions. Consider the program which reads an array of 'n' integers and sorts the array in ascending order. The following would be the program for the same.

```
/* Functions and Numerical Arrays – Sorting an array of 'n' integers in ascending order */
#include<stdio.h>
main()
{
int a[100],n;
void read_arr(int[],int);
void print_arr(int[],int);
```

```
void ascend(int[],int);
clrscr();
printf("\n\nEnter how many elements you want to enter : ");
scanf("%d",&n);
printf("\n\nEnter the elements : ");
read_arr(a,n);
printf("\n\nThe elements you have entered are : ");
print_arr(a,n); /* First time print_arr() gets called */
ascend(a,n);
printf("\n\nThe array in ascending order is : ");
print_arr(a,n); /* Second time print_arr() gets called */
getch();
}
void read_arr(int x[], int n)
{
int i;
for(i=0;i<n;i++)
scanf("%d",&x[i]);
}
void print_arr(int x[], int n)
{
int i;
for(i=0;i<n;i++)
printf("%d\t",x[i]);
}
void ascend(int a[],int n)
{
int i,j,temp;
for(i=0;i<n;i++)
for(j=i+1;j<n-1;j++)
if(a[i]>a[j])
{
temp=a[i];
```

```
a[i]=a[j];
a[j]=temp;
}
}
```

We see that we have one more function ascend(), whose prototype is very much similar to the other two functions read_arr() and print_arr(). This function performs the sorting process and once this function gets called from the function main(), the array would have been sorted in ascending order and when we call the function print_arr() for the second time, the sorted array gets printed. This can be observed from the Figure 7.12.

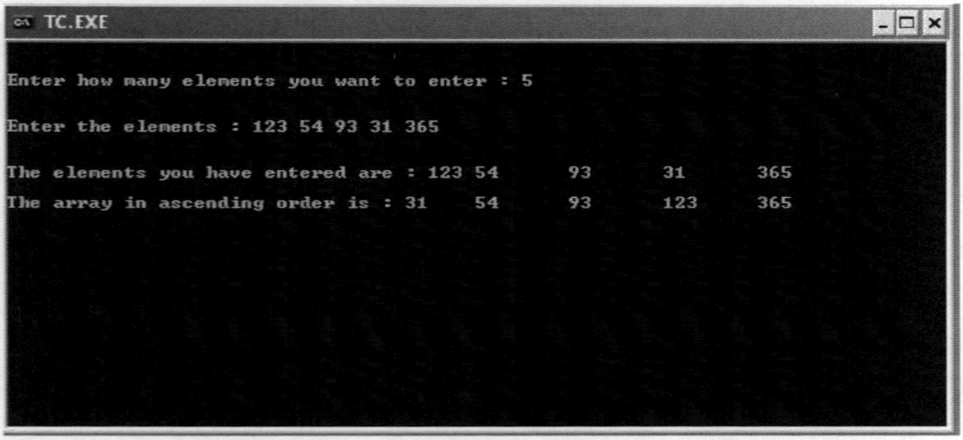

Figure 7.12 Sorting an array in ascending order

The memory picture of the above program is shown in the Figure 7.13. There is just one addition to the memory map shown in the Figure 7.11, the function ascend().

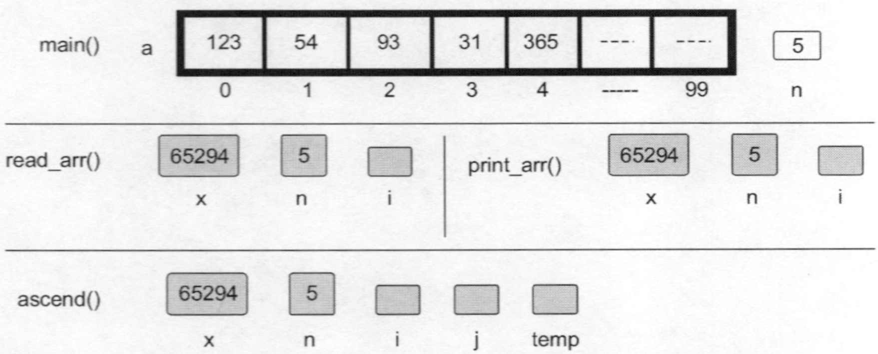

Figure 7.13 Memory picture—sorting an array in ascending order

With this idea of working with single-dimensional array, we proceed to discuss on how the two-dimensional arrays work with functions in the following section.

7.5.2 Functions and Two-Dimensional Numerical Arrays

Two-dimensional numerical arrays work in a similar manner with the functions as the single-dimensional array. Let us consider the program to read and find the transpose of an mXn matrix. The following is the program coding.

```
/* Reading and finding the transpose of an mn matrix */
#include<stdio.h>
main()
{
int a[10][10],m,n;
void read_arr(int[][],int,int);
void print_arr(int[][],int,int);
void transpose(int[][],int,int);
clrscr();
printf("\n\nEnter the order of the matrix :  ");
scanf("%d%d",&m,&n);
printf("\n\nEnter the array elements :  ");
read_arr(a,m,n);
printf("\n\nThe array you have entered is\n");
print_arr(a,m,n);
printf("\n\nThe transpose of the matrix is\n");
transpose(a,m,n);
getch();
}
void read_arr(int x[][10],int m,int n)
{
int i,j;
for(i=0;i<m;i++)
for(j=0;j<n;j++)
scanf("%d",&x[i][j]);
}
```

```
void print_arr(int x[][10],int m,int n)
{
int i,j;
for(i=0;i<m;i++)
{
for(j=0;j<n;j++)
printf("%d\t",x[i][j]);
printf("\n");
}
}
void transpose(int x[][10],int m,int n)
{
int i,j;
for(j=0;j<n;j++)
{
for(i=0;i<m;i++)
printf("%d\t",x[i][j]);
printf("\n");
}
}
```

We have three functions namely read_arr(), print_arr() and transpose(), to read, display and print the transpose of an mXn matrix. We also see that when we call the function we call it with the base address of the 2-D array 'a' and with the values of 'm' and 'n'. When we define the function, we observe that the size of the second dimension alone is collected in the header and that of the first dimension is left out. This is because once the base address of the entire 2-D array is known when the control comes to the function and once the second dimension's size is also known, the memory will be allocated accordingly in the function. I would suggest you check this fact by using the following statement in the program. This would print the base address of the 2-D array.

```
printf("%u\n",x);
```

Also, try to figure out yourself the memory map of the above program. Figure 7.14 shows a sample execution of the above program.

7.6 FUNCTIONS AND CHARACTER ARRAYS

Character arrays, as discussed in Chapter 4, differ from the numerical arrays only with respect to the NULL character that is used to terminate the array. Apart from this, there is no difference in accessing and manipulating the individual elements of the array.

Let us straight away discuss a program with the character array. Consider the following program in which we reverse the string that is read. We use four functions, readfn(), writefn(), lenfn() and revfn() for reading, displaying, finding length and reversing the given string. The function for finding the length is important since the function revfn() must know the length before it reverses the input string.

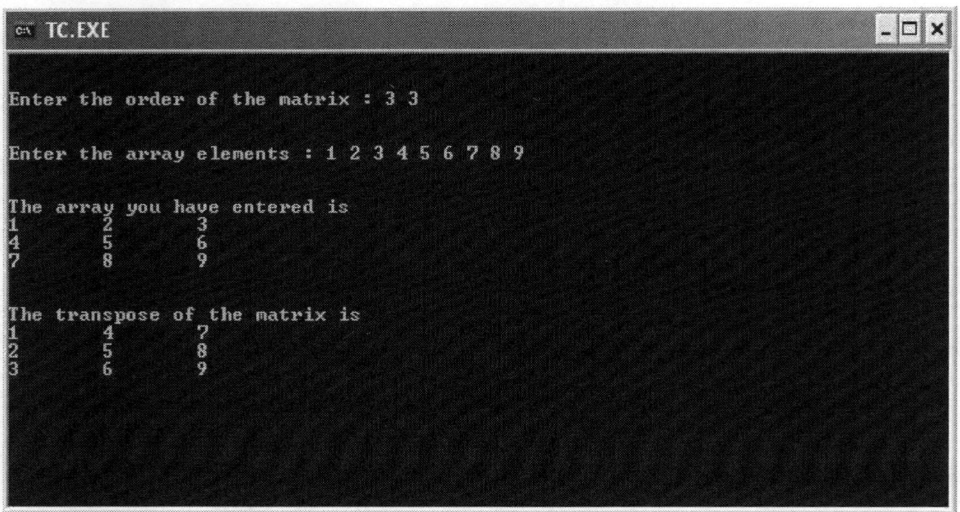

Figure 7.14 Transpose of the matrix of order mXn

```
/* Read and reverse a string using functions */
#include<stdio.h>
main()
{
char a[100];
int l;
void readfn(char[]);
void writefn(char[]);
int lenfn(char[]);
void revfn(char[],int);
```

```
clrscr();
printf("\n\n\tEnter the string :   ");
readfn(a);
printf("\n\n\tThe string entered is :   ");
writefn(a);
l=lenfn(a);
printf("\n\n\tThe length of the string is : %d",l);
printf("\n\n\n\tThe reversed string is :  ");
revfn(a,l);
getch();
}
void readfn(char x[])
{
int i;
for(i=0;(x[i]=getchar())!='$';i++);
x[i]='\0';
}
void writefn(char x[])
{
puts(x);
}
int lenfn(char x[])
{
int i;
for(i=0;x[i];i++);
return i;
}
void revfn(char x[],int l)
{
int i;
for(i=l;i>=0;i--)
putchar(x[i]);
}
```

Figure 7.15 shows the output of the above program and Figure 7.16 shows the memory picture of the program. If we analyse the program, we can understand that when the functions readfn(), writefn() and lenfn() are called, the base address of the array 'a' is passed from the function main(). When the function revfn() is called, along with the base address we pass the value of the length of the string. For more clarity, go through the memory picture of the program. The base address of the array is found to be 65386 in my execution of the program and it may differ when you execute the same program. This address could be obtained by using the following printf() any one of the user-defined functions used in the program.

```
printf("\n%u",x);
```

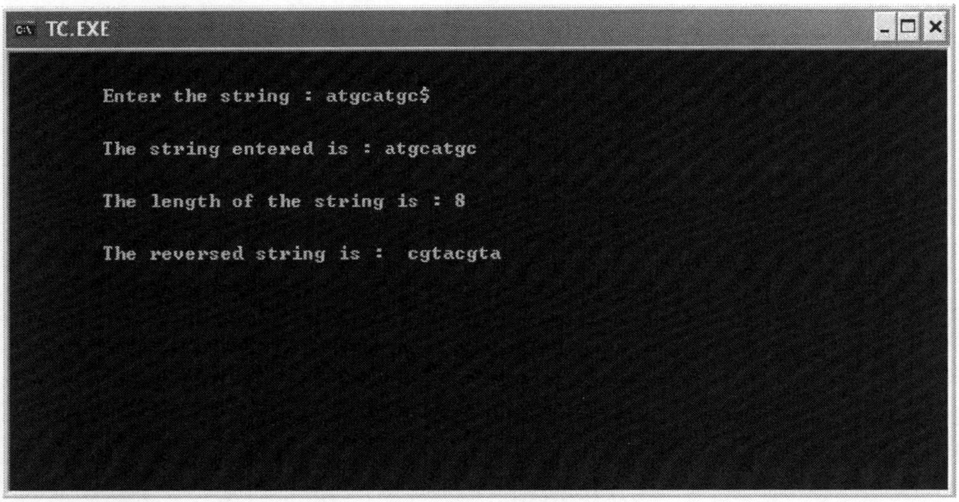

Figure 7.15 Sample run—reversing a string

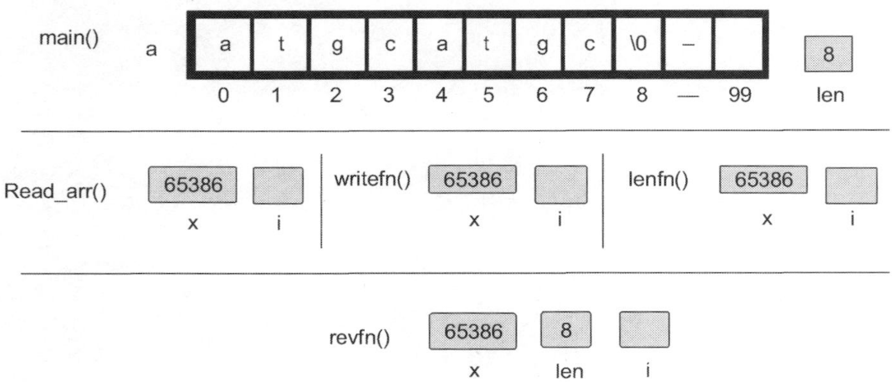

Figure 7.16 Memory picture—reversing a string

Let us discuss one more program using functions, in which we would find the number of times a particular pattern exists in the given string. The following is the program coding for the same.

```
/* Check for the occurence of a pattern in the given string/text */
#include<stdio.h>
main()
{
char a[100],b[100];
int count;
void readfn(char[]);
void writefn(char[]);
int check(char[],char[]);
clrscr();
printf("\n\n\tEnter the text :  ");
readfn(a);
fflush(stdin);
printf("\n\tEnter the pattern to check :  ");
readfn(b);
fflush(stdin);
printf("\n\n\tThe string entered is :   ");
writefn(a);
count=check(a,b);
printf("\n\nThe given pattern \"%s\" occurs in the given text \"%s\" \"%d\"times",b,a,count);
getch();
}
void readfn(char x[])
{
int i;
for(i=0;(x[i]=getchar())!='$';i++);
x[i]='\0';
}
void writefn(char x[])
{
puts(x);
```

```
}
int check(char x[],char y[])
{
int i,j,k,l=0;
for(i=0;x[i];i++)
{
for(j=0,k=i;y[j]==x[k]&&y[j];k++,j++);
if(y[j]=='\0')
l++;
}
return l;
}
```

In the program above, we are using a new function fflush(stdin). This function is essentially needed when one manipulates with multiple character arrays. In this program we use two character arrays 'a' and 'b'. Normally when we give the input strings through the keyboard, we give an enter or a space or a tab character between the two inputs (default delimiters that we had discussed in Chapter 2). Every device attached with the computer has its own memory and this memory is called as a 'buffer' or a temporary memory. When we operate with more than one character array, any one of the abovesaid delimiter character will appear in between the two input strings and hence when we check the existence of a pattern with the given text, only the 'enter' character (which is in the array 'b') gets checked with the given text in the array 'a' and hence you will get the output much against your anticipation. I would suggest, you please check with this by removing the function fflush(stdin) given between the two inputs and after the second input. You will not get the right answer. Figure 7.17 shows the output when the above-quoted function is ignored from the program (just remark it to ignore it from the program and do not delete it). Observe the last line of the Figure 7.17. You can see that it is spread over two lines. The reason is very simple. When we ignore the fflush(stdin) function, the 'enter' character that is available in the keyboard buffer (the 'enter' character that is given after the first text) will be appended automatically with the first input text. Hence we are getting the last line spread over two lines. The output is also wrong, since it is obvious from the input text atgcatcg that the test string at occurs two times, whereas the output indicates that it occurs '0' times. The reason is that when we ignore the function fflush(stdin) from the program the 'enter' character that is given after the first text (atgcatcg) is checked for its occurrence and not the string 'at'. Now include the function fflush(stdin) and repeat the execution with the same inputs. You will get the correct result in this execution. This is because, the usage of the function fflush(stdin) will clear off the keyboard buffer and when the input text (atgcatcg) is checked for the occurrence of the pattern 'at', it gives a right result. Figure 7.18 shows the execution of the program with the function fflush(stdin)included in the program.

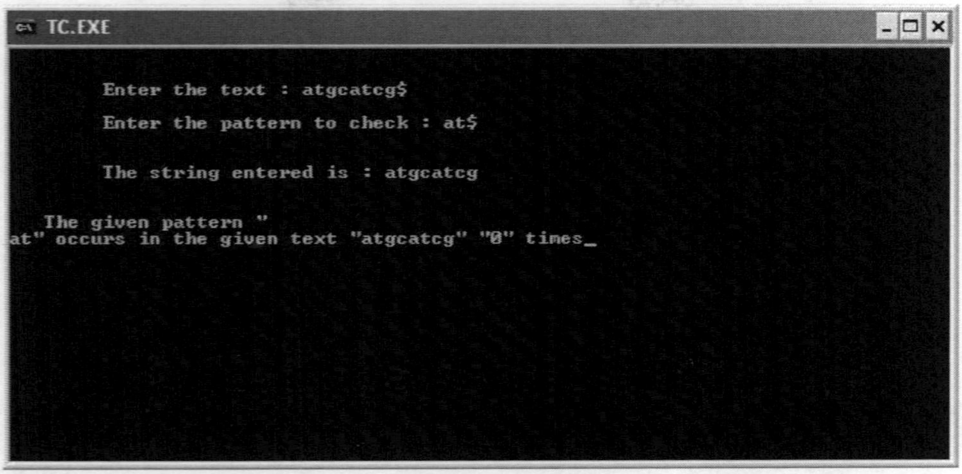

Figure 7.17 Pattern matching—O/p when fflush(stdin) not included

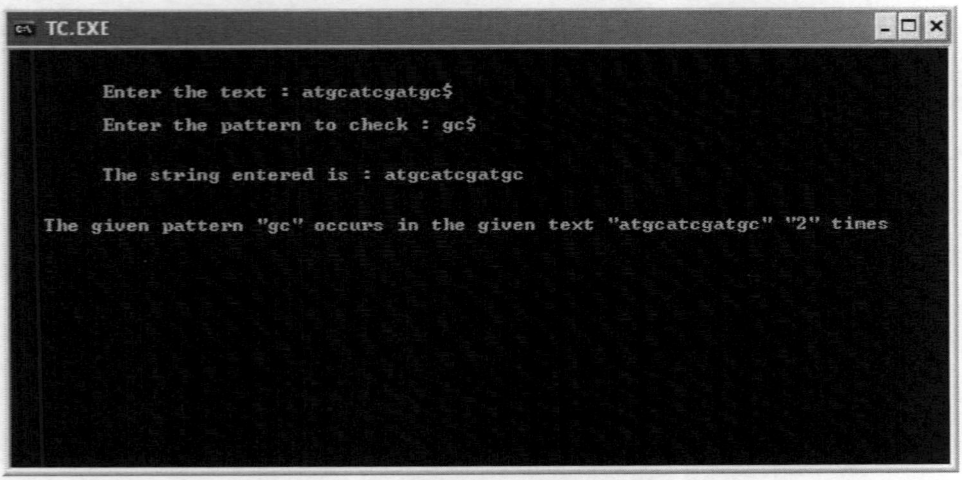

Figure 7.18 Pattern matching—O/p when fflush(stdin) included

So, we can conclude that whenever we operate with multiple character arrays, we must use the function fflush(stdin) to clear off the keyboard buffer and get the anticipated result.

When you run the program for different sets of inputs, you might find a discrepancy. For example, test the program with the inputs "hello world" and "hell". As such the word "hell" does not occur in the input text, but it is an integral part of the word "hello" and hence when you execute the above program with the inputs "hello world" and "hell", you will get the result that the word is occurring once in the given text. So, our logic not only checks the exact occurrence of a pattern in the given input text, but also the integral part of the input text. The output shown in Figure 7.19 justifies this. I leave it as an exercise to the learners of the book to identify the most suitable logic that would work with any input text and test pattern.

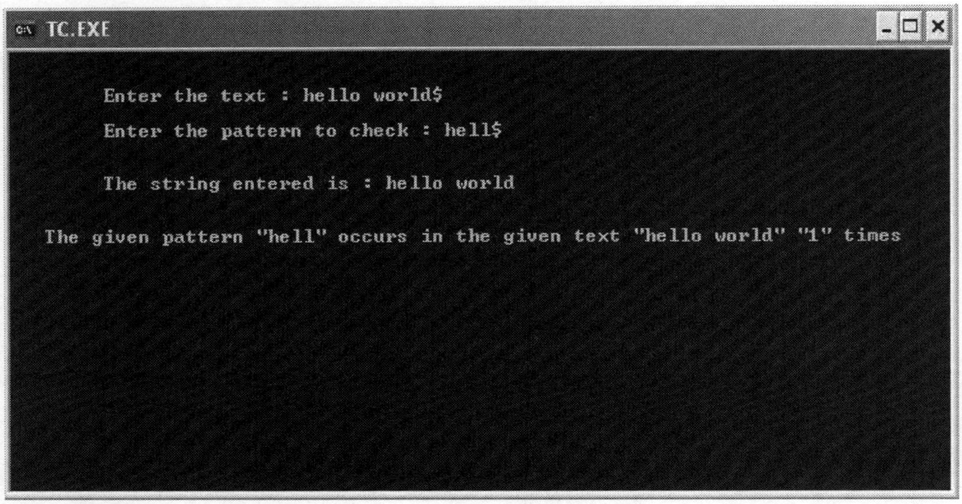

Figure 7.19 Part of the input text is also matched with given pattern

I would appreciate if the learners figure out the memory picture for the above program, which will surely increase the gravity of their understanding.

With this we conclude the discussion of how the character arrays work with the functions. I am sure the programs that I have discussed would be of immense use for the learners. Try to solve the exercises given at the end of this book pertaining to this chapter so that you can work more confidently with the functions. What you have understood till now in functions would be of great significance when you learn the concepts in the forthcoming chapters.

7.7 FUNCTIONS AND POINTERS

We have been discussing functions earlier in this chapter and in our discussion we did not include pointers while we wrote simple functions. We also know that any function at a given time can return only one value or no value at all. No function can return more than one value after processing. This simply means that using the *return* statement, one can make a function only one value after the process. But think of those situations when a programmer wants to get more than one value from the function. Is this achievable? Does C have a provision to do this? Yes. Of course, it has.

When we write functions with pointers we would possibly get more than one value returned from the function and that too without explicitly using the *return* statement. This method of calling functions and getting the result is called as **call by value**. This is the most effective method especially when the functions are used with arrays. We shall see the concept of using the arrays with functions and pointers later in this chapter. Let us see how the functions work with the pointers. Consider the program where we would like to swap two values of type integer. We would be writing a simple function swap() to achieve this task. The following is the program code for the same.

```
/* Swapping two values of type integer */
#include<stdio.h>
main()
{
int a,b;
void swap(int*,int*);
clrscr();
printf("\n\tEnter the values to swap : ");
scanf("%d%d",&a,&b);
printf("\n\tThe values of a and b before swapping : %d\t%d",a,b);
swap(&a,&b);
printf("\n\n\tThe values of a and b after swapping : %d\t%d",a,b);
getch();
}
void swap(int *x,int *y)
{
int temp;
temp=*x;
*x=*y;
*y=temp;
}
```

We see from the program that we have a function **swap()** that swaps the values of two integers. We see that no values are returned from the function. But still the values would be swapped. How is this possible? If you observe the function prototype **void swap(int*,int*)** and the function call **swap(&,a,&b)**, we pass the address of the variables a and b to the function swap() and hence whatever changes are made to the variables in the function, they are reflected in the function main(). Earlier when we used call by value method of accessing functions, we just copied the values of the variables that we passed from the function main() to the called function. In other words, we created dummy variables for copying the values that are passed. When we made changes in the copies, they are not reflected in the original variables. In the above program we have used **call by reference** method of accessing functions.

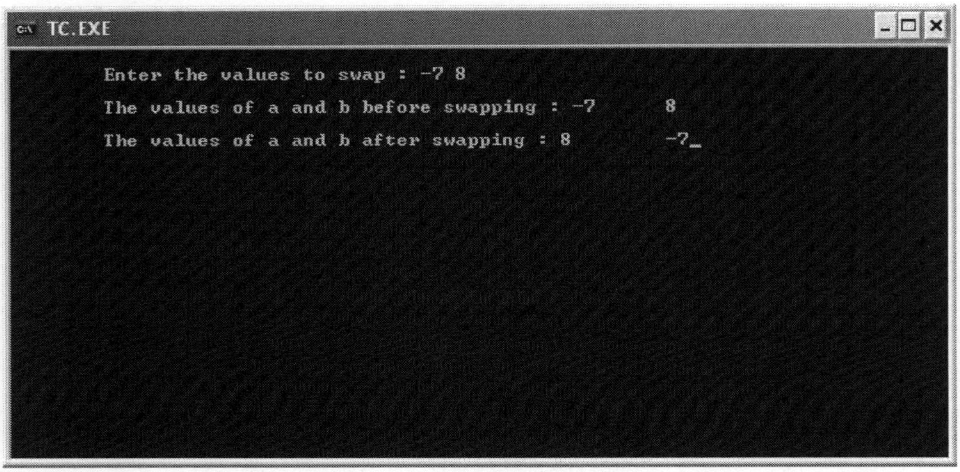

Figure 7.20 Swapping two integers using functions and pointers

The output of the above program is shown in the Figure 7.20. The memory picture shown in the Figure 7.21 explains how the call by reference method of function is effectively used.

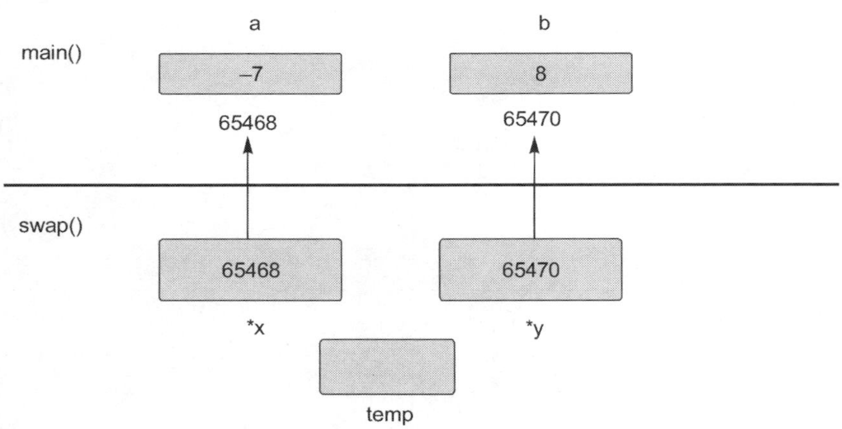

Figure 7.21 Memory picture for swapping program

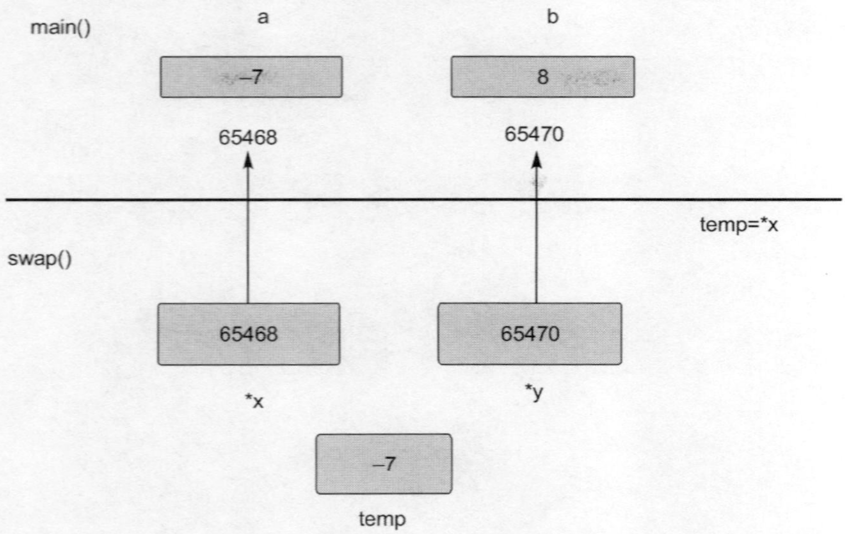

Figure 7.22 Execution of the code temp=*x in the function swap()

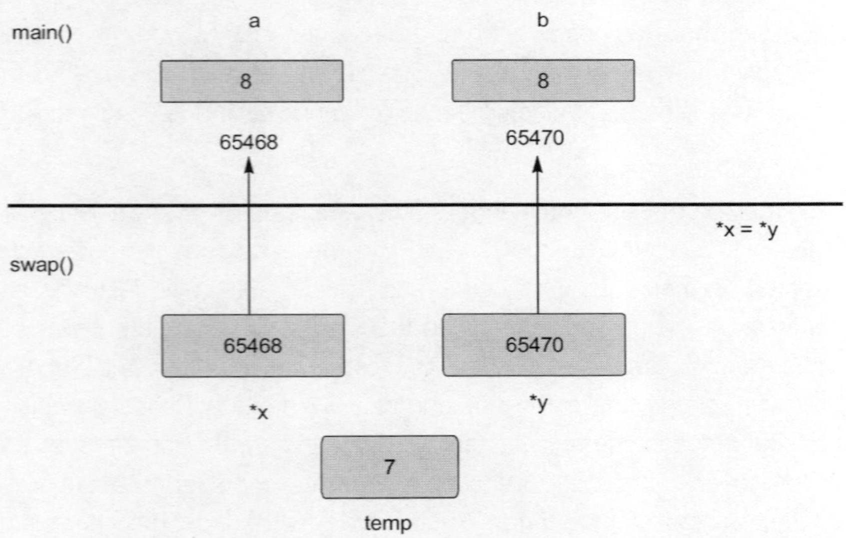

Figure 7.23 Execution of the code *x=*y in the function swap()

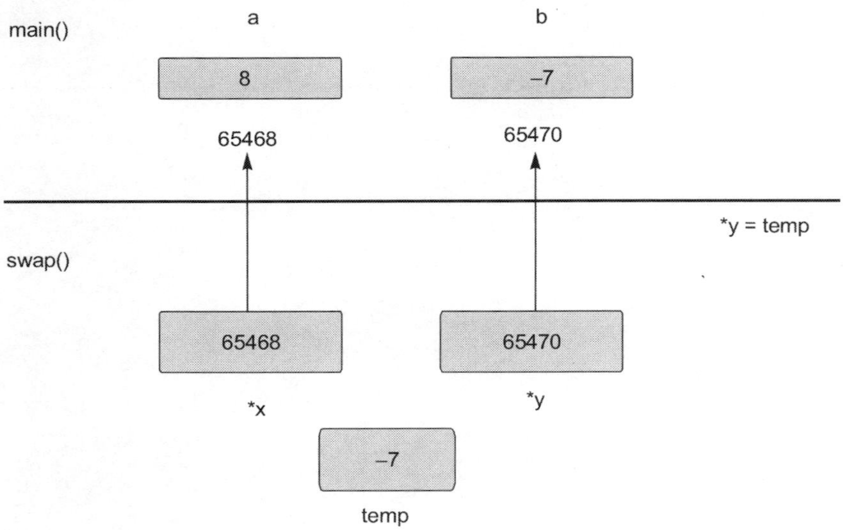

Figure 7.24 Execution of the code *y=temp in the function swap()

Analyse the function swap() closely. Since the pointer variables x and y in the function swap() have the addresses of the variables a and b in the function main(), when we process the pointer variables in the function swap(), they make changes in the variables in the function main(). The execution of the three lines of coding in the functions swap() is shown in Figure 7.22, 7.23 and 7.24.

From the memory pictures shown in Figure 7.22, 7.23 and 7.24, we can easily observe that the changes are made in the called function swap() and these changes get reflected in the calling function, the main(). Earlier, this was not the case, since there is no relationship between the variables in the calling function and the called function, except that the variables in the called function just act as a copy (dummy) of those in the calling function. But in the above program, the addresses of the variables in the calling function are available with the variables in the called function and hence they become the pointers for them and so, we can say that a relationship is established. When the code temp=*x is encountered by the compiler, it is converted internally to temp=*65468 which is equivalent to temp=-7. Thus we are able to get the value of a variable in the function main() to the function swap() by call by reference method. The other two lines of coding work in a similar way. The line of coding *x=*y is converted internally as *65468=*65470. Due to this conversion, we get the value of 'b' (8) in the address *65468, which is the address of 'a'. The line of coding *y=temp is internally taken as *65470=-7 and hence the value of 'a' occupies 'b'. Thus at the end of these three lines, we get the values of a and b in the function main() swapped.

Let us discuss another program for a better understanding of the concept. Consider the program in which we are finding the factorial of the given unsigned integer value using functions and pointers. The following is the program coding for the same.

```
/* Factorial of a given unsigned integer */
#include<stdio.h>
main()
{
unsigned n;
long unsigned f=1;
void fact(unsigned,long unsigned*);
clrscr();
printf("\n\tEnter the number for which factorial has to be found out : ");
scanf("%u",&n);
if(n==0)
{
printf("\n\tThe factorial of %u is %lu",n,f);
getch();
return;
}
fact(n,&f);
printf("\n\tThe factorial of %u is %lu",n,f);
getch();
}
void fact(unsigned n, long unsigned *pf)
{
int i;
for(i=n;i>1;i--)
*pf*=i;
}
```

We see that the value of the number for which the factorial value has to be found out is obtained as an unsigned number (since factorial for negative values is meaningless). Also, when the value for 'n' is given as '0', the function fact need not be called at all, since the factorial value for '0' is defined as '1'. Hence we check this in the beginning itself and terminate the program in the function main() itself, if the input is '1'. The variable that is supposed to hold the result of the factorial is 'f' and it is initialized as '1' and when we call the function fact(), we pass the value of 'n' and the address of 'f', since the variable 'f' alone is getting manipulated in the whole program. Once we pass the address of 'f' it is collected in the pointer variable 'pf' and as we manipulate 'pf' in the function swap(), the change is done in the original variable 'f' and hence when we print the result in the function main(), the correct value of factorial for the given number is printed. Observe also that the variable 'f' is declared as **long unsigned**, since the result of the factorial will be a bigger number in the case of input exceeding '7' and an ordinary integer cannot hold that big value, as it will exceed the range of it. I would suggest you to go through the list of data types and their permissible range of values for more clarity on this point. Figure 7.25 shows a sample run of the program (for the inputs '0' and another valid value for 'n').

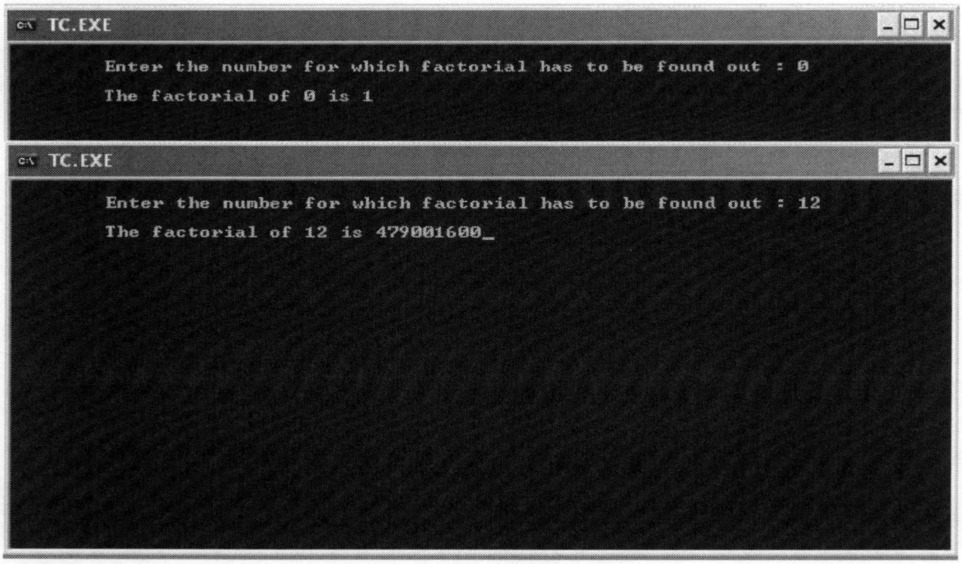

Figure 7.25 Finding the factorial of the given unsigned integer

The memory picture for the above program is shown in Figure 7.26. From the memory picture, we see that since the variable 'pf' in the function swap() holds the address of the variable 'f' in the function main(), the changes that we make with the variable 'pf' reflects in the variable 'f'. Initially the variable 'f' had the value '1' and when it is getting manipulated through 'pf', its value changes according to the input value for 'n'. Thus when we give 12 as the input for 'n', the result of the factorial, which is 47,90,01,600, is correctly shown as in the displayed figure above. The program works fine till the input value of 32, after which the range of long unsigned integer is exceeded and hence we will start getting different values, which are not anticipated.

I would suggest you to practice all the programs that are given at the end of the book pertaining to this chapter, to gain efficiency in working with functions and pointers.

7.8 FUNCTIONS, POINTERS AND ONE-DIMENSIONAL NUMERICAL ARRAYS

After having gained a thorough knowledge of how pointers are used in association with the functions, we are ready to advance ourselves to the next idea of using pointers and arrays with functions. To begin with we will discuss how pointers work with one-dimensional numerical arrays when they are used in functions. We know how pointers work with arrays from the previous chapter and how functions work with numerical arrays in this chapter in the section 7.5.1. We will combine those ideas in this section and will try to understand the way all the three work together. For that let us consider a program where we find the biggest and smallest of 'n' integers and swap their positions in the array and display the new array. The following is the program coding for the same.

```
/* Finding the biggest and smallest of 'n' integers and swap their positions using functions */
#include<stdio.h>
main()
{
int a[100],small,big,n,*posbig,*possmall;
void readfn(int*,int);
void writefn(int*,int);
void bigfn(int*,int*,int**,int);
void smallfn(int*,int*,int**,int);
void swapfn(int**,int**);
clrscr();
printf("\n\tEnter the number of elements in the array : ");
scanf("%d",&n);
printf("\n\n\tEnter the array elements : ");
readfn(a,n);
printf("\n\n\tThe array elements are : ");
writefn(a,n);
bigfn(a,&big,&posbig,n);
printf("\n\n\tThe biggest of the %d integers is %d",n,big);
```

```
smallfn(a,&small,&possmall,n);
printf("\n\n\tThe smallest of the %d integers is %d",n,small);
swapfn(&posbig,&possmall);
printf("\n\n\tThe array after swapping the biggest and smallest integers is:\n\n\t");
writefn(a,n);
getch();
}
/* Reading the elements of the array */
void readfn(int *pa,int n)
{
int *pt=pa;
for(;pa<(pt+n);pa++)
scanf("%d",pa);
}
/* Printing the elements of the array */
void writefn(int *pa,int n)
{
int *pt=pa;
for(;pa<(pt+n);pa++)
printf("%d\t",*pa);
}
/* Finding the biggest of the 'n' elements and its position */
void bigfn(int *pa,int *pbig,int **posbig,int n)
{
int *pt=pa;
*pbig=*pa;
*posbig=pa;
for(;pa<(pt+n);pa++)
if(*pa>*pbig)
{
*pbig=*pa;
```

```
*posbig=pa;
}
}
/* Finding the smallest of the 'n' elements and its position */
void smallfn(int *pa,int *psmall,int **possmall,int n)
{
int *pt=pa;
*psmall=*pa;
*possmall=pa;
for(;pa<(pt+n);pa++)
if(*pa<*psmall)
{
*psmall=*pa;
*possmall=pa;
}
}
/* Swapping the biggest and smallest using their positions */
void swapfn(int **posbig, int **possmall)
{
int temp;
temp=**posbig;
**posbig=**possmall;
**possmall=temp;
}
```

We see from the program that we have used five user-defined functions, to read the array, to display it, to find the biggest of its elements with its address and the smallest of its elements with its position and finally swap their positions using the addresses obtained. Even though the program seems to be complicated, the logic used is quite easy to understand. Here we use not just pointers, we use pointers to pointers since we need the addresses of the biggest and smallest numbers once they are found out from the respective functions.

Note the prototype of the functions readfn() and writefn(). We pass the base address of the array (by mentioning the name of the array) and the value of the number of elements (which is obtained and stored in 'n'). Thus we have a mixed type of function call, i.e., we use both the call by value and call by reference in the same function.

Note that the prototype of the functions bigfn() and smallfn() are similar and we are passing the base address of the array from the calling function, addresses of the variables that hold the biggest and smallest values (big and small), the addresses of the pointers posbig and possmall (in order to collect the addresses of biggest and smallest of 'n' integers in the array) and the value of 'n'.

Finally the prototype of the function swapfn() is simple, where we pass the addresses of the pointers from the function main() and once we manipulate the values in those addresses, the biggest and smallest values in the array get interchanged and the array when reprinted, shows the biggest number in the smallest number's position and vice versa.

The output of the above program code is shown in Figure 7.26 and the memory pictures of the same are shown in Figure 7.27 through Figure 7.31.

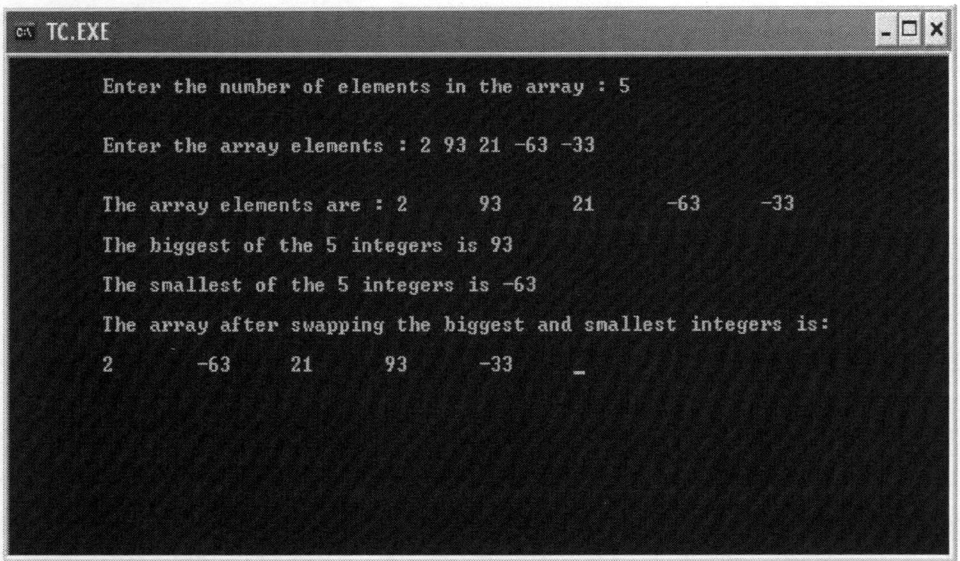

Figure 7.26 Finding the biggest and smallest and swapping them

From Figure 7.26, we can observe that the biggest number in the array is 93 and the smallest number in the array is –63 and their positions (not addresses) in the array are 2 and 4 respectively. Once we find out the addresses of these two elements, we pass them to the function swapfn(), which swaps them.

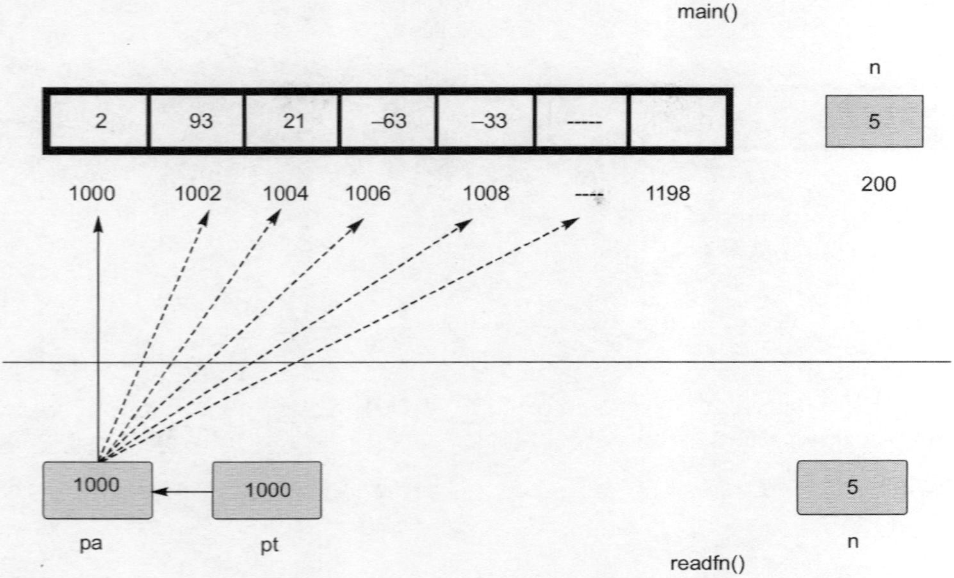

Figure 7.27 Reading the array elements using readfn()

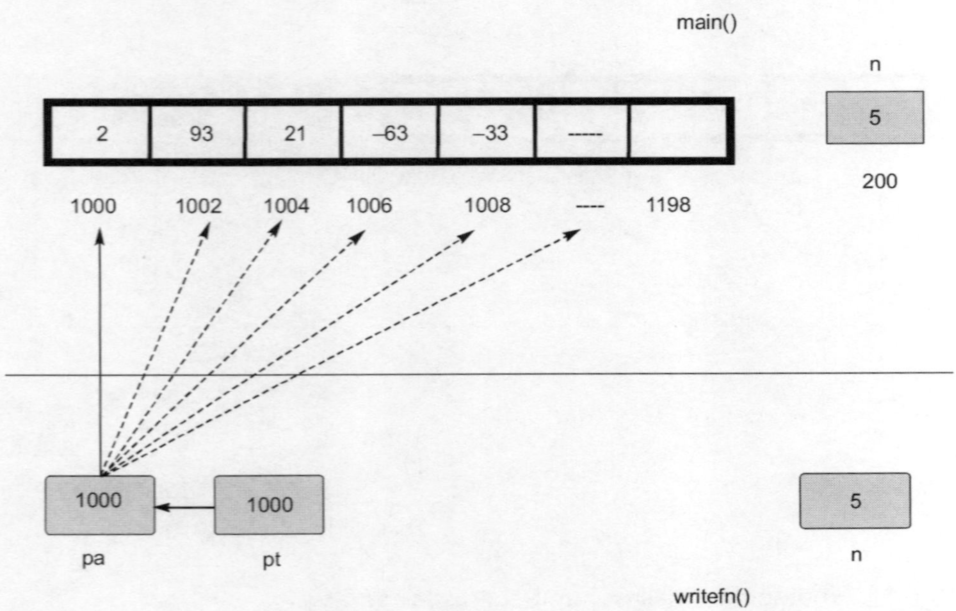

Figure 7.28 Displaying the array elements using writefn()

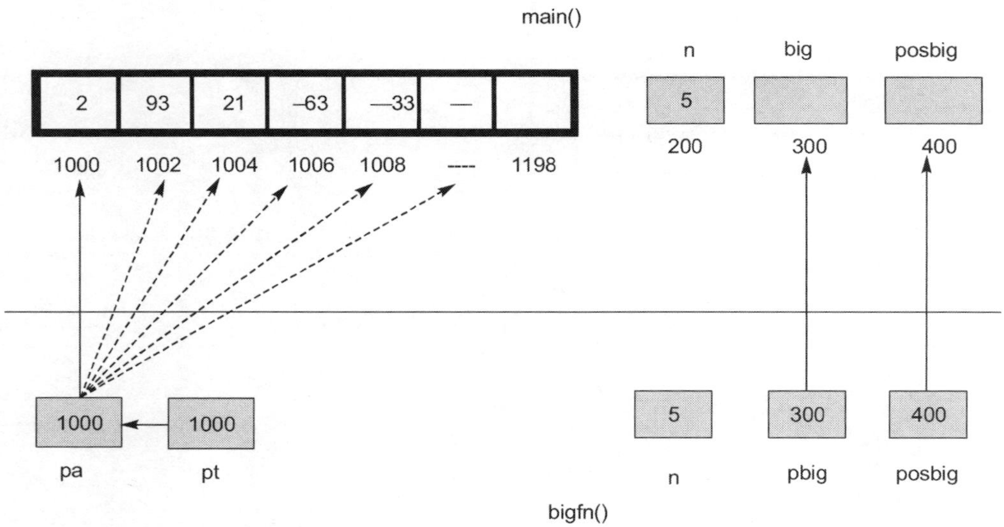

Figure 7.29 Finding the biggest number and its address

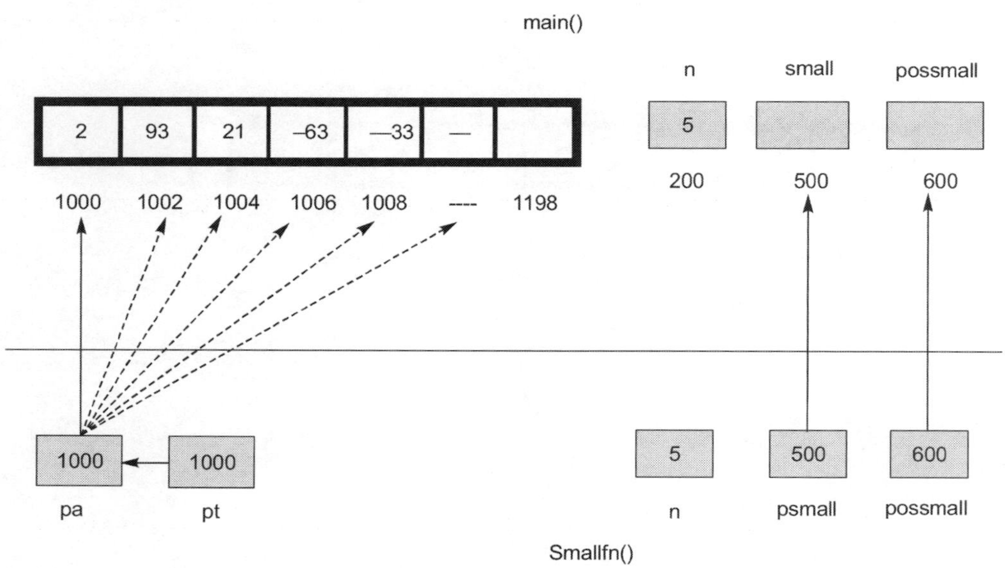

Figure 7.30 Finding the smallest number and its address

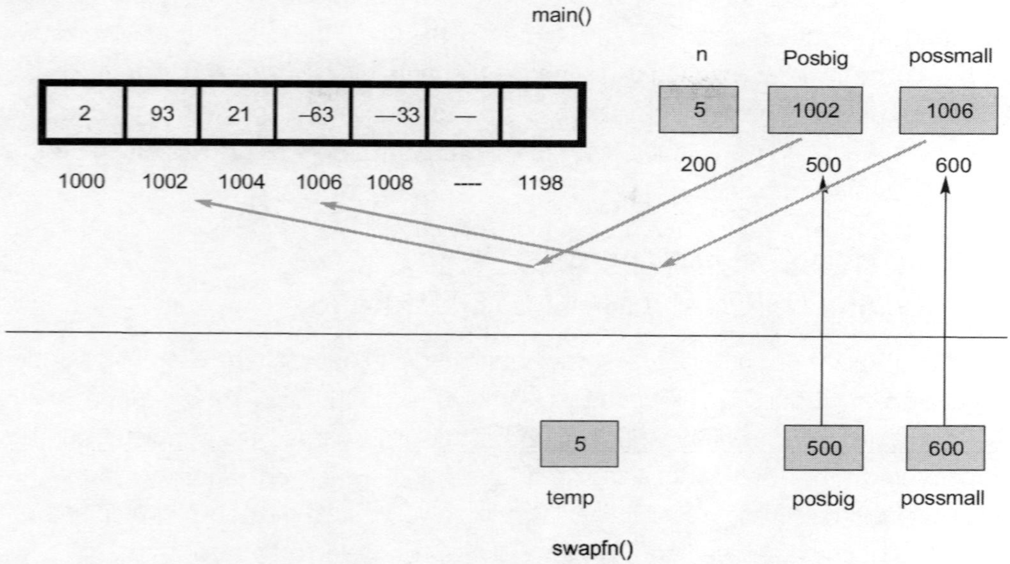

Figure 7.31 Swapping the biggest and smallest numbers

From Figure 7.31, we see that the addresses of the variables posbig and possmall are passed from the function main() and they are collected in the pointer variables posbig and possmall in the function swapfn(). We also note that the variables posbig and possmall in the function main() are pointers and they contain the addresses of the biggest and smallest numbers updated through the functions bigfn() and smallfn(). Hence the variables posbig and possmall in the function swapfn() are not just pointers, they are pointers to pointers. This is indicated with grey arrows in the figure. So, once the addresses of posbig and possmall are received in the respective pointers of the function swapfn(), the biggest and smallest numbers are swapped through the variable **temp** in the function swapfn(). The following are the equivalent lines of coding for those in the function swapfn().

```
temp=**posbig;              temp=*(*(400))=*(1002)=93
**posbig=**possmall;        *(*400))=*(*(600))=*(1006)=-63
**possmall=temp;           *(*(600))=temp=93
```

Thus at the end execution of these three lines, the value at the address of biggest number would be swapped with that at the address of the smallest number. In other words, the address of the biggest number now would have become that of the smallest number and vice versa. The pointer variable **pt** is used in all the functions (except the function swapfn()) so as not to lose the base address of the array 'a' received in 'pa'. Since 'pa' in all the functions is manipulated in the loops, we would lose the base address of the array 'a'.

I am sure the above program, explanation, screenshots and the memory pictures would have been a great deal of understanding for the learners of the book. In the discussion, we have not just covered the call by reference method of calling the function. We have covered how the base address of the array is passed on to the functions and how the concept of 'pointer to pointer' is getting implemented. I would advice the learners, to go through the entire discussion again to get the concepts in a more lucid manner.

7.9 FUNCTIONS, POINTERS AND ONE-DIMENSIONAL CHARACTER ARRAYS

We have already discussed how pointers work with character arrays in the chapter 6 (Section 6.4.2) and we have also discussed how pointers work with functions in this chapter Section 7.7). This section would be just a combination of both and above all, we had just completed our discussion on how pointers work with functions and arrays. So, I do believe that understanding and working the programs in the discussion that follows would be very much easier. We would straight away start with a program where we find whether a given string is a palindrome or not, to explain the concept. The following would be the program for that.

```
/* Checking whether the given string is a palindrome or not */
#include<stdio.h>
main()
{
char a[100];
int flag=1,len=0;
void readfn(char*);
void writefn(char*);
void lenfn(char*,int*);
void palinfn(char*,int*,int);
clrscr();
printf("\n\tEnter the string to check : ");
readfn(a);
printf("\n\n\tThe string you have entered is : ");
writefn(a);
lenfn(a,&len);
```

```
printf("\n\n\tThe length of the string is : %d",len);
if(len%2==0||len==1)
{
printf("\n\n\tThe string \"%s\" is not compatible for the test",a);
getch();
return;
}
palinfn(a,&flag,len);
if(flag)
printf("\n\n\tThe string %s is a palindrome",a);
else
printf("\n\n\tThe string \"%s\" is not a palindrome",a);
getch();
}
/* Reading the string */
void readfn(char *pa)
{
for(;(*pa=getchar())!='$';pa++);
*pa='\0';
}
/* Displaying the string */
void writefn(char *pa)
{
for(;*pa;pa++)
putchar(*pa);
}
/* Finding the length of the string */
void lenfn(char *pa,int *plen)
{
```

```
for(;*pa;pa++)
(*plen)++;
}
/* Checking for the palindromic nature */
void palinfn(char *pa, int *pflag,int len)
{
char *pt,*pj,*pk;
pt=pa;
pa=pa+len/2;
pj=pa-1;
pk=pa+1;
for(;pj>=pt&&*pk;pk++,pj--)
if(*pj!=*pk)
{
*pflag=0;
break;
}
}
```

We have four functions in the program, readfn(), writefn(), lenfn() and palinfn(). The string to be checked is read and displayed using the functions readfn() and writefn(), the length is found out using the function lenfn() and it is checked for its palindromic character using the function palinfn(). We see that for the functions readfn() and writefn(), the base address of the character array 'a' is passed from the calling function main(). For the function lenfn(), we pass the address of the variable 'len' (whose initial value is zero) along with the base address of the array 'a'. When we do so, the length gets updated in the calling function and hence when we print the same in the calling function, it is getting printed correctly. The length of the string is essential for checking the palindromic nature of the input string, as obvious from the program coding. We use a variable 'flag' (treated as a Boolean variable even though declared as an integer), whose initial value is '1'. When we call the function palinfn(), the base address of the array 'a', address of the variable 'flag' and the value of the length of the string, which is available in the variable 'len' are passed to it.

We can also see that when the function palinfn() is coded, we use a character pointer 'pt' in which we store the base address of the array that is received in the pointer 'pa', since we would not lose it when we code the condition part of the *for* loop. We see that in the function main(), once the length of the string is found out, immediately we check whether the length is odd or even and if it is even, we skip the call to the function palinfn(), since most of the palindromic strings are of odd length. Thus we make the program work only for strings which contain odd number of characters, including the spaces. Again while the string has odd number of characters or the length of the string is an odd number, we skip the value '1', which again is not valid for the palindrome test, as no palindromes would be having the length as '1'.

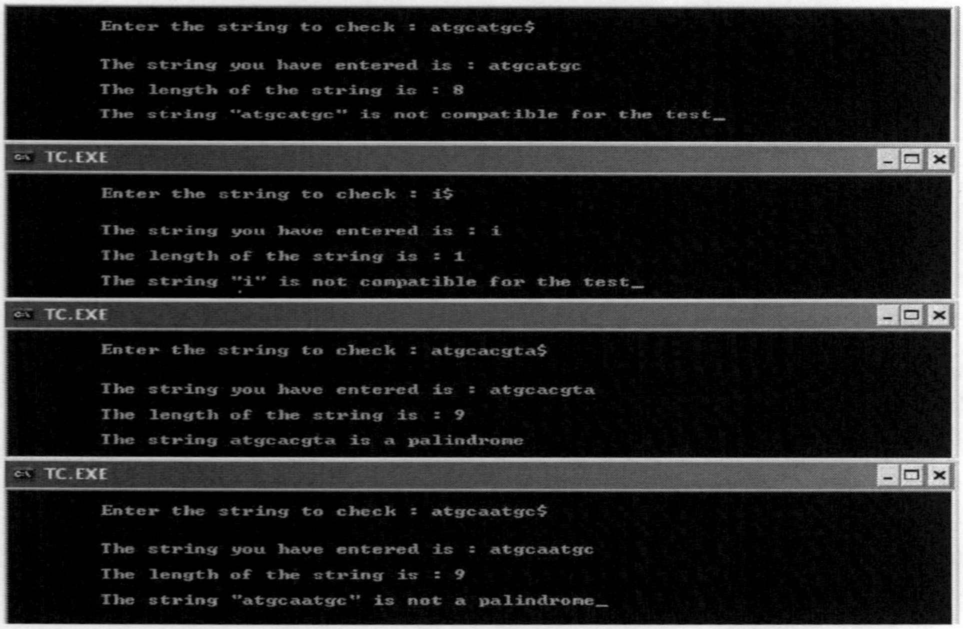

Figure 7.32 Checking the palindromic nature of a string

Figure 7.32 shows the sample run of the program for four different kinds of inputs (For a string whose length is odd, for a string whose length is '1', for a string whose length is odd and which is a palindrome, and for a string whose length is odd and which is not a palindrome).

I hope the program and output screenshot is clear to you. For more clarity, do go through the memory pictures given for each function call, in Figures 7.33–7.36 and the explanation given at the end. I leave it to the readers to figure out logic and come out with a program, when the length of the string is even and it is a palindrome (as in the case of the strings such as **peep**, **deed**, etc.).

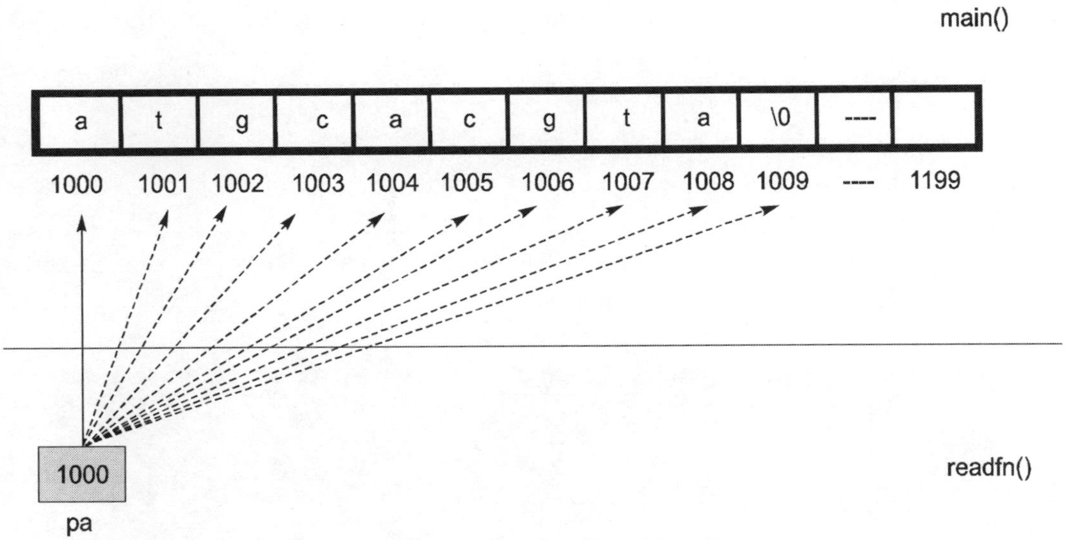

Figure 7.33 Reading the string through the function readfn()

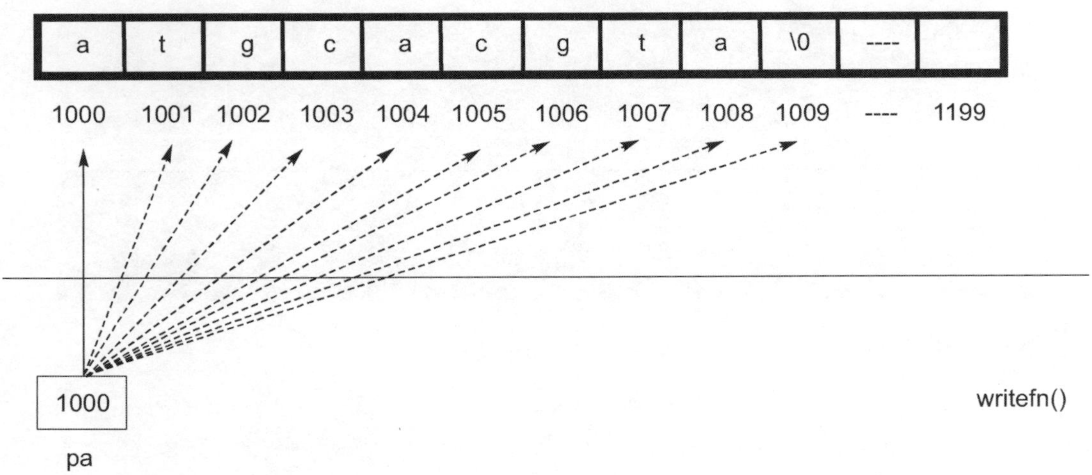

Figure 7.34 Displaying the string through the function writefn()

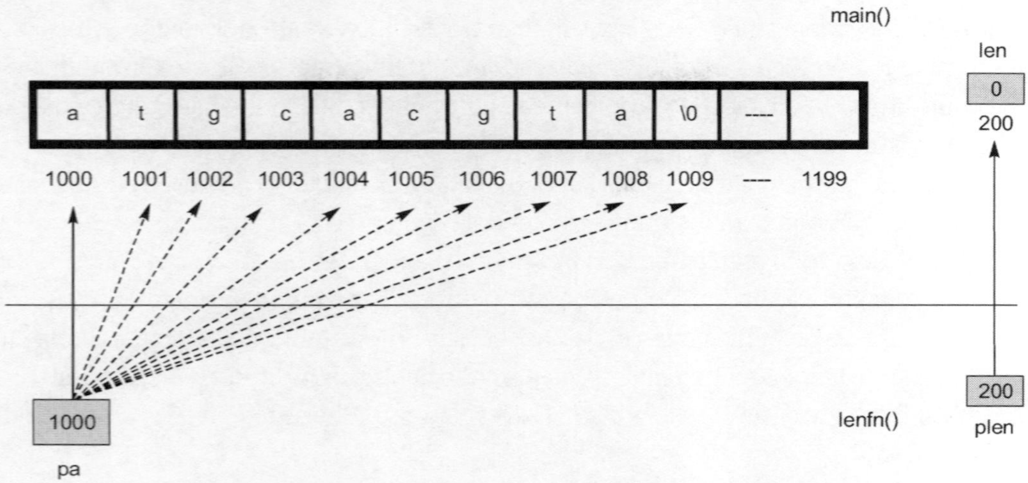

Figure 7.35 Finding the length of the string using the function lenfn()

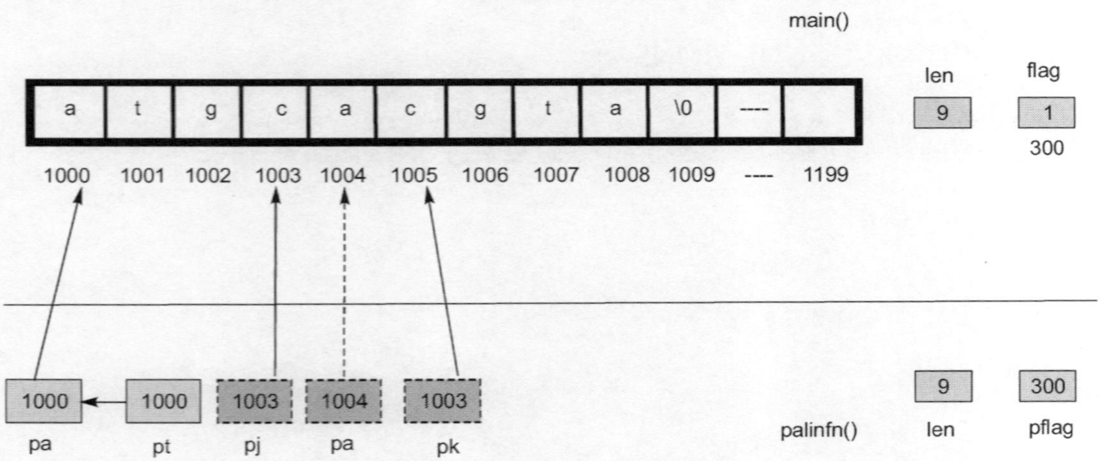

Figure 7.36 Checking the nature of the string using function palinfn()

We could see that the base address is passed while we read and display the array. Once we read the string, we store NULL character at the end of the array. So, when we read the array back and display the string, we check for the occurrence of NULL character. For the function lenfn(), we pass the base address of the array along with the address of the variable 'len', whose value is '0' from the function main(). As we read the array in the function lenfn(), we update the variable 'len' in the function main() through the pointer 'plen' in the function lenfn(). When we check the nature of the string (whether it is a palindrome or not), we pass the base address of the array along with the address of the variable 'flag' (Boolean variable) whose initial value is '1' and the value of length obtained using the earlier function. The base address of the array is collected in the pointer 'pa' and the address of the variable 'flag' is collected in the pointer 'pflag'. We have three more pointers declared in the function palinfn(), namely 'pt' for storing the base address so that we won't lose it, 'pj' for positioning one character towards left of the middle character of the given string and 'pk' for positioning one character towards right of the middle character of the given string. As we go with the coding, 'pj' is decremented and 'pk' is incremented and as we do so, we check for the equality of the characters pointed by them and if so, we proceed. Even if one mismatch occurs, then we stop by setting the value of the variable 'flag' to '0' (this is done through the pointer to flag 'pflag') and break the loop. In the function main(), we print the nature of the string (whether palindrome or not) based on the value of the variable 'flag'. If the value is '1', the string given is a palindrome and if it is '0', it is not.

Thus we are able to associate functions and pointers with arrays and build an efficient relationship among these three. You may try out how the multi-dimensional arrays work with functions and pointers. The discussion of it, however, would be beyond the scope of this book.

7.10 RECURSIVE FUNCTIONS

Recursive functions are a significant feature of C language. Any function that calls itself several times is called as a recursive function. Consider the following simple code.

```
main()
{
printf("Good Morning");
main();
}
```

What would be the result of executing the above program code? Yes, you are right. We will get the message **Good Morning** infinite number of times and the reason for this is the function main() is being called from within its coding. So there would be no termination for the execution of the program, unless the stack overflows and you get an error message as shown in the Figure 7.37.

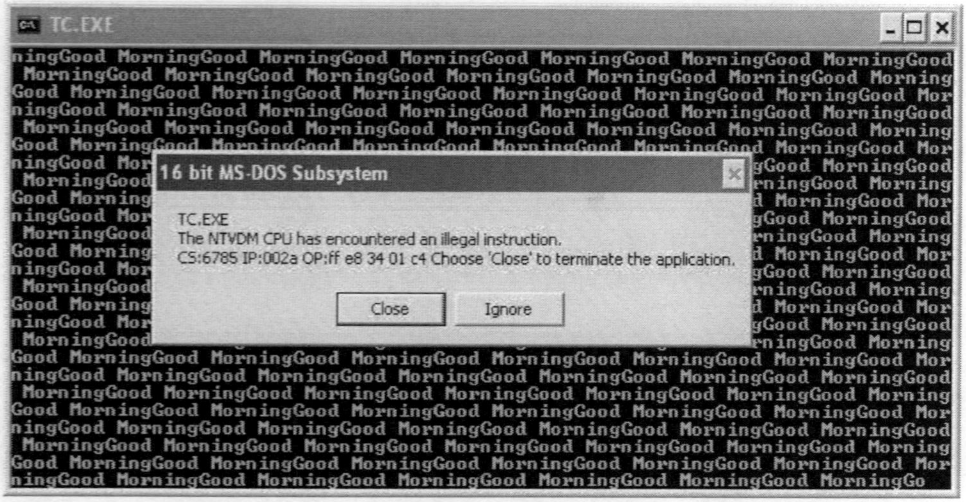

Figure 7.37 Error message displayed for the recursive program

Recursive functions are used in many applications, when we want to pass a string through a cycle and iterate it and print it as many times we would like to without really using the idea of the loops. Unlike the above program, any recursive function shall have criteria for terminating it. Otherwise, it will never terminate and you will get error as shown in the figure. The key advantage of using recursive functions is that, as said earlier, they simply replace loops as the recursive functions themselves are iterative in nature. The disadvantage of using recursive functions is that computationally they are very expensive since more number of interrupts will occur during the program execution.

But, in most of the applications involving calculations and string processing, recursion is very useful. The most important thing one should keep in mind before using recursive logic is that, the function has to terminate at some point of time once a proper condition is given in the function coding. Be aware of the fact that what can be done with recursion, could be done away even without it. So use recursive logic in your programs only when their need is essential. Let us consider a program where we find out the factorial of a given unsigned integer using recursive function. The following is the program code for that.

```
/* Factorial of a given number using recursion */
#include<stdio.h>
#include<conio.h>
void main()
```

```
        {
        unsigned long f;
        int n;
    unsigned long fact(int);
        clrscr();
        printf("\n\tEnter the number : ");
        scanf("%d",&n);
        f=fact(n);
        printf("\n\tThe factorial of the number %d is %lu",n,f);
        getch();
        }
    unsigned long fact(int x)
        {
        unsigned static long f1;
        if(x<=1)
        return 1;
        else
        f1=x*fact(x-1);
        return f1;
      }
```

We see that in the function fact(), we have the condition for termination as if(x==1). When the condition stated above is true, the control comes back to the calling function and prints the value of the factorial. If we give an input '5' for the variable 'n' in the function main(), it is processed in the function fact() recursively and finally when the control returns to the function main(), it would have the factorial of '5', which is '120' and it would be printed then. We also note from the function fact() that, the variable 'f1' is declared as unsigned static long. When we declare something as static, the initial value of that variable would be '0' and it will retain the values between the function calls. Figure 7.38 shows the factorial value for the input '12'. The flow of the control inside the function fact() while it is executed is shown in the Figure 7.39. For the sake of convenience, we have taken the value of 'n' as '5'.

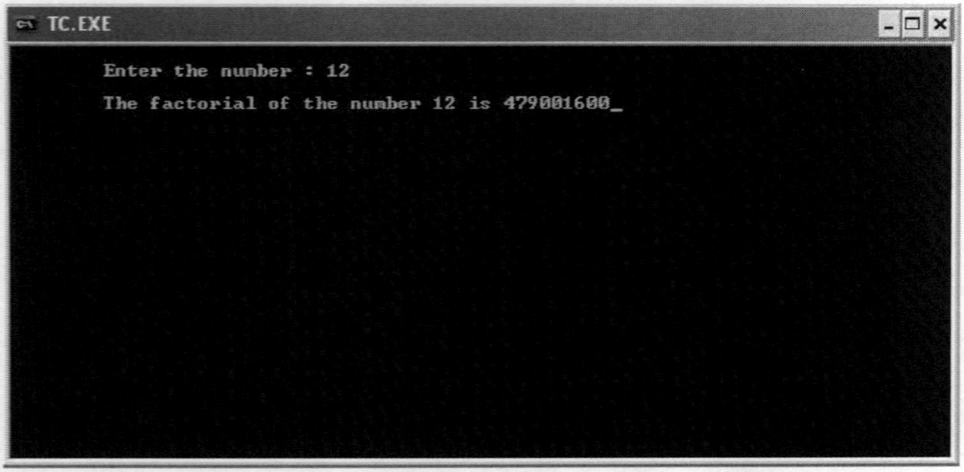

Figure 7.38 Factorial of '12' using recursion

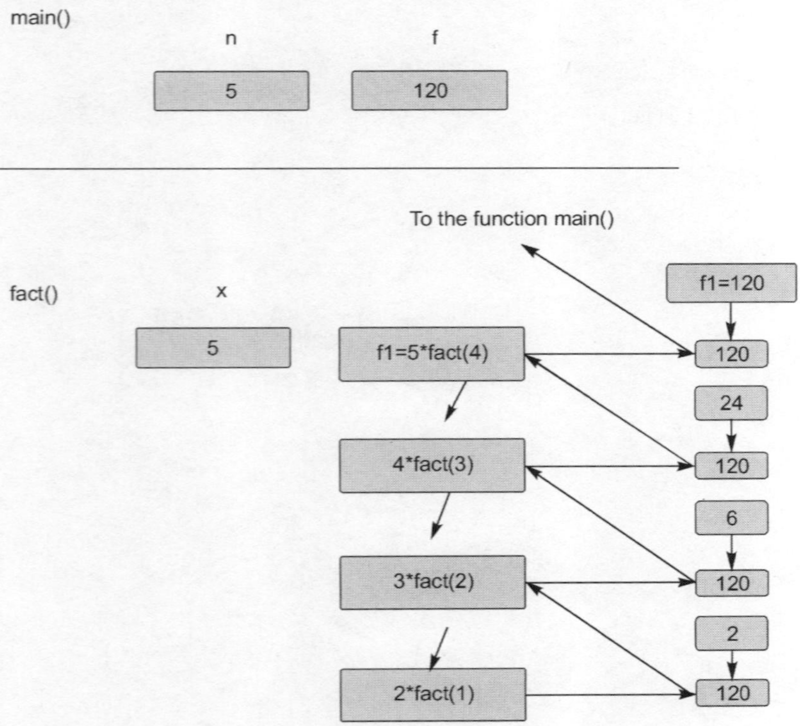

Figure 7.39 Factorial of '5' using recursion—flow of control

We note that when the value of 'x' reaches '1', a value of '1' is returned to the calling function. Thus the expression 2*fact(1) will evaluate to 2, which is returned to its calling function, which is recursively the function fact() and it is multiplied with the value '3' and the result '6' is returned to its calling function which is again fact() and this goes on and on and finally the control comes back to the calling function main() with the final result '120' and this gets printed in the output.

Let us work out another program with string, where we would print the reversed form of the given input string. The following is the program code for that.

```
/* Reversing a string using recursion */
#include<stdio.h>
#include<conio.h>
int i;
char a[100];
void main()
{
void rev(int);
clrscr();
printf("Enter the string : ");
for(i=0;(a[i]=getchar())!='$';i++);
a[i]='\0';
i--;
printf("\nReversed string : ");
rev(i);
getch();
}
void rev(int j)
{
if(j<0)
return;
else
{
printf("%c",a[j]);
j--;
rev(j);
}
}
```

The output of the above program is shown in the Figure 7.40.

Figure 7.40 Reversing a string using recursion

I would suggest you to make out the flow of control inside the function rev() and hence draw the memory picture. With this program we would wind up the discussion of recursive functions and their usage.

For recursion, additional work has to be done in the form of saving return addresses, so that the correct statement is executed next when the *return* is executed. Compared to iteration, recursion is more efficient.

SUMMARY

This entire chapter has dealt with the concept of functions with a special attention to the user-defined functions. We started with explaining the reasons for having user-defined functions and we discussed the categories of the functions. We also discussed the methods of accessing the functions namely call by value and call by reference. Following that we discussed what function prototyping is and how significant is prototyping in C language. Different types of function headers were discussed in brief and we wrote some simple functions and validated them with C programs.

Based on our understanding of the basic usage of functions, we started discussing the way the functions work with one-dimensional and two-dimensional numerical arrays. We had an elaborate discussion on how the numerical arrays are passed on to the functions and how they are accessed and manipulated in the functions. We continued the discussion with the one-dimensional character arrays and we did a few programs to understand how the character arrays work with the functions.

We then followed the discussion of functions and pointers and what is the importance of using pointers with functions. We discussed a few programs that demonstrated the importance of passing addresses to the functions and thereby return more than one value from the functions. This section has provided the basis for the subsequent sections, where we discussed how the pointers and functions work with the numerical and character arrays. We discussed elaborately the way the pointers and functions work with the arrays. We had screenshots and memory pictures in the sections for a better understanding of the concepts and the technical details involved in a program.

We had a discussion on recursive functions and their applications in the concluding part of this chapter. We illustrated two programs for getting the concept in a much better way, with necessary screenshots and memory pictures.

I am sure the concepts that are provided in this chapter would have been a great amount of significance to the learners and obviously the concepts would be a good foundation for those who aspire to develop the best algorithms.

CRACK THE NUT

Identifying the output for the following programs:

1. small(int,int);

```
main()
{
Int a,b;
scanf("%d%d",&a,&b);
small(&a,&b);
}
small(int p,int q)
{
```

```
if(p==r)
printf("The numbers are equal");
else
if(p<r)
printf("The smaller number is : %d",p);
else
printf("The smaller number is : %d",q);
}
```

2. main()

```
{
int a,b,diff;
printf("Enter the integers :  ");
scanf("%d%d",&a,&b);
diff=sub(a,b);
printf("The difference is %d",diff);
int sub(int p, int q)
return p-q;
}
```

3. main()

```
{
printf("The following message is displayed");
message();
}
message();
{
printf("Good Morning");
}
```

4. main()

```
{
Int n,sqr;
Printf("Enter the number : ");
scanf("%d",&n);
sqr=sqrvalue(&n);
printf("The square of %d is %d",n,sqr);
}
int sqrvalue(int a)
{
return a*a;
}
```

5. main()

```
{
int x,y,z,sum ;
printf("Enter the numbers : ");
scanf("%d%d%d",&x,&y,&z);
sum=add(x,y,z);
printf("The total is %d",sum);
}
int add(int p, int q, int r)
{
return(p,q,r);
}
```

6. main()

```
{
int num=10;
fn(num);
printf("%d",num);
```

```
}
fn(int num)
{
return ++num;
}
```

7. main()

```
{
printf("\nUVCE");
main();
}
```

8. void main(int x, int y)

```
{
printf("%d",x) ;
}
```

9. main()

```
{
char x=67, y='c';
fn(x,y);
}
fn(x,y)
{
printf("x=%d, y%d", x, y);
}
```

10. main()

```
{
int m=50, n=60, x;
x=fn(m,n) ;
printf("x=%d",x);
}
```

```
fn(int m, int n)
{
int y, z;
y=m+n;
z=m*n;
return (z,y,z+y);
}
```

11. main(){

```
int total;
total=sum(2);
printf("%d",total);
}
int sum(int i){
static int even=0;
if(i<=20){
even=even+i;
sum(i+2);
}
return even;
}
```

12. int max(int x,int y){return x>y?x:y;}

```
void main(){
int m;
m=max(max(4,max(11,6)),max(10,5));
printf(«%d»,m);
getch();
}
```

13. main(){

```
int x,num=4;   clrscr();
```

```
x=call(num);
printf("%d",x);   getch();}
int call(int num){
static int x=1,y;
if(num>0)
{
x=x*num;
y=call(num-1)+call(num-2);
}
return x;
}
```

14. main(){

```
int x,num=5;   clrscr();
x=call(num);
printf("%d",x);   getch();}i
nt call(int num)
{   static int x=1,y;
if(num>0){
x=x+num;
y=call(num-2)+call(num-3);
}
return x;
}
```

15. main(){

```
int x,num=1;
clrscr();   x=call(num);
printf("%d",x);   getch();}
int call(int num){
static int x=0;
```

```
if(num<4)
{
x=2*x-num;        call(num+1);
call(num+2);  }  return x;
}
```

8

FILES AND COMMAND LINE ARGUMENTS

8.1 INTRODUCTION TO FILES

Files are places for reading data from or writing data to. This includes disk files and its devices such as the printer or the monitor of a computer. C treats all information which enters or leaves a program as though it were a stream of bytes: a file. The most commonly used file streams are stdin (the keyboard) and stdout (the screen), but more sophisticated programs need to be able to read or write to files which are found on a disk or to the printer, etc.

An operating system allows a program to see files in the outside world by providing a number of channels or portals (inlets and outlets) to work through. In order to examine the contents of a file or to write information to a file, a program has to open one of these portals. The reason for this slightly indirect method of working is that channels/portals hide operating-system-dependent details of filing from the programmer. Think of it as a protocol. A program which writes information does no more than pass that information to one of these portals and the operating system's filing subsystem does the rest. A program which reads data simply reads values from its file portal and does not have to worry about how they got there. This is extremely simple to work in practice.

The standard library in C has many file I/O functions. The file-handling functions, like the other functions, are easy to use and above all they are powerful and complete. This chapter elaborates the various file operations and file manipulations that are important for any person who uses C to develop an application. The following is the sequence of I/O operations while using files.

Open the file

Read from or write to the file

Close the file

A program opens a file by calling a standard library function and is returned a file pointer by the operating system, which allows a program to address that particular file and to distinguish it from all others.

C provides two levels of file handling; these can be called high level and low level. High level files are all treated as text files. In fact, the data which go into the files are exactly what would be seen on the screen, character by character, except that they are stored in a file instead. This is true whether a file is meant store characters, integers or floating point types. Any file, which is written to by high level file handling functions, ends up as a text file, which could be edited by any text editor.

High level text files are also read back as character files, in the same way that input is acquired from the keyboard. This all means that high level file functions are identical in concept to keyboard/ screen input/output.

The alternative to these high level functions is obviously low level functions. These are more efficient, in principle, at filing data as they can store data in large lumps, in raw memory format, without converting to text files first. Low level input/output functions have the disadvantage that they are less programmer-friendly than the high level ones, but they are likely to work faster.

8.2 FILE POSITIONS

When data are read from a file, the operating system keeps track of the current position of a program within that file so that it only needs to make a standard library call to 'read the next part of the file' and the operating system obliges by reading some more and advancing its position within the file, until it reaches the end. Each single character which is read causes the position in a file to be advanced by one.

Although the operating system does a great deal of task regarding file positions, a program can control the way in which that position changes, if need be. In most cases it is not necessary and it should be avoided, since complex movements within a file can cause complex movements of a disk drive mechanism which in turn can lead to wear on disks and the occurrence of errors.

8.3 HIGH-LEVEL FILE HANDLING FUNCTIONS

Most of the high level input/output functions which deal with files are easily recognizable in that they start with the letter f. Some of these functions will appear strikingly familiar. For instance:

```
fprintf()
fscanf()
fgets()
fputs()
```

These are all generalized file handling versions of the standard input/output library. They work with generalized files, as opposed to the specific files stdin and stdout which printf() and scanf() use. The file versions differ only in that they need an extra piece of information: the file pointer to a particular portal. This is passed as an extra parameter to the functions. They process data in an identical way to their standard I/O counterparts. Other filing functions will not look so familiar. For example:

```
fopen()
fclose()
getc()
ungetc();
putc()
fgetc()
fputc()
feof()
```

Before any work can be done with high level files, these functions need to be explained in detail. This is exactly what is done in the following sections.

8.4 OPENING A FILE

Consider the following program coding:

```
/* Demonstrating the file operations in C */
#include <stdio.h>
main( )
{
FILE *fp;
char name[100];
```

```
int index;
clrscr();
fp = fopen("sample.txt","w"); /* Open the file for writing */
/* Some lines of coding go here */
strcpy(name,"Welcome to the World of C Programming");
for (index = 1;index <= 10;index++)
fprintf(fp,"%s Line number %d\n",name,index);
fclose(fp); /* Close the file before ending the program */
printf("\nThe line is written specified number of times in the file you have specified!!!");
getch();
}
```

We begin as before with the "#include" statement for "stdio.h", then define some variables for use in the example including a rather strange-looking new type. The type FILE is used for a file variable and is defined in the "stdio.h" file. It is used to define a file pointer for use in file operations. The definition of C contains the requirement for a pointer to a FILE, and as usual, the name can be any valid variable name (follow the variable-naming conventions here).

Before we can write to a file, we must open it. What this really means is that we must tell the system that we want to write to a file and what the filename is. We do this with the function fopen() as illustrated in the program. The file pointer, fp in our case, points to the file and two arguments are required in the parentheses, the filename first, followed by the mode of accessing the file.

The first argument is the filename, which is any valid filename and can be expressed in upper or lower case letters, or even mixed if you so desire. It is enclosed in double quotes. For this example we have chosen the name sample.txt. This file should not exist on your disk at this time. If you have a file with this name, you should change its name or move it because when we execute this program, its contents will be erased. If you don't have a file by this name, that is good because we will create one and put some data into it.

The second argument is the file attribute and can be any of three letters, "r", "w", or "a", and must be lower case. When an "r" is used, the file is opened for reading, a "w" is used to indicate a file to be used for writing, and an "a" indicates that you desire to append additional data to the data already in an existing file.

Opening a file for **reading** requires that the file already exist. If it does not exist, the file pointer will be set to NULL and the same can be checked by the program. When a file is opened for **writing**, it will be created if it does not already exist and it will be reset if it does resulting in deletion of any data already there. When a file is opened for **appending**, it will be created if it does not already exist and it will be initially empty. If it does exist, the data input point will be the end of the present data so that any new data will be added to any data that already exists in the file.

8.5 OUTPUTTING TO A FILE

The job of actually outputting to the file is nearly identical to the outputting we have already done to the standard output device. The only real differences are the new function names and the addition of the file pointer as one of the function arguments. In the programs that access files, the function **fprintf()** replaces our familiar function printf() and the file pointer defined earlier is the first argument within the parentheses. The remainder of the statement looks like, and in fact is identical to, the function printf().

8.6 CLOSING A FILE

To close a file, you simply use the function **fclose()** with the file pointer in the parentheses. Actually, in this simple program, it is not necessary to close the file because the system will close all open files before returning to DOS. It would be good programming practice for you to get in the habit of closing all files in spite of the fact that they will be closed automatically, because that would act as a reminder to you of what files are open at the end of each program. You can open a file for writing, close it, and reopen it for reading, then close it, and open it again for appending, etc. Each time you open it, you could use the same file pointer, or you could use a different one. The file pointer is simply a tool that you use to point to a file and you decide what file it will point to.

8.7 WRITING SIMPLE PROGRAMS IN FILES

Compile and run the above program. You will only get the message that is shown in the Figure 8.1. To get the contents of the actual file, look at your directory for a file named **sample.txt** and use the DOS command **type** to view its contents. That is where your output will be. Compare the output with that specified in the program. It should agree. If you desire to choose the DOS prompt from within C editor, you may select the option **File** from the editor (by pressing Alt+f) and then select the option **OS shell** (by pressing the letter O from the keyboard) as shown in Figure 8.2. When you view the contents it shall be the one as shown in the Figure 8.3. Do not erase the file named sample.txt as we will use it later in this chapter.

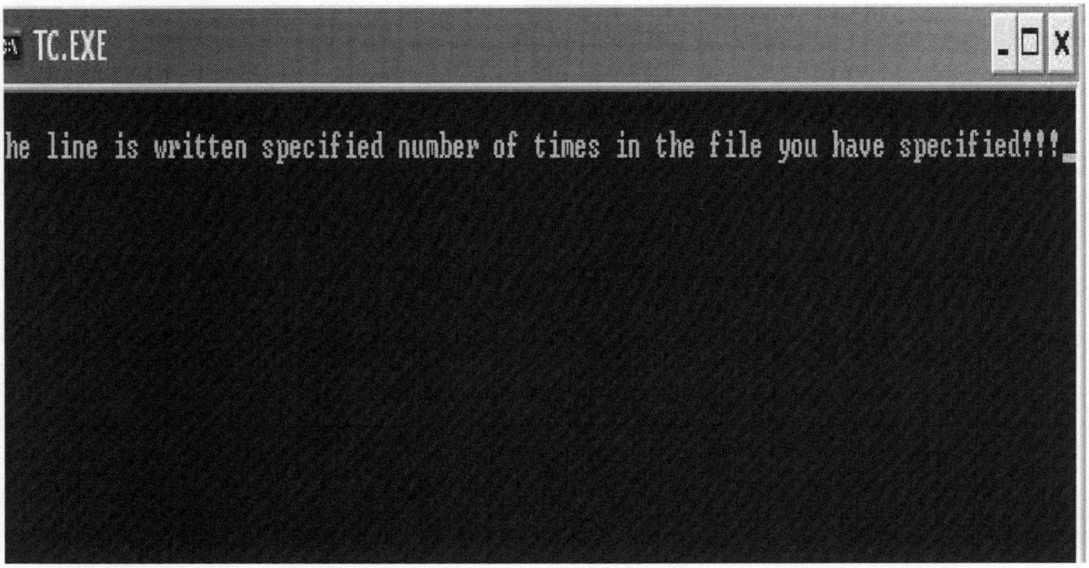

Figure 8.1 Output for the program that writes data to the file

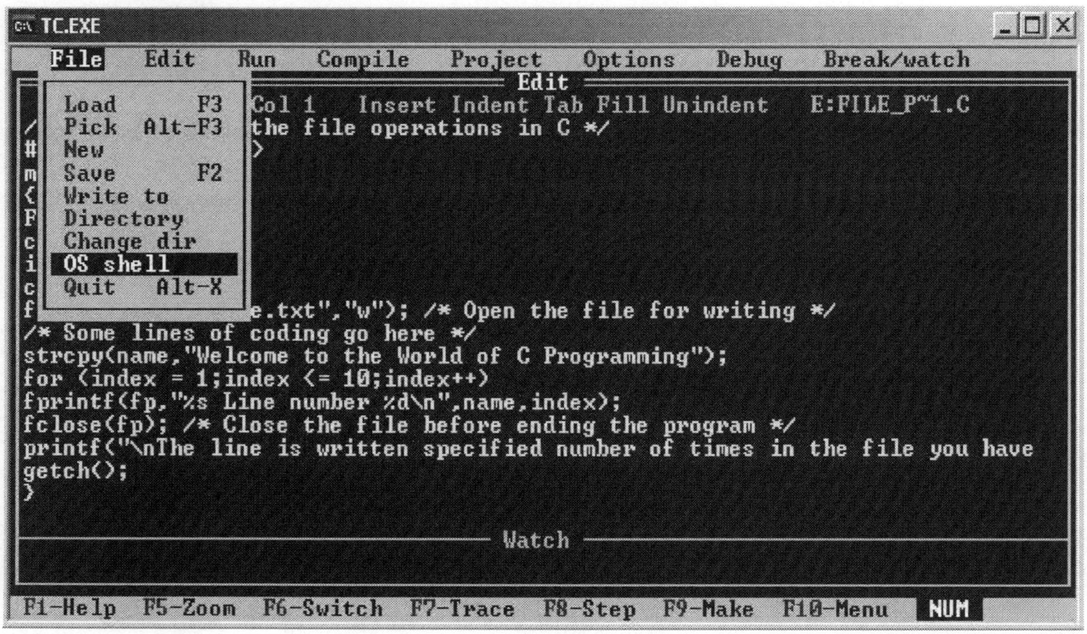

Figure 8.2 Getting into the DOS prompt from within the C editor

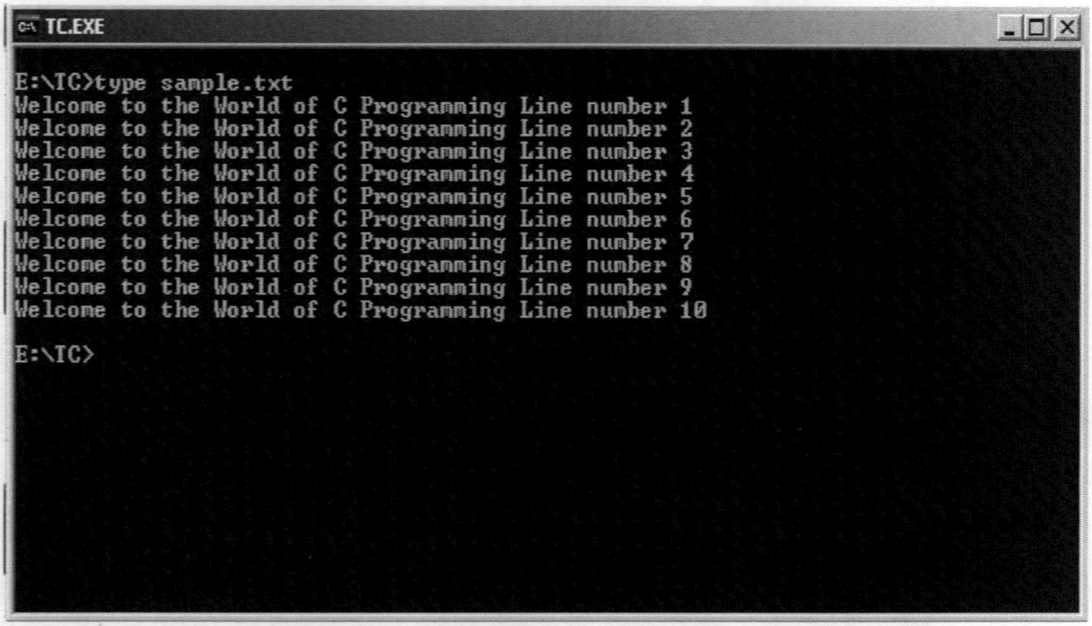

Figure 8.3 Contents of the file sample.txt in the directory

Consider the following program coding:

```
/* Appending contents to an existing file */
#include <stdio.h>
main()
{
FILE *fp;
char name[100];
int index,count; clrscr();
strcpy(name," Know the wonders of C by exploring it!!!");
fp = fopen("SAMPLE.TXT","a"); /* Open for appending */
for (count = 1;count <= 10;count++)
{
for (index = 0;name[index];index++)
putc(name[index],fp); /* Output a single character to the file */
putc('\n',fp); /* Output a linefeed to the file */
}
```

```
fclose(fp);
printf("\nThe given contents is appended successfully");
getch(); }
```

Figure 8.4 shows the output for the above program. We view the contents of the file **sample.txt** in which the appending is carried out.

Figure 8.4 Appended contents of the file sample.txt

The program begins with the "include" statement, then defines some variables including a file pointer. We have named the file pointer as **fp**. We then define a string of characters to use in the output function using the function strcpy(). We are ready to open the file for appending and we do so in the function fopen(), except this time we use the upper cases for the filename and its extension. This is done simply to illustrate that DOS does not care about the case of the filename. Notice that the file will be opened for appending so we will add to the lines inserted during the last program. The program is actually coded with two nested *for* loops. The outer loop is simply a count to ten so that we will go through the inner loop ten times. The inner loop calls the function putc() repeatedly until a character in the array **name** is detected to be a zero.

The part of the program we are interested in is the function putc(). It outputs one character at a time, the character being the first argument in the parentheses and the file pointer being the second and last argument. Why the designer of C made the pointer first in the function fprintf(), and last in the function putc() is a good question for which there may be no answer. It seems like this would have been a good place to have used some consistency. When the array name is exhausted, a new line is needed because a new line was not included in the definition above. A single **putc** is then executed which outputs the "\n" character to return the carriage and do a linefeed. When the outer loop has been executed ten times, the program closes the file and terminates. Compile and run this program but once again there will be no output to the monitor. Following execution of the program, view the contents of the file using the DOS command **type** and you will see that the 10 new lines were added to the end of the 10 that already existed. If you run it again, yet another 10 lines would be added.

Now that we have been able to write contents to the file and append the same, we are ready to read the contents that we have written and then appended. For that we may have a coding like the one shown below.

```
/* Program to read the contents of a file – One character at a time */
#include <stdio.h>
main( )
{
FILE *fp;
int c;
clrscr();
fp= fopen("sample.txt","r");
if (fp == NULL)
{
printf("File doesn't exist\n");
getch();
return;
}
do
{
c = getc(fp); /* Get one character from the file */
putchar(c); /* Display it on the monitor */
} while (c != EOF); /* Repeat until EOF (end of file) */
fclose(fp);
getch();
}
```

This program begins with the familiar "include", some data definitions, and the file opening statement which should require no explanation except for the fact that an "r" is used here because we want to read it. In this program, we check to see that the file exists, and if it does, we execute the main body of the program. If it doesn't, we print a message and quit. If the file does not exist, the system will set the pointer equal to NULL which we can test.

The main body of the program is the *do-while* loop in which a single character is read from the file and output to the monitor until an EOF (end of file) is detected from the input file. The file is then closed and the program is terminated. Figure 8.5 shows the output for the above program coding (see that we have read the file sample.txt and the contents of it are displayed on the screen). We could use any loop other than a *do-while* and I would appreciate the learners if they use the other two loops to achieve the same result.

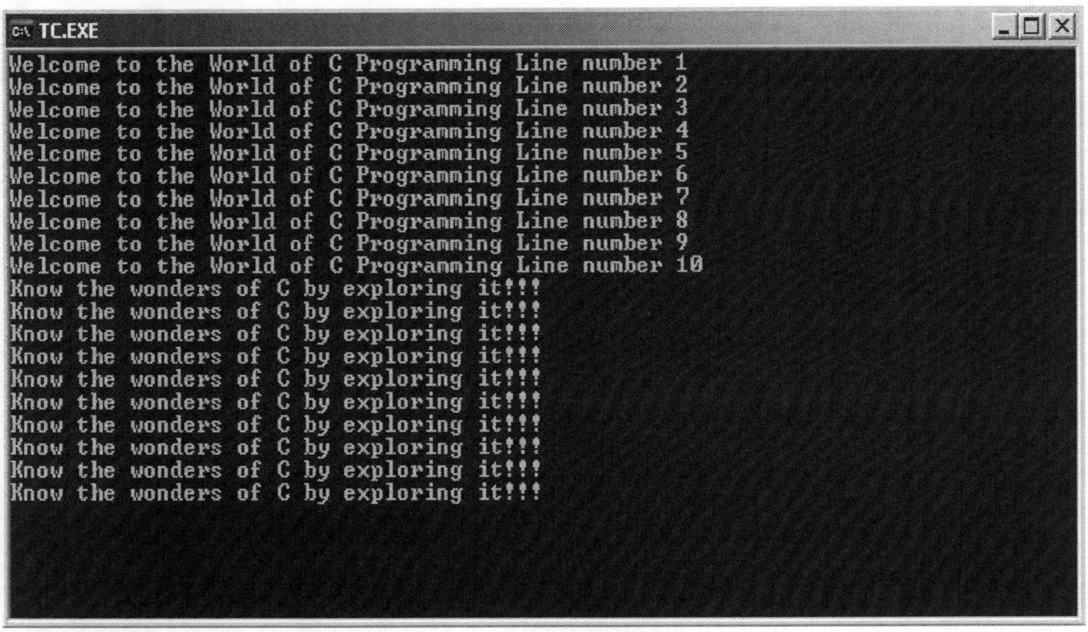

Figure 8.5 Reading the contents of sample.txt character by character

At this point, we have the potential for one of the most common and most perplexing problems of programming in C. The variable returned from the function getc() is a character (based on the prototype), so we could use a "char" variable for this purpose. There is a problem with that however, because on some, if not most, implementations of C, the EOF returns a minus one which a "char" type variable is not capable of containing. A "char" type variable can only have the values of zero to 255, so it will return a 255 for a minus one on those compilers that use a minus one for EOF. This is a very frustrating problem to try to find because no diagnostic is given. The program simply can never find the EOF and will therefore never terminate the loop. This is easy to prevent, always use an "int" type variable for use in returning an EOF. You can tell what your compiler uses for EOF by looking at the "stdio.h" file where EOF is defined. That is the standard place to define such values.

After you compile and run this program and are satisfied with the results, it would be a good exercise to change the name of "sample.txt" and run the program again to see that the NULL test actually works as stated. The output when the file name is changed is shown in the Figure 8.6.

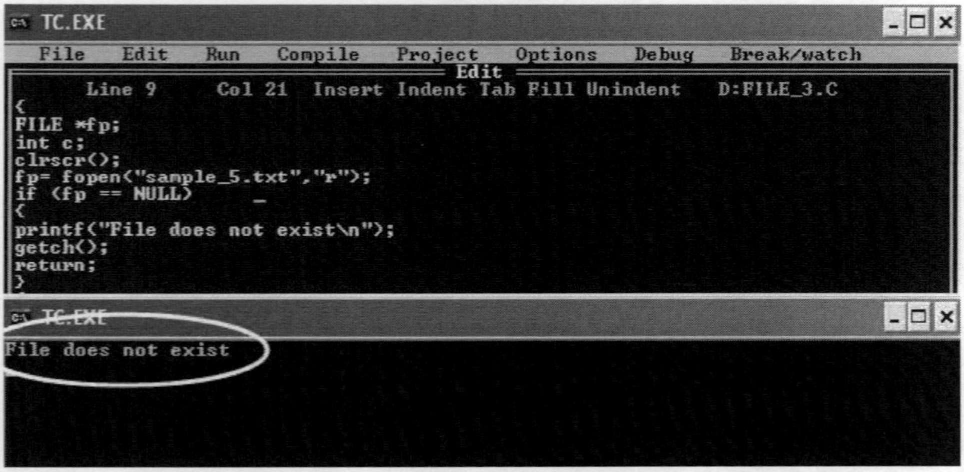

Figure 8.6 Reading the contents of the file sample_5.txt

In the above program, we have read one character at a time from the file and displayed the same on the monitor. But if we want to read one word at a time, we could do it and the following would be the modified coding for the same.

```
/* Program to read the contents of a file - One word at a time */
#include <stdio.h>
main( )
{
FILE *fp;
int c;
char word[100];
clrscr();
fp= fopen("Gandhi.txt","r");
if (fp == NULL)
```

```
{
printf("File does not exist\n");
getch();
return;
}
do
{
c = fscanf(fp,"%s",word); /* Get one word from the file */
printf("%s\n",word); /* Display it on the monitor */
} while (c != EOF); /* Repeat until EOF (end of file) */
fclose(fp);
getch();
}
```

The output of the above program is shown in the Figure 8.7 below.

Figure 8.7 Reading the contents of sample.txt word by word

Inspection of the program output (from Figure 8.7) will reveal that when we read data in and detect the EOF, we print out something before we check for the EOF resulting in an extra line of printout. What we usually print out is the same thing printed on the prior pass through the loop because it is still in the buffer of the array **word**. We therefore must check for EOF before we execute the function printf(). Compile and execute the above program and observe the output. There will be additional lines displayed because of the function printf() before checking for EOF. Thus in the following program coding we use NULL instead of checking EOF.

In the earlier two examples, we read the file character by character and then word by word. Now we will try reading the contents line by line. The following would be coding required for the same.

```
/* Program to read the contents of a file - One line at a time */
#include <stdio.h>
main( )
{
FILE *fp;
char *c;
char word[100];
clrscr();
fp= fopen("Gandhi.txt","r");
if (fp == NULL)
{
printf("File does not exist\n");
getch();
return;
}
do
{
c = fgets(word,100,fp); /* Get one line from the file */
if(c!=NULL)
printf("%s",word); /* Display it on the monitor */
} while (c != NULL); /* Repeat until NULL */
fclose(fp);
getch();
}
```

Figure 8.8 is the output of the above coding and we see that there are no additional lines as in Figure 8.7. Due to the presence of additional lines, we get a partial output in Figure 8.7 and this is rectified in the above program and hence we get the complete contents of the file.

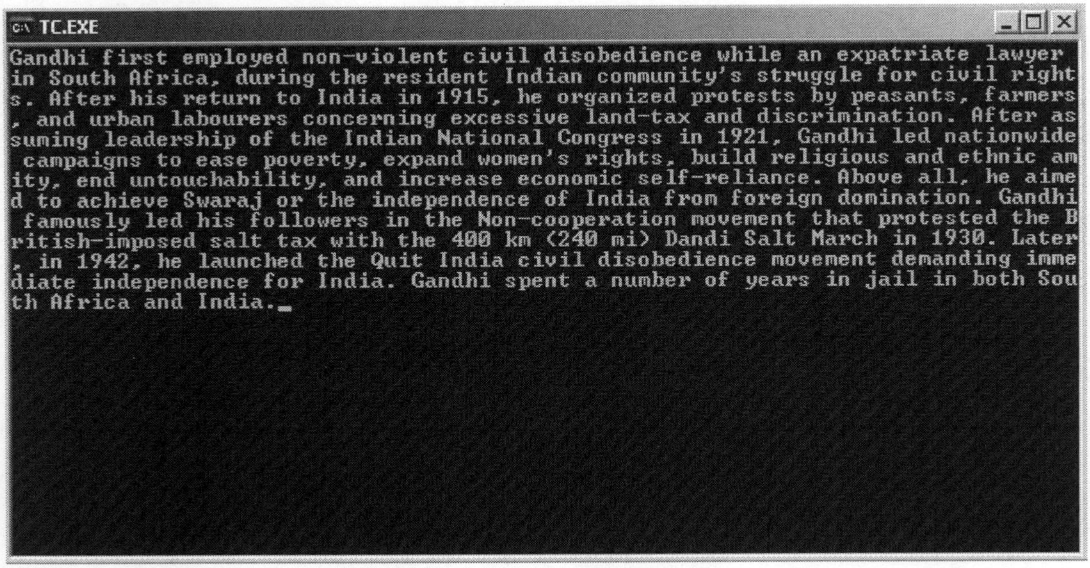

Figure 8.8 Reading the contents of sample.txt line by line

8.7.1 How to use a Variable File Name?

So far we have been opening the file by specifying the name with extension directly as one of the arguments to the function fopen(). What if we want to open the file by giving the name with extension through the keyboard? This is possible by just doing a minor modification in the above program code. I have presented the new coding below and it should be clear to you in all the aspects, since we have dealt with them already.

```
/* Program to read the contents of a file - One line at a time – Getting File Name from the keyboard using the
function scanf() */
#include <stdio.h>
main()
{
FILE *fp;
char *c;
char word[100], filename[100];
clrscr();
printf("\nEnter the file name -> " );
scanf("%s",filename); /* Read the name of the desired file to be opened */
printf("\n\n" );
```

```
fp= fopen(filename,"r");
if (fp == NULL)
{
printf("File does not exist\n");
getch();
return;
}
do
{
c = fgets(word,100,fp); /* Get one line from the file */
if(c!=NULL)
printf("%s",word); /* Display it on the monitor */
} while (c != NULL); /* Repeat until NULL */
fclose(fp); getch();
}
```

If you compile and execute this program, it will request the file name from you. Enter the desired file to be opened. The file you desire could be even a .c file and not necessarily .txt file. The output for the above program is shown in Figure 8.9.

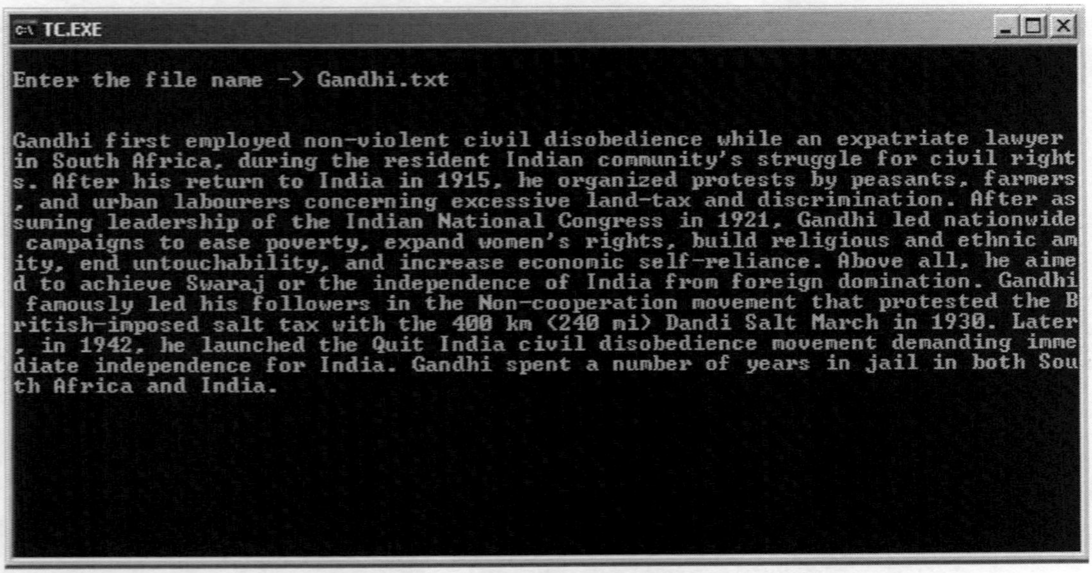

Figure 8.9 Reading the file whose name is given dynamically

There is another point the learners must observe and understand here. When we enter the name of the file, we are supposed to give the complete path of the file if we are not displaying the contents of that file which is available in the working directory of C. In the output shown above, we have chosen to display a .c file which is available in the directory specified from the prompt and it is not the one in the working directory of C.

8.8 COMMAND LINE ARGUMENTS—WHAT AND WHY?

When we use files, if we incorporate the idea of command lines, the program would be more efficient. What are these command line arguments and what is a command line after all?

If you have observed the example programs that we have used in all the Chapters, we have used the function main() invariably in all the programs and as mentioned in Chapter 2, it is the heart-throb of any C program. We know from the concept of functions that a function could be called by passing values or addresses as arguments to it (from Chapter 7 on functions). Have we ever passed any arguments to the function main()? The answer is very clear. We have never used such idea in any of our programs. Is it possible for us to pass arguments to the function main() as we do with the other functions (both built-in and user-defined)? The answer is yes. If you recall the function prototyping discussion, we had used two terms namely **caller** and **callee**. The **caller** function is that which calls some other function from within its code and the **callee** function is that which is called from another function. Invariably all the functions do have a **caller** for them. So, who is the caller for the function main()? The operating system is the caller for the function main() and one can pass arguments from the operating system to the function main() and make the program work. In our discussion, we have used DOS as the platform and hence we would say DOS is the **caller** for the function main(). Hence if we want to pass arguments to the function main(), we could do so from the DOS prompt, which is called as the **Command Line**. I am sure that by now you would have understood what command line arguments are. In simple words, the arguments that are passed from the command line are called the command line arguments and they are passed to the function main(). They are significant in programming in the sense that the user can execute such programs from the command line and he/she need not get into the coding of the program while validating the program. The following is a simple example of a program where we have used the concept of command line arguments. Here we read the contents of the file and display the same on the monitor.

```
/* Program to read the contents of a file - Command Line Arguments */
#include <stdio.h>
main(int argc, char *argv[])
{
```

```
FILE *fp;
char *c;
char word[100], filename[100];
/*clrscr();*/
if(argc<2)
{
printf("\nUsage Error");
getch();
return; }
fp= fopen(argv[1],"r");
if(fp==NULL)
{
printf("File does not exist\n");
getch();
return;
}
/* Printing the addresses of the arguments */
printf("\n\n\nAddress of the first argument : %u\nAddress of the second argument :
%u\n\n\n",argv[0],argv[1]);
do
{
c = fgets(word,100,fp); /* Get one line from the file */
if(c!=NULL)
printf("%s",word); /* Display it on the monitor */
}while (c!=NULL); /* Repeat until NULL */
fclose(fp);
getch();
}
```

We see that we pass the arguments to the function main() from the DOS prompt and hence get the display of the contents of the file that we want to view. This program shall be compiled first using the key F9 as usual, but the execution shall not be done by pressing the key combination ctrl+F9 as in the earlier cases. Choose the option **File** from the menu and choose the option OS shell. Then type the .exe file name of the presently executed file (invariably .exe file name shall be the same as the original .c source file) as the first argument and the second argument shall be the file name or the path of the filename whose contents are going to be displayed. The command that is to be given from the DOS prompt or the OS shell is the one as given below for the above program.

```
d:\progra~1\tc\bin>file_7 d:\swap_fun.c
```

Assume that the program is saved as file_7.c and the file whose contents are to be viewed is swap_fun.c and it is stored in the D drive of the system. When the .c file is compiled to success, we would get the corresponding .exe file and hence we are ready to execute the program from the command line. Once we press the enter key after the commands are given, we would get the display of the contents of the file swap_fun.c on our screen. Figure 8.10 shows the output along with the command line command. We can display any type of file compatible for the display using these command line parameters.

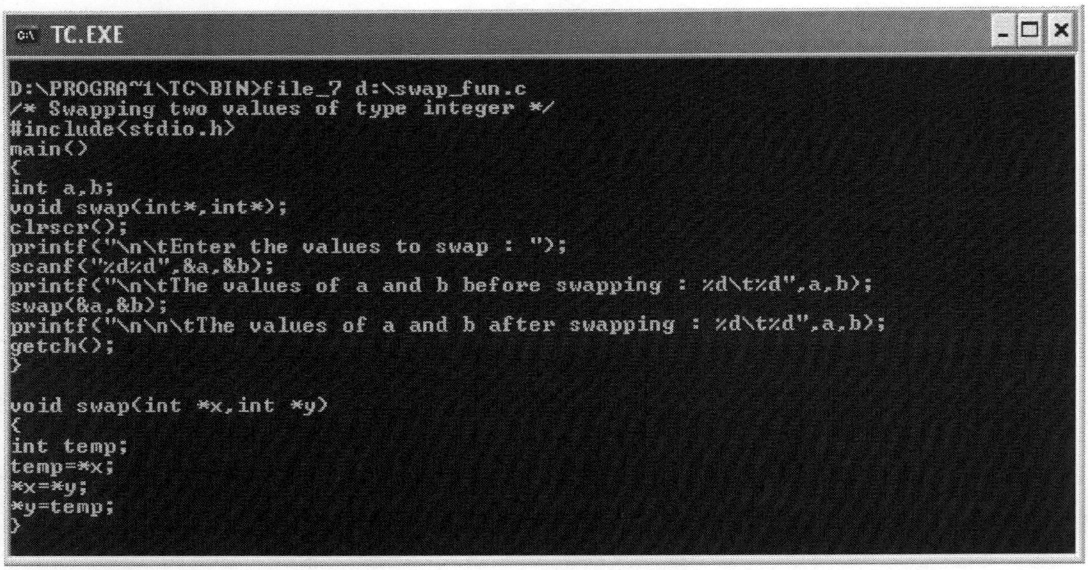

```
TC.EXE
D:\PROGRA~1\TC\BIN>file_7 d:\swap_fun.c
/* Swapping two values of type integer */
#include<stdio.h>
main()
{
int a,b;
void swap(int*,int*);
clrscr();
printf("\n\tEnter the values to swap : ");
scanf("%d%d",&a,&b);
printf("\n\tThe values of a and b before swapping : %d\t%d",a,b);
swap(&a,&b);
printf("\n\n\tThe values of a and b after swapping : %d\t%d",a,b);
getch();
}

void swap(int *x,int *y)
{
int temp;
temp=*x;
*x=*y;
*y=temp;
}
```

Figure 8.10 Displaying the contents of a file through command line

The programmers and those who execute the above program must keep in mind the fact that the command line execution is possible with the .exe file only from the path where it is located and not necessarily the path of the present working directory. Let me explain this idea. We might have installed the C compiler in D drive inside the sub-folder tc. But for our convenience, we might have created two sub-folders namely source_programs and bin inside the folder tc for storing all the C programs and their corresponding .exe files respectively. So, the present working directory in our case is D:\tc. All the source files would be stored in D:\tc\source_programs and all the corresponding

.exe files would be stored in D:\tc\bin. Thus when we choose the option **OS Shell** from the **File** menu, we would be in the path D:\tc\source_programs. We cannot execute the program from here, since the .exe file is in the path D:\tc\bin. Hence, before we execute the program, we must get into the path D:\tc\bin. This is illustrated in the Figure 8.11. If we do not follow the above procedure, then we would get the error, which is also illustrated in Figure 8.11.

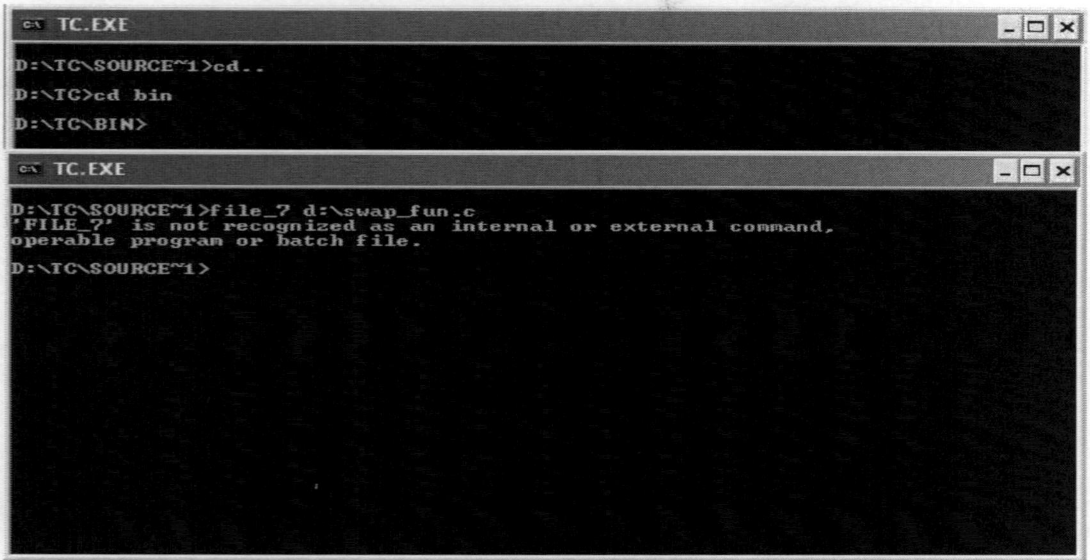

Figure 8.11 Executing the program through the command line

Thus we can conclude that command line parameters could be executed provided we have the corresponding .exe files in the path from where we pass the parameters. I am sure the learners are now ready to work with more of such programs. We will do some programs in the next section to understand the way the command line arguments work. But before we proceed, one must have a clear understanding of the two arguments **argc** and **argv** and hence the way the command line arguments work.

The argument **argc** is the **argument counter** and **argv** is the **argument vector**. The argument counter keeps a check on minimum required number of arguments that the user must pass through the command line. If you observe the program closely, we have checked this minimum requirement through *if* condition and if the minimum number of arguments required is not supplied from the command line, we immediately flash the error and terminate the program. Since the number of arguments is always counted as a whole number, its type is an integer.

The argument **argv** as mentioned is the argument vector that keeps the base addresses of the arguments that are passed through the command line. That is the reason why it is declared as an

array of pointers. You must be wondering why the base addresses of the arguments are stored and what must be the significance of doing so. If you again recall the method of passing the arguments from the command line, those arguments are nothing but the names of different files and all these files are located within the computer's hard disk and hence they must possess base address. The file names are nothing but strings and when they are accessed through their names, it is enough if we knew the base address of these strings (which will invariably be the base address of the corresponding files). When we collect these addresses and store them in a separate array, the array becomes the array of pointers. In our case, the array is that of **character pointers** and hence the declaration **char *argv[]**. The size of the array is not specified, which means that it would be dynamic. Thus argv[0] would contain the base address of the first argument, argv[1] the address of the second argument and so on. In the program coding, we do not need the argv[0] at all since it is this argument that drives the program and gets the output for us. One can print the addresses of the arguments inside the function main() to ensure the above explanation. Figure 8.12 shows the output that contains the **addresses of the arguments** that we pass from the command line.

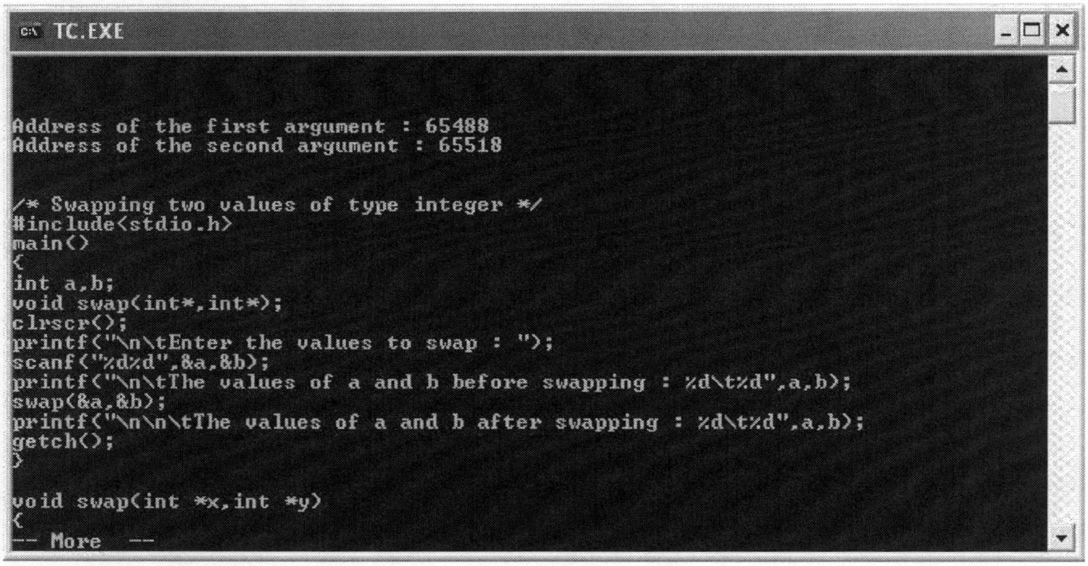

Figure 8.12 Addresses of the command line arguments printed

Another point the user must understand here is that the names of the arguments **argc** and **argv** could be changed while programming and there is absolutely no issue in doing so. I would appreciate if the programmers adhere to the same naming convention, as it would not defeat the spirit of the language.

8.9 SIMPLE PROGRAMS USING COMMAND LINE PARAMETERS

Let us write simple programs using the ideas that we have learnt in the earlier sections with respect to the command line arguments.

8.9.1 File Copying

To start with let us write a simple program for copying the content of one file to another. This aspect is called file copy and this is achieved through the copy command in DOS. The following is the coding for the same.

```
/* Program to copy data from one file to another */
#include<stdio.h>
main(int argc,char *argv[])
{
        char a;
        FILE *fp1,*fp2;
        if(argc<3)
            {
            printf("Usage Error");
            return;
            }
        fp1=fopen(argv[1],"r");
        fp2=fopen(argv[2],"w");
        if(fp1==NULL||fp2==NULL)
            {
            printf("\n\nFile Opening Error");
            return;
            }
        a=getc(fp1);
        while(a!=EOF)
            {
            putc(a,fp2);
            a=getc(fp1);
            }
```

```
        fclose(fp1);
        fclose(fp2);
    }
```

Let us execute the above program and copy the file Gandhi.txt, which is available in the directory SOURCE in the drive D: and the target file new.txt would be created in the directory TARGET in the same drive. While executing the program, the programmers are advised to be cautious about the file-naming convention. The DOS (Disk Operating System) follows an eight-letter naming convention for the files and hence when the name of the file in the SOURCE directory has more than eight characters, the programmers shall make sure that only seven characters are mentioned through the command line. For example, if the directory SOURCE has the file name sample_one.txt and if the users specify the full name of the file as it is, then File Opening Error will be displayed while executing the program. Instead, if the file name is specified as sample~1.txt from the command line, the anticipated output will be generated. The generation of the error as well as the output is illustrated in Figure 8.13. The name of the above program is specified as p182.c and it is the **first of the three arguments** passed through the command line. The name of

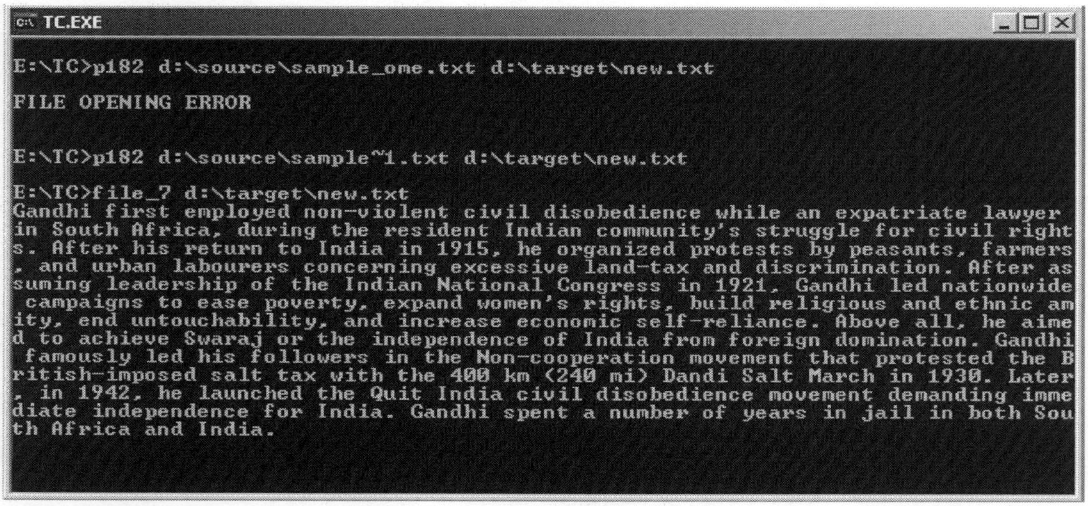

Figure 8.13 File copying through command line

the source file is sample_one.txt and it is the **second argument** and the name of the target file is new.txt, which is the **third argument**. In the program coding, we check the minimum number of arguments that must be passed through the command line and the minimum number required is three. We also see that the target file is read and its contents are displayed using the program to read and display the contents of a file that we had written in the previous section. The name of that program is file_7.c which was the **first argument** passed through the command line.

8.9.2 File Merging

Let us write a program to merge the contents of two files together and display the contents of the merged file. For that we will take the file Gandhi.txt and Gandhi_1.txt and call the merged file as merged.txt. Both the files are available in the working directory (we call that the Present Working Directory). The following is the program code for the merging process. The output is shown in Figure 8.14. The contents of the merged file are viewed through the program, file_7.c.

```
\/* Program to merge the contents of two files */
#include<stdio.h>
main(int argc,char *argv[])
        {
        char a,b;
        FILE *fp1,*fp2,*fp3;

        if(argc!=4)
            {
            printf("USAGE ERROR");
            return; }
        fp1=fopen(argv[1],"r");
        fp2=fopen(argv[2],"r");
        fp3=fopen(argv[3],"w");
        if(fp1==NULL||fp2==NULL||fp3==NULL)
                {
                printf("FILE OPENING ERROR");
                return;
                }
        for(;(a=getc(fp1))!=EOF;)
        putc(a,fp3);
        for(;(b=getc(fp2))!=EOF;)
        putc(b,fp3);
        fclose(fp1);
        fclose(fp2);
        fclose(fp3);
}
```

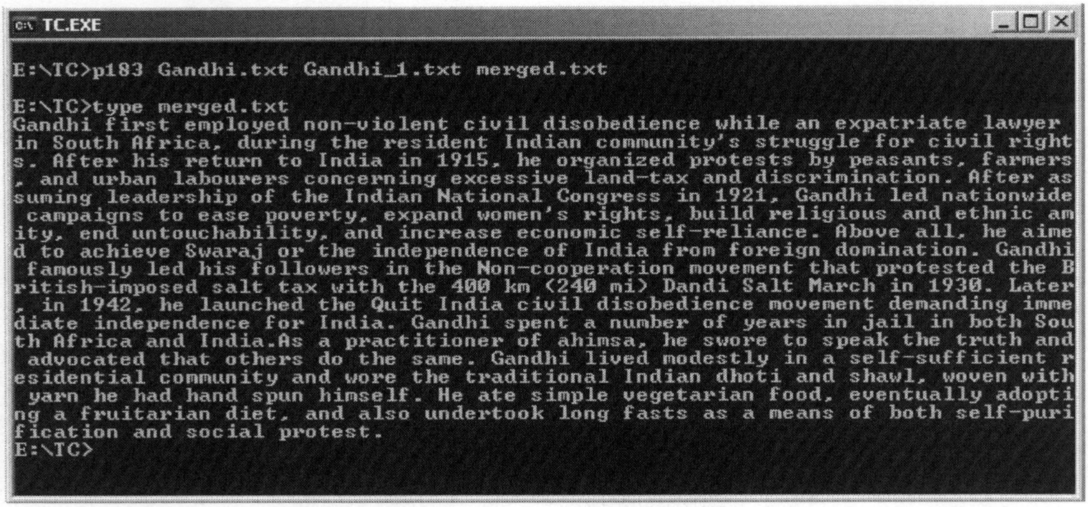

Figure 8.14 File merging through command line

8.9.3 Case Conversion

Let us write a program to convert the contents of a file in lowercase to uppercase. The coding for the same is given below.

```
/* Program to copy lowercase content in one file as uppercase */
#include<stdio.h>
main(int argc,char *argv[])
{
        char a;
        FILE *fp1,*fp2;
        if(argc!=3)
            {
            printf("USAGE ERROR");
            return;
            }
        fp1=fopen(argv[1],"r");
        fp2=fopen(argv[2],"w");
        if(fp1==NULL||fp2==NULL)
```

```
                {
                printf("FILE OPENING ERROR");
                return;
                }
        for(;(a=getc(fp1))!=EOF;)
                {
                if(a>='a'&&a<='z')
                a-=32;
                putc(a,fp2);
                }
        fclose(fp1);
        fclose(fp2);
}
```

The above program is executed with the file Gandhi.txt in the working directory and the contents of the file that are in lowercase are copied to another file ucase.txt in the same directory. The output of the program is shown in the Figure 8.15. This kind of program could be useful when we want to convert the lowercase contents of a file to uppercase which could be used for some other purpose. We can have the otherwise logic too, wherein we would check for the existence of uppercase characters and convert the same to lowercase. We just have to make minor changes in the above program to achieve this.

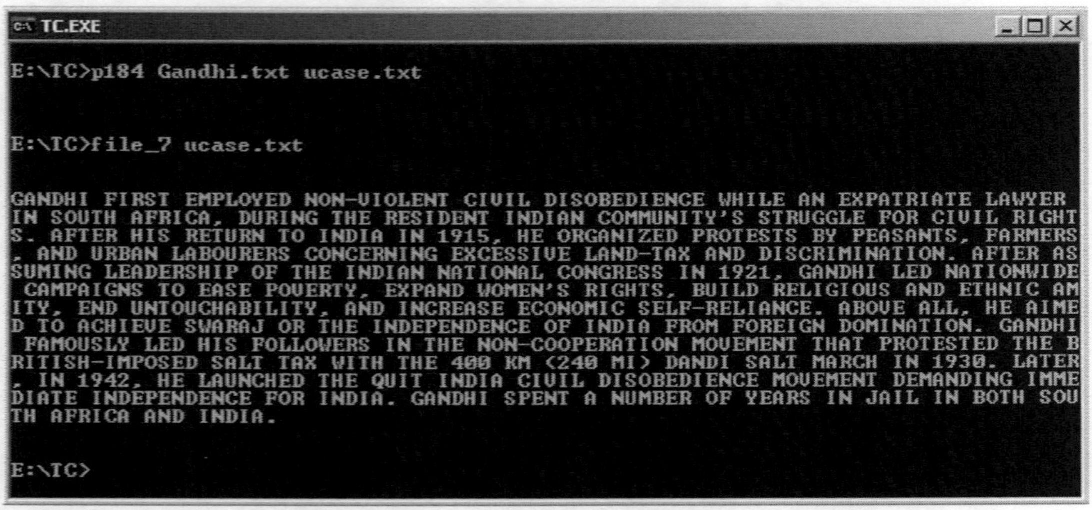

Figure 8.15 Case conversion

8.9.4 Pattern Matching

```
/* Program to get the pattern through command line and check the number of times it is used in the particular
file */
#include<stdio.h>
main(int argc,char *argv[])
                {
                char a[1000];
                int i,j,k,w=0;
                FILE *fp;
                if(argc!=3)
                        {
                        printf("USAGE ERROR");
                        return;
                        }
                fp=fopen(argv[1],"r");
                if(fp==NULL)
                        {
                        printf("FILE OPENING ERROR");
                        return;
                        }
                for(i=0;(a[i]=getc(fp))!=EOF;i++)
                             {
                             if(a[i]=='\n')
                                     {
                                     a[i]='\0';
                                     for(i=0;a[i];i++)
                                             {
                for(k=i,j=0;argv[2][j]==a[k]&&argv[2][j]!='\0';k++,j++);
                                             if(argv[2][j]=='\0')
                                             w++;
                                             }
                                     i=-1;
                                     }
                             }
```

```
printf("The pattern you have entered occurs %d time(s)",w);
        fclose(fp);
        }
```

The above program is used for checking the number of occurrence of the pattern **printf** in the file **p184.c**. The output is shown in the Figure 8.16. We can check the number of occurrences of any big pattern in the given file. In the next program, we go one step ahead, where we display the lines in which the given pattern occurs, along with the line number. The program coding is shown below followed by its output in the Figure 8.17.

8.9.5 Displaying the Lines in which the Pattern Occurs

```
/* Program to get the pattern through the command line argument and search for that pattern in the file and
display the lines in which the pattern occurs along with the line number */
#include<stdio.h>
main(int argc,char *argv[])
        {
        char a[1000];          int i,j,k,l=0;          FILE *fp;
        if(argc!=3)
                {
                printf("USAGE ERROR");
                return;
                }
        fp=fopen(argv[1],"r");
        if(fp==NULL)
                {
                printf("FILE OPENING ERROR");
                return;
                }
        for(i=0;(a[i]=getc(fp))!=EOF;i++)
                {
                if(a[i]=='\n')
                        {
                        l++;
                        a[i]='\0';
```

```
for(i=0;a[i];i++)
{
for(k=i,j=0;argv[2][j]==a[k]&&argv[2][j];k++,j++);
{
if(argv[2][j]=='\0')
{
printf("%d\t",l);
puts(a);
}
} }
i=-1;
}
}
fclose(fp);
}
```

Figure 8.16 Pattern count

Figure 8.17 Lines in which the pattern occurs

8.9.6 Displaying the Lines in which the Pattern does not Occur

The following program coding illustrates how to get the display of the lines in which the given pattern does not occur. The output shown in the Figure 8.18 is partial and we could see that those lines which do not have the pattern printf are displayed with their line numbers.

```
/* To print the lines in which the pattern does not exist */
#include<stdio.h>
main(int argc,char *argv[])
    {
    char a[1000];
    int i,j,k,l,f;
    FILE *fp;
    if(argc!=3)
        {
        printf("USAGE ERROR");
        return;
        }
    fp=fopen(argv[1],"r");
    if(fp==NULL)
        {
        printf("FILE OPENING ERROR");
        return;
        }
    l=0;
    for(i=0;(a[i]=getc(fp))!=EOF;i++)
        {
        if(a[i]=='\n')
            {
            f=0;
            l++;
```

```
a[i]='\0';
for(i=0;a[i];i++)
        {
for(j=0,k=i;argv[2][j]==a[k]&&argv[2][j]!='\0';k++,j++);
                {
                if(argv[2][j]=='\0')
                        {
                        f=1;
                        continue;
                        }}      }
i=-1;
if(f==0)  {
   printf("%d\t",l);
   puts(a);
   }    }      }            fclose(fp);              }
```

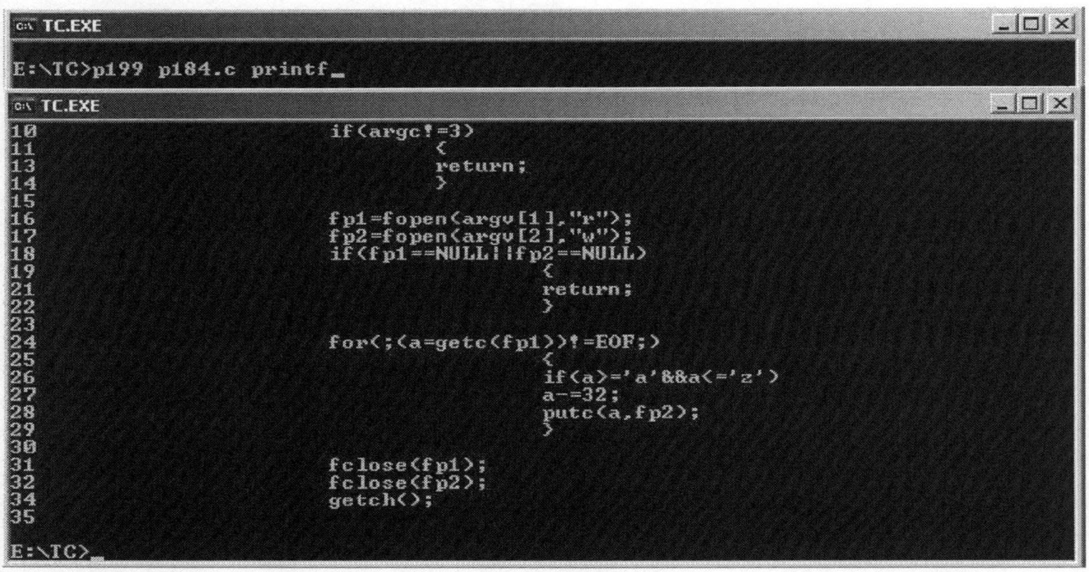

Figure 8.18 Lines in which the pattern does not occur

I am sure the programs that are discussed above using the command line parameters would be an eye opener for the learners of this book and I would also suggest them to do more such exercise to get full confidence in using the command line arguments efficiently.

SUMMARY

We began this chapter with an introduction to files and the various operations that are possible with them. We had a brief discussion on the file positions and high-level file handling functions. With sufficient illustrations, we understood the concepts of opening and closing a file, writing into a file and reading from a file. We also discussed how to append contents in the existing file (without overwriting the existing contents). We then discussed how to get the file name as a variable name from the keyboard and open the same for the display on the monitor.

We then came to an important aspect of the files, namely the command line arguments, which are considered to be the key idea in efficiently handling files. We had a brief introduction to command line arguments and how significant it is in accessing the files using it. We followed the discussion with some simple programs in which we checked for the number of occurrences of a pattern, the lines in which the pattern occurs with the line number and the lines in which the pattern does not occur.

CRACK THE NUT

Identify the output for the following programs:

1. main()

```
{
FILE *fp;
char c;
clrscr();
fp=fopen("myfile.txt","r");l
fclose(myfile);
}
```

2. main()

```
{
FILE fptr;
clrscr();
fptr=fopen("file_1.txt","r");
if(fptr=NULL)
{
puts("file opening error");
exit();
}
fclose(fptr);
}
```

3. main()

```
{
FILE *fp;
char str[30], str2[8];
fp=fopen("myfile.txt","r");
printf("Enter a string :  ");
gets(str);
fputs(str,fp);
fputs("\n",fp);
fclose(fp);
}
```

4. main()

```
{
FILE *fp;
char str[30], str2[8];
fp=fopen("myfile.txt","r");
while(fgets(str,fp)!=NULL)
```

```
puts(str); fclose(fp);
}
```

5. main()

```
{
FILE *fp;
char var[20];
fp=fopen("sample.c","r");
while(fgets(var,10,fp))!=EOF)
printf("%c",var); }
```

6. #include<dos.h>

```
main()
{
printf("%d",_argc);
}
```

7. main(int argc,char *argv[])

```
{
printf("%s",strrev(argv[1]));
}
```

8. main(int count,char *argv[])

```
{
int i=0;
for(i=0;argv[i];i++)
printf("\n%s",argv[i]);
}
```

9. #include"dos.h"

```
#include"string.h"
main()
{
char str[15];
```

```
int i=0;
strcpy(str,_argv[1]);
for(i=0;i<=strlen(str);i++){
if(str[i]>=97&&str[i]<=122)
str[i]=str[i]-32;
}
```

10. #include<stdio.h>

```
int main(int argc, char **argv)
{
printf("%s\n", argv[argc-1]);
return 0;
}
```

9 INTRODUCTION TO DATA STRUCTURES

9.1 WHY SHOULD YOU LEARN DATA STRUCTURES?

In computing terms, a data structure is something that facilitates the computer to store and retrieve the data efficiently. In any programmer's opinion, there are *three* important and necessary things that are essential to write efficient and interesting programs, namely, *loops, representation of data* (through variables and pointers) and *functions/methods* and *recursive programs*. By now, I am sure, you would have almost mastered 75–100% of the above concepts. When we try to differentiate a data type and a data structure, the difference is as simple as anything. A data *type* is a well-defined collection of data and it will have well-defined set operations on it. For example, take the case of a variable pertaining to an integer data type. We can quote a fixed range of values for the variable and can almost perform most of the operations that are valid in C. On the other hand, a data *structure* is an implementation of an *abstract data type*. The abstract data type may not be having a specific range of values and it may be restricted to a few set of operations, which would be defined from time to time.

Clearly, this is not the first instance in this book that we are actually starting to learn about data structures. We have already been acquainted with the terminology and we have done an initial study on the concept in the form of arrays and structures. Rightly saying, arrays are the simplest form of the data structures and structures are the next advanced level. On further extending the idea, we will land up into linked lists and binary trees (simply termed as B-trees). I would advise all the readers here to recall the concepts of arrays and structures and their linkage with the pointers (from the Chapters 4, 5 and 6). This will facilitate you to have a fair idea of them and you will not be surprised when I later talk about the related concepts such as linked lists, binary trees (which are just an extended form of the arrays, structures and pointers).

There arises a question why should I study and know in detail the data structures . The answers are umpteen.

They play a central and crucial role in modern computer science. In fact, you cannot imagine any real-life application in computing science in which data structure is not forming one of the ingredients. In very simple words, data structures add spice to a program. Consider your Google, or your mail server. Can you imagine them work without the data structures? These routine applications are obvious examples for data structures acting as building blocks in getting efficient algorithms. I am sure, you, as a reader and a beginner in data structures, would not resist learning them completely and thoroughly.

9.2 DIFFERENT DATA STRUCTURES IN USE

There are quite a few data structures actually in use depending on the nature of the applications and programs developed and also on the nature of the creator of the program. Broadly one can classify the available data structures as linear and non-linear. As a professional programmer, one would be expected to master all those and develop efficient applications using them.

As I had mentioned in the previous sub-chapter, we have already looked into the preliminaries and pre-requisites for getting into the core data structure ideas, in the form of arrays and structures. I presume that all the readers are aware of these two essential ideas and if not, I would advise them to go through the respective chapters once again, before they go forward.

The next level in learning a data structure starts with the *stacks* and *queues*. Stacks are the data structures in which the order of *last-in, first-out* is followed and queues are those which maintain the order of *first-in, first-out*. Though queues are seemingly fairer compared to stacks, both have significant applications in programs as a data structure.

We have something called as dynamic data structures that grow and shrink during the execution. The idea of dynamic memory allocation follows from this data structure. We can even call a file as a data structure, as it has the records and they in turn have the attributes in terms of tables.

We do have linked lists, in which we have a sequence of data records and in each of them we have what is referred to as a *field*. The field would contain a reference (otherwise termed as a *link*) to the next record in the sequence. Linked lists are the simplest among the existing data structures and serve as the catalysts for implementing the other important data structures such as stacks, queues, hash tables and many more. We do have a separate discussion on liked lists that follows from here.

9.3 LINKED LISTS

The term list refers to linear collections of data items. Data processing frequently involves storing and processing data organized into lists. One way to store data elements is by means of arrays. Another way of storing a list in memory is to have each element in the list to contain a field called a link or pointer, which contains the address of the next element in the list.

For writing a program for linked lists and the operations that are possible with them, we use self-referential structures, which we had seen at the introductory level in Chapter 5. Before we proceed to write programs to explain the operations on linked lists, let us have an idea about the advantages and disadvantages of linked lists.

Advantages

- Overflow can never occur unless the memory is actually full.
- Insertions and deletions are easier than for contiguous lists.
- With large records, moving pointers is easier and faster than moving the items themselves.
- These lists are dynamic.

Disadvantages

- The pointers used in the lists require extra space.
- Linked lists do not allow random access.
- Time must be spent traversing and changing the pointers.
- Programming is typically trickier with pointers.

The core idea behind using a linked list is that the data are stored somewhere in the memory, but still bound together. This fact makes a list much different from arrays. The elements of the array are stored in contiguous memory locations and those of the linked list are dispersed, still connected through the links.

Linked lists often confuse the beginners even though they are good in using the pointers and structures. Those who have gained expertise in both are exceptional to this.

9.3.1 Singly Linked Lists and Operations on it

A singly linked list or one-way list is a linear collection of data elements called nodes, where the linear order is given by means of pointers, which may vary in number. A simple way to represent a linear list is to be expanding each node to contain a link to the next node. This representation is called one-way chain or singly-linked linear list as shown in the Figure 9.1.

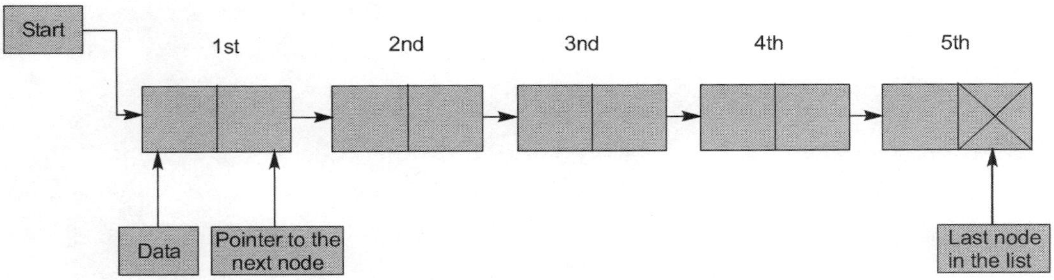

Figure 9.1 Representation of a singly linked list

Each node is divided into two parts. The first part represents the information of the element and the second part contains the address of the next node. The last node does not have a successor node and no actual address is stored in the pointer field and it is left empty. Usually a NULL value is stored as the address in the last node. The operations that are possible with the linked list are insertion, deletion, searching, traversing and modification.

The following is the program for the various operations on a singly-linked list which uses pointers and structures.

```
/* Operations on Singly-Linked List */
#include<stdio.h>
#include<alloc.h>
/* Structure Declaration */
struct node
        {
        int data;
        struct node *next; /* Self-referential structure */
        };
/* Declaration of pointers to the structure node */
struct node *first=0,*last=0;
```

```
main()
{
int ch;
clrscr();
do
{
gotoxy(15,2);
printf("Operations on a Singly Linked List");
gotoxy(15,3);
printf("**********************************");
gotoxy(15,6);
printf("1.Append");
gotoxy(15,8);
printf("2.Deletion");
gotoxy(15,10);
printf("3.Display");
gotoxy(15,12);
printf("4.Search");
gotoxy(15,14);
printf("5.Insertion");
gotoxy(15,16);
printf("6.Quit");
gotoxy(15,18);
printf("Enter your choice :  ");
scanf("%d",&ch);
switch(ch)
        {
        case 1:
                append();
                break;
        case 2:
                delete();
                break;
```

```
        case 3:
                display();
                break;
        case 4:
                search();
                break;
        case 5:
                insert();
                break;
        case 6:
                clrscr();
                gotoxy(25,12);
                printf("Good Bye....Have a Nice Day!!!");
                sleep(2);
                break;
        default:
                clrscr();
                gotoxy(12,12);
                printf("Invalid option....");
                sleep(2);
                gotoxy(12,12);
                printf("Please enter the right option while the menu loads...");
                sleep(2);
                clrscr();
        }
}while(ch!=6);
}
/* Insert Operation */
insert()
{
struct node *temp,*pp,*cp;
int inse,before;
temp=(struct node*)malloc(sizeof(struct node));
```

```
clrscr();
gotoxy(12,3);
printf("Enter the element to insert :  ");
scanf("%d",&temp->data);
gotoxy(12,5);
printf("Enter before which element you want to insert :  ");
scanf("%d",&before);
if(first->data==before)
        {
        temp->next=first;
        first=temp;
        }
        else
        {
        for(pp=cp=first;cp!=NULL;pp=cp,cp=cp->next)
                {
                if(cp->data==before)
                        {
                        pp->next=temp;
                        temp->next=cp;
                        }
                }
        }
        gotoxy(12,7);
        printf("The element inserted successfully....");
        sleep(1);
        gotoxy(12,9);
        printf("Please wait while the menu loads...");
        sleep(2);
        clrscr();
}
/* Append Operation */
append()
```

```
        {
        struct node *temp;
        temp=(struct node*)malloc(sizeof(struct node));
        clrscr();
        gotoxy(12,3);
        printf("Enter element to append :  ");
        scanf("%d",&temp->data);
        gotoxy(12,5);
        printf("Element appended successfully.....");
        sleep(1);
        gotoxy(12,7);
        printf("Please wait while the menu loads.....");
        sleep(2);
        clrscr();
        if(first==0)
                first=last=temp;
        else
                last->next=temp;
        last=temp;
        last->next=0;
        }
/* Delete Operation */
delete()
        {
        struct node *pp,*cp;
        int dee;
        clrscr();
        gotoxy(12,3);
        printf("Enter the element to delete :  ");
        scanf("%d",&dee);
        gotoxy(12,5);
        printf("Element deleted successfully...");
        sleep(1);
```

```
        gotoxy(12,7);
        printf("Please wait while the menu loads...");
        sleep(2);
        clrscr();
        if(first->data==dee)
                first=first->next;
        for(pp=cp=first;cp!=NULL;pp=cp,cp=cp->next)
        {
                if(cp->data==dee)
                pp->next=cp->next;
        }
        if(cp==last)
        {
        last=pp;
        last->next=0;
        }
}
/* Display Operation */
display()
        {
        struct node *temp;
        clrscr();

        if(first==0)
        {
        gotoxy(12,3);
        printf("The list is empty");
        }
        gotoxy(12,3);
      printf("0054he singly linked list is :  ");
        for(temp=first;temp!=NULL;temp=temp->next)
        printf("%d->",temp->data);
        getch();
```

```
        clrscr();
    }
    /* Search Operation */
    search()
        {
        struct node *temp;
        int sear;
        clrscr();
        gotoxy(12,3);
        printf("Enter the element to search :  ");
        scanf("%d",&sear);
        for(temp=first;temp!=NULL;temp=temp->next)
            {
            if(temp->data==sear)
                {
                gotoxy(12,5);
                printf("Element Found....");
                break;
                }
    }
            if(temp==NULL)
    {
            gotoxy(12,5);
            printf("Element Not Found....Try another element...");
    }
            sleep(2);
            clrscr();
    }
```

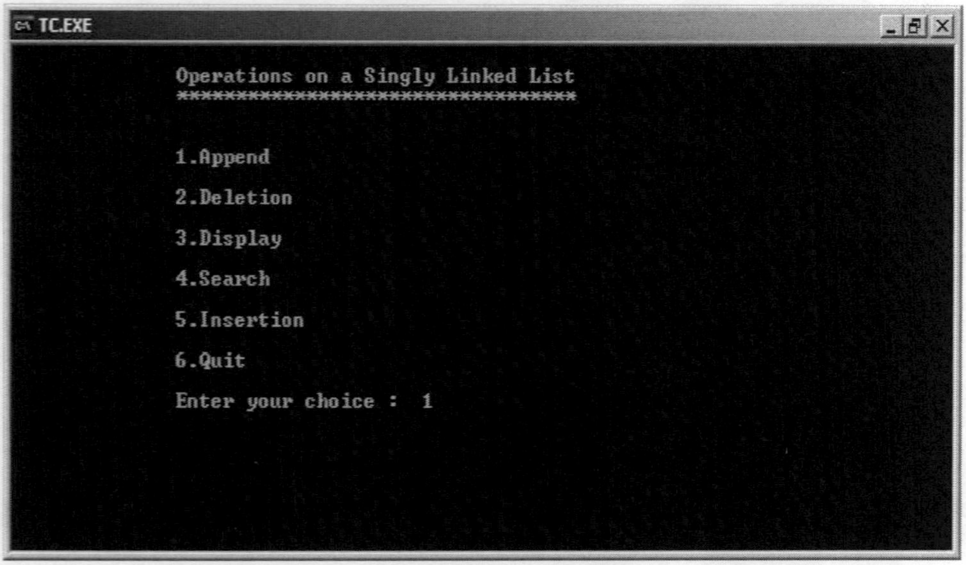

Figure 9.2 Menu of oprations on the singly linked list

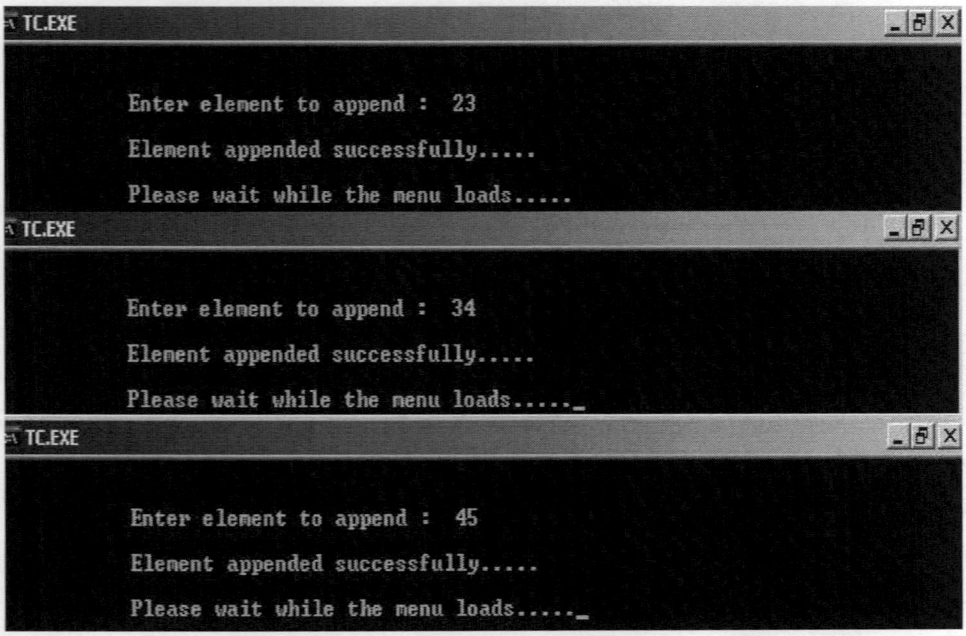

Figure 9.3 Three elements appended to the list

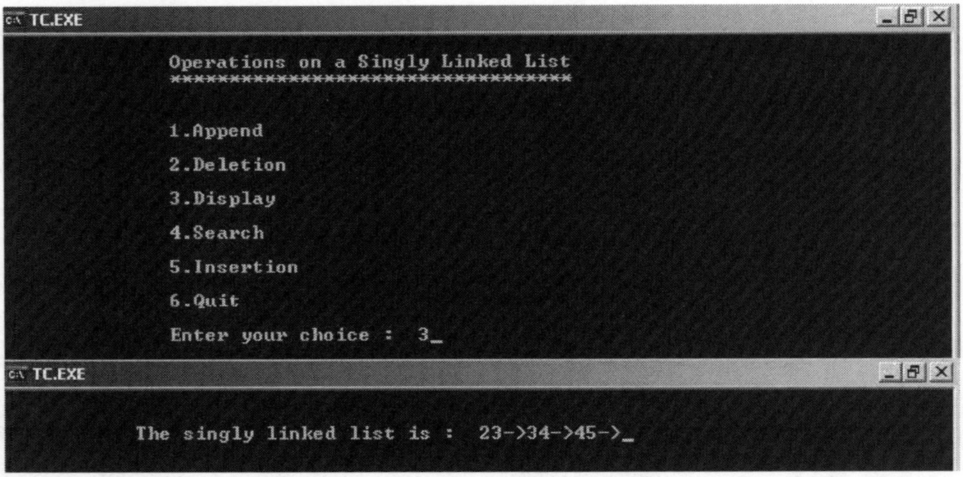

Figure 9.4 Display option in action

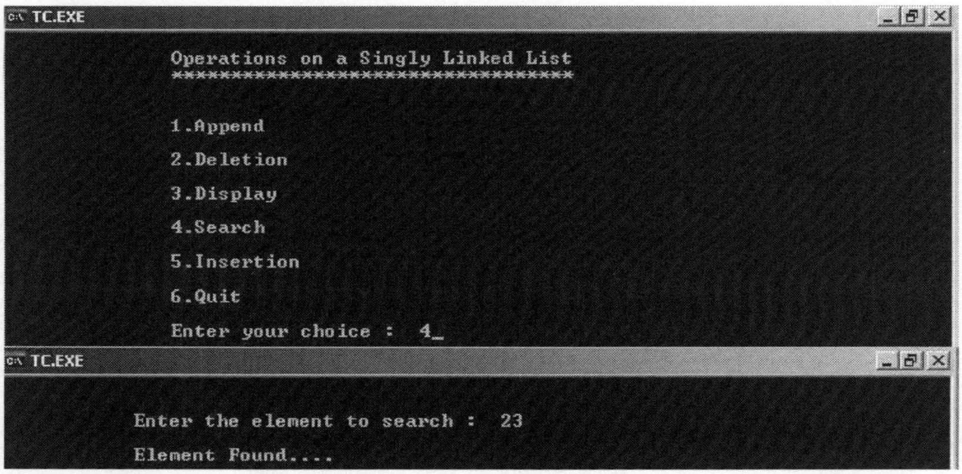

Figure 9.5 Search option in action (element found)

If you closely observe the program, we have not only used pointers and structures, but also user-defined functions, which is an additional and added feature of C language. Functions are discussed in the chapter 7 and once you read the chapter 7 and understand how user-defined functions are declared, defined and used in C programming, it would be definitely easy to understand the execution of the above program. The concepts of structures and using a pointer with it have been already discussed. The figures 9.2 through 9.8 show the sample execution of the above program coding and it clearly explains the variety of operations that are possible with a singly linked list.

Please note that, in the coding that is given above, we have used the built-in function gotoxy() in order to make a very interactive style of execution.In programs such as this, this way of coding is highly appreciated, since it facilitates the programmer, who is the end-user, to thoroughly understand and conceive the ideas.

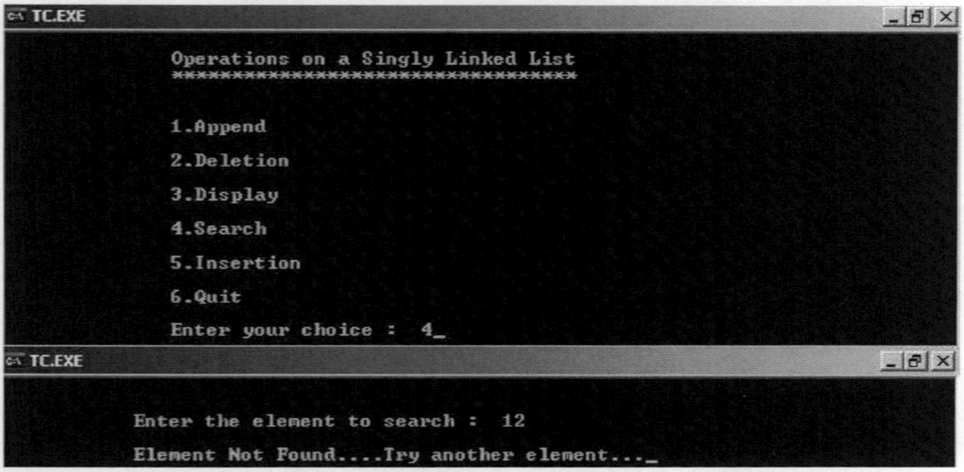

Figure 9.6 Search option in action (element not found)

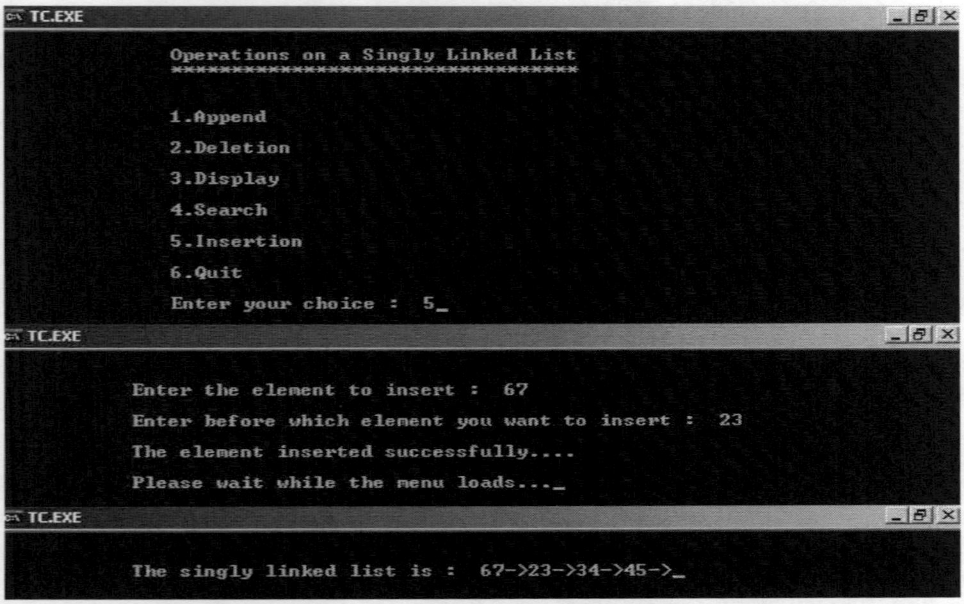

Figure 9.7 Insert option in action

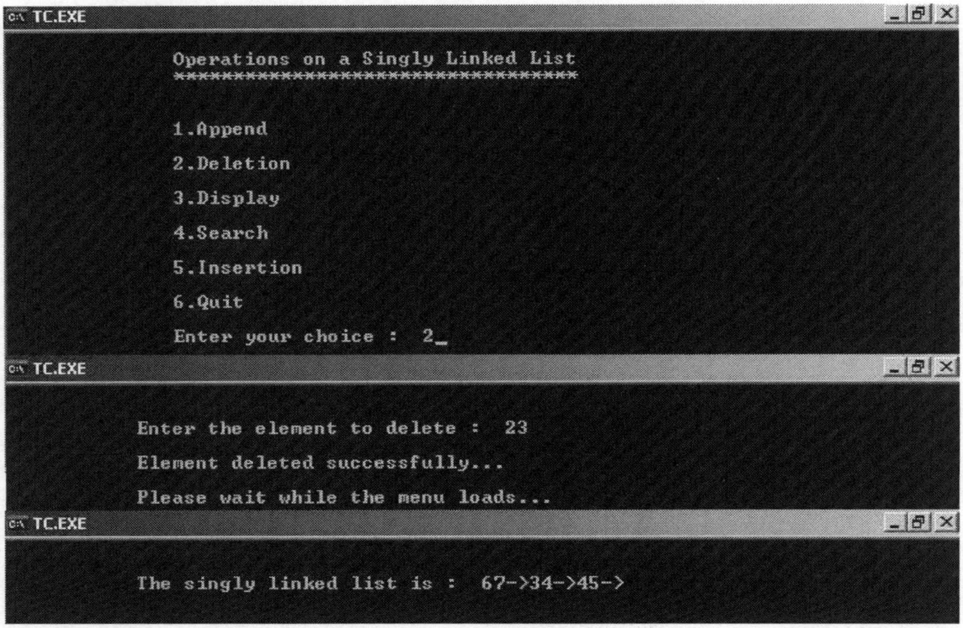

Figure 9.8 Delete option in action

9.3.2 Doubly-Linked Lists

Doubly-linked list (abbreviated as DLL) is a more sophisticated kind of linked list. Unlike the singly-linked list, in DLL, each node has two links: one pointing to the previous node and one to the next node. The previous link of the first node and the *next* link of the last node, both point to a NULL. It can be viewed as two singly linked lists formed from the same data items, in opposite orders. The two links allow walking along the list in either direction in a balanced way. Compared to a singly linked list, manipulating a doubly linked list is a bit tedious. A typical and simple representation of a doubly linked list is shown in the Figure 9.9.

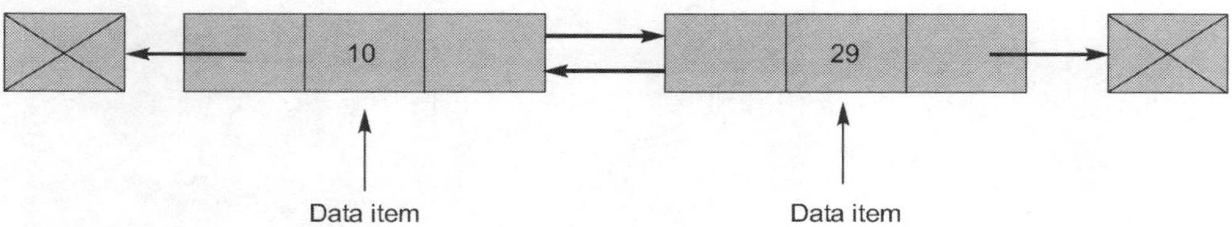

Figure 9.9 A doubly-linked list

I would leave it to the readers to design a suitable code for manipulating a doubly-linked list. I am sure that you will be tinkering with your mind to figure out a good and efficient program, and that would really be a good exercise for you to recall all those that you had learned earlier.

9.3.3 Circular Lists

Circular linked list is another special type of linked list in which all the nodes are linked in a continuous circle without any pause or a break. A circular list can be a singly linked or a doubly linked list. Since the links are arranged in a circular form, there is no way that a circular list can have a NULL. The elements of a circular list can be added at the back and removed from the front at any given point of time.

Another key advantage of a circular list is that we can traverse the full list beginning at any particular node (again due to the arrangement of the elements in circular form). This avoids the necessity of storing the first node and the last node (like the singly linked list and the doubly linked list). But care shall be taken in case we have an empty list and the need for a NULL is realized in such a list. The addition and deletion of the nodes will be easier when we use such a representation. An empty list will always be a special case. Figure 9.10 shows a simple circular list.

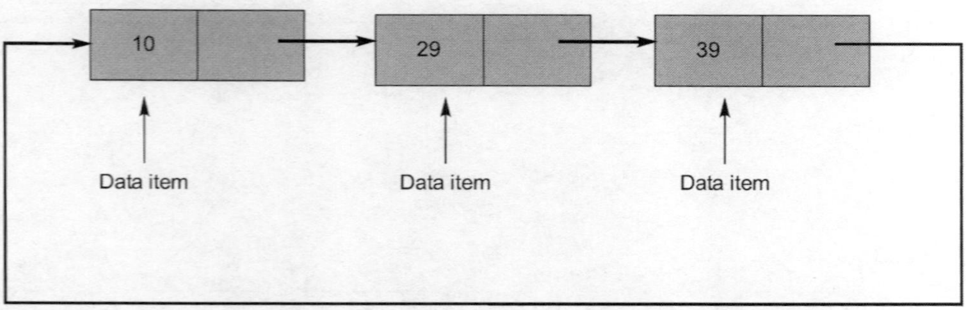

Figure 9.10 A circular list

When we want to represent structures that are naturally circular in nature, it would be wise to use a circular list. This, as I had mentioned earlier, would facilitate us in traversing the list starting at any point. Through a single pointer, we can easily traverse the first and the last records of the list. The implementation of the circular linked list is beyond the scope of this book, though, I would request the readers to make an exercise out of this and implement the same.

9.4 STACK AND IMPLEMENTATION OF ITS OPERATIONS

Stack is a very common data structure used in almost in all of the available programming languages. There are many ways to explain the meaning of a stack. But, for me, the best ever way, which I am sure, is easier for even laymen is, to consider a pile of plates or books. Now, think of something that should be possible with the considered situation. The minimum set of actions that we can do with the pile of plates or books are that, we can add one more plate or a book on the top of the pile, we can take one off from the pile, we can check whether the pile is empty and we can also check whether the pile is full to its specified limits.

These minimal set of operations are technically termed as push (places one element on the top), *pop* (removes one element from the top), *IsEmpty* (checks whether the stack is empty) *and IsFull* (checks whether the stack overflows).

Readers, would you mind in making a physical analogy of the stack? If not, try to draw the stack in terms of what you have understood. I am sure you would be coming out with a vertical diagram, where the term *top* will really be on the top of the diagram. If you still do not follow what I have just said, follow the figures 9.11 through 9.13. I am sure these figures are more or less self-explanatory.

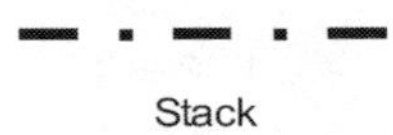

Figure 9.11 Empty stack

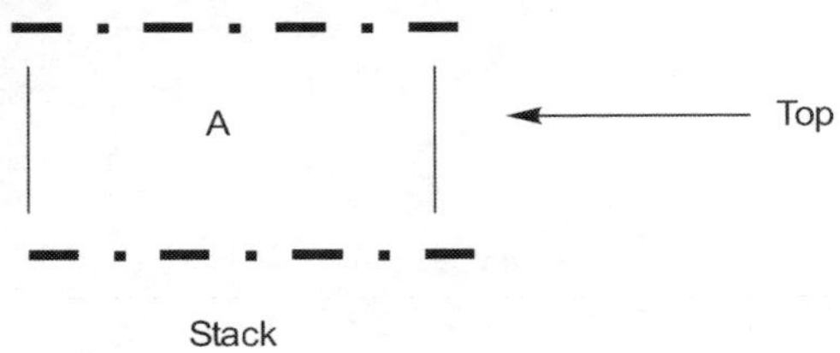

Figure 9.12 Stack with one element

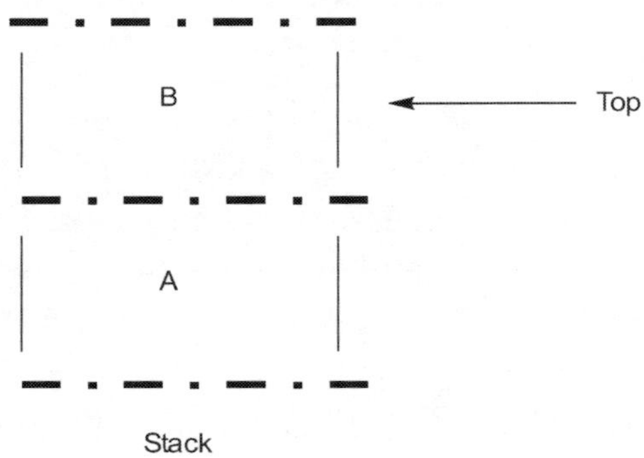

Figure 9.13 Stack with two elements

Figure 9.14 Stack from which the element **B** is popped out

Figure 9.11 corresponds to an empty stack, figure 9.12 has one element and Figure 9.13 has two elements in it. Obviously, we had performed the operation *push* on the stack to get transformed from the Figure 9.11 to 9.12 and from 9.12 to 9.13. Assume that we want to remove one of these two elements from the stack. Then we have to start from the top of the stack and use the operation *pop*. First time when we apply the *pop*, we will get the element B removed, since it is the element at the top of the stack (Figure 9.14) and subsequent usage of *pop* would remove the second element A from the stack, after which the stack becomes empty.

Assume that we want to push another element C in the stack. That will be pushed and placed above the element A. The same is shown in the Figure 9.15.

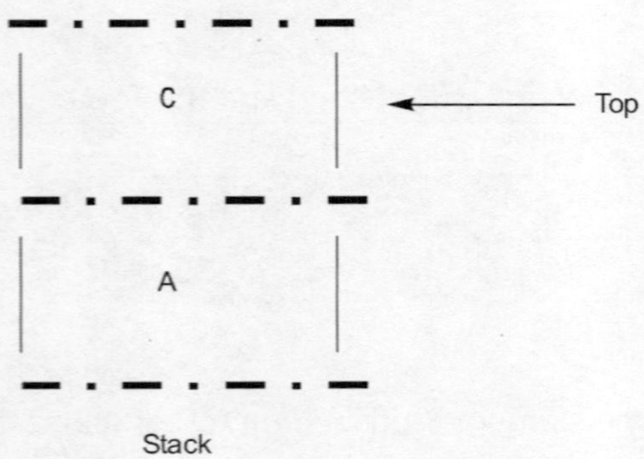

Figure 9.15 Stack in which the element **C** is pushed

We observe from the Figures 9.14 and 9.15 that the last element that was pushed earlier is the first one to be popped out and any element that would be pushed is likely to be placed on the top of the stack. This enables us to conclude that a stack works in Last-In-First-Out (LIFO) methodology. With this idea, let me proceed in implementing the *push* and *pop* operations performed on the stack, through a C program. To begin with, I would use the array implementation and then I would migrate to linked list implementation of the stack. There are two reasons for me choosing the array for implementing stack. One, we have already used the arrays and two, a stack holds a bunch of items of similar type and an array too is similar in this aspect.

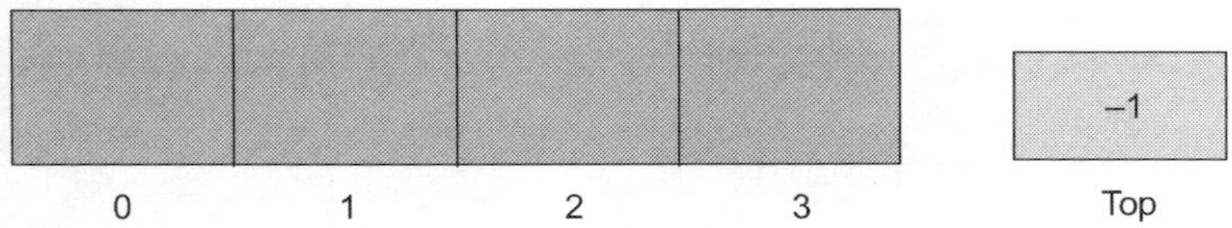

Figure 9.16 Empty stack

Consider an empty stack (in our case, it is an empty array) such as the one shown in the Figure 9.16. The top of the stack is initially assigned a value '–1' since the array starts is counted from '0'.

Now, let us push two elements A and B into the stack. After pushing the elements using the *push* operation, we have the structure of the array modified as shown in the Figures 9.17 and 9.18. Please note that the top of the stack value is also updated to '1'. Initially it was '0', then after the first element has been pushed, it became '0' and after the second one is pushed, it had become '1'.

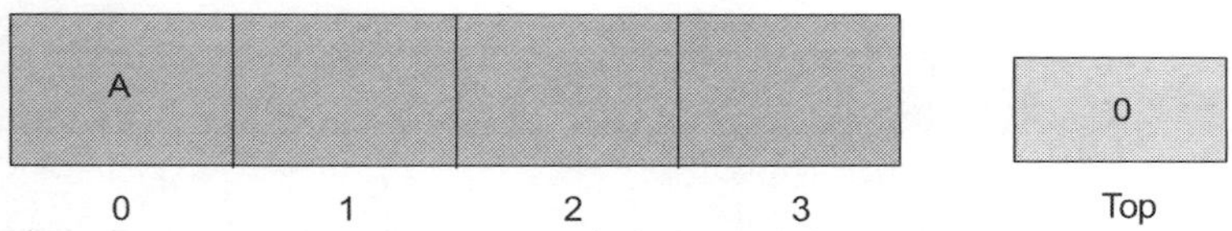

Figure 9.17 Stack with the element **A** pushed and the Top being '0'

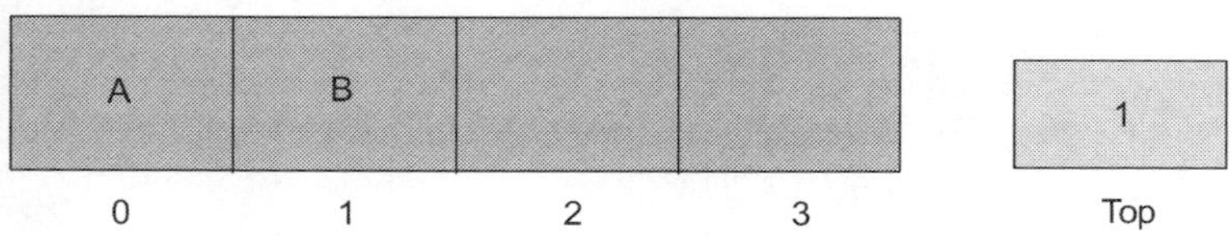

Figure 9.18 Stack with the element **B** pushed on top of **A** and the Top being '1'

Now if we want to pop the element B and push the element C, we have to apply *pop* first on the array followed by the *push*. The figures 9.19 and 9.20 show the corresponding status of the array after doing so.

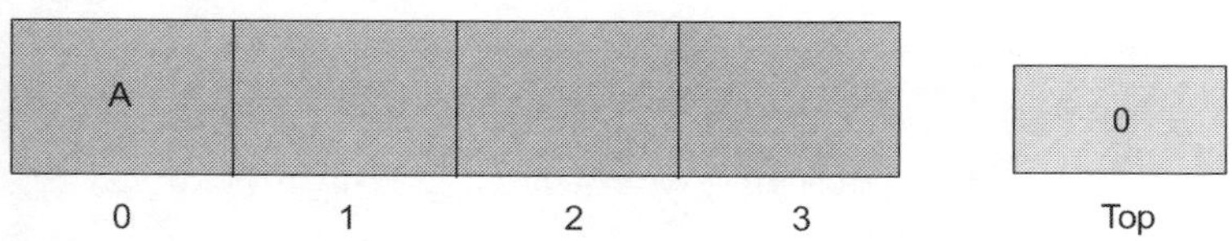

Figure 9.19 Stack with the element B popped and the Top becoming '0'

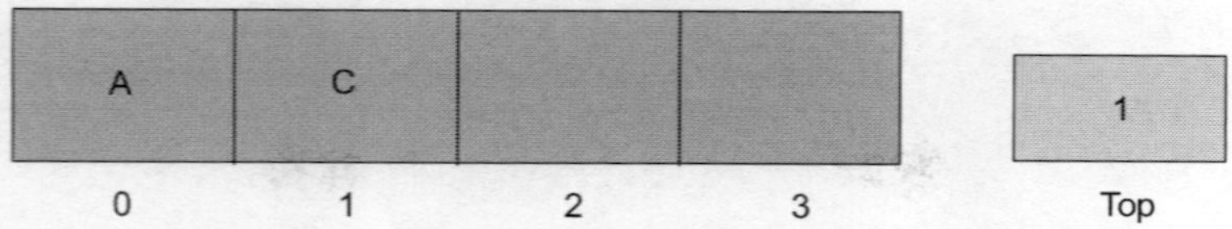

Figure 9.20 Stack with the element **C** pushed and the Top becoming '1'

Now that you had understood and visualized the way the program would work, I am giving the program itself in the listing below.

```
/* Inclusion of header files */
#include<stdio.h>
#include<conio.h>
#include<dos.h>
/* Function Prototypes */
void push(int);
int pop(void);
void display(void);
/* Function main() */
main()
{
int choice,val;
clrscr();
do
{
gotoxy(12,4);
printf("Operations on Stack");
gotoxy(12,5);
printf("*******************");
gotoxy(12,6);
printf("1. Push");
gotoxy(12,7);
printf("2. Pop");
gotoxy(12,8);
```

```
printf("3. Display");
gotoxy(12,9);
printf("4. Exit");
gotoxy(12,10);
printf("Enter Your Choice : ");
scanf("%d",&choice);
switch(choice)
{
case 1:
        sleep(1);
        clrscr();
        gotoxy(12,12);
        printf("Enter the element to be Pushed : ");
        scanf("%d",&val);
        push(val);
        sleep(1);
        clrscr();
        break;
case 2:
        sleep(1);
        clrscr();
        val=pop();
        if(val!=-1)
        {
        gotoxy(12,12);
        printf("Popped Element : %d",val);
        }
        sleep(1);
        clrscr();
        break;
```

```
case 3:
        clrscr();
        gotoxy(12,12);
        display();
        getch();
        clrscr();
        break;
case 4:
        gotoxy(12,12);
        printf("Good Bye!!!!....Have a Nice Day....");
        sleep(2);
        break;
default:
        clrscr();
        gotoxy(12,12);
        printf("Wrong Choice....wait till menu is loaded again...");
        sleep(2);
        clrscr();
        break;
}
}while(choice!=4);
return 0;
}
/* Size of the stack is defined and subsequently an array confining to the same is declared */
#define maxsize 100
int stack[maxsize];
int stacktop=0;
/* User-defined function push() */
void push(int val)
{
if(stacktop<maxsize)
stack[stacktop++]=val;
```

```
else
{
clrscr();
gotoxy(12,12);
printf("Stack Overflow");
}
/* User-defined function pop() */
int pop()
{
int a;
if(stacktop>0)
{
stacktop—;
a=stack[stacktop];
return a;
}
else
{
gotoxy(12,12);
printf("Stack is Empty");
return -1;
}
}
/* User-defined function display() */
void display()
{
int i=0;
if(stacktop>0)
{
printf("Elements of the stack are : ");
while(i<stacktop)
printf("%d ",stack[i++]);
```

```
}
else
{
gotoxy(12,12);
printf("Stack is Empty");
} }
```

The screenshots for different executions of the above program are shown in the Figures 9.21 through 9.25. I am sure the screenshots are self-explanatory. I would advise you to go through the program and follow the screenshots for a better understanding of the execution of the program. This would facilitate you in understanding the concept more clearly when we implement the same using linked list.

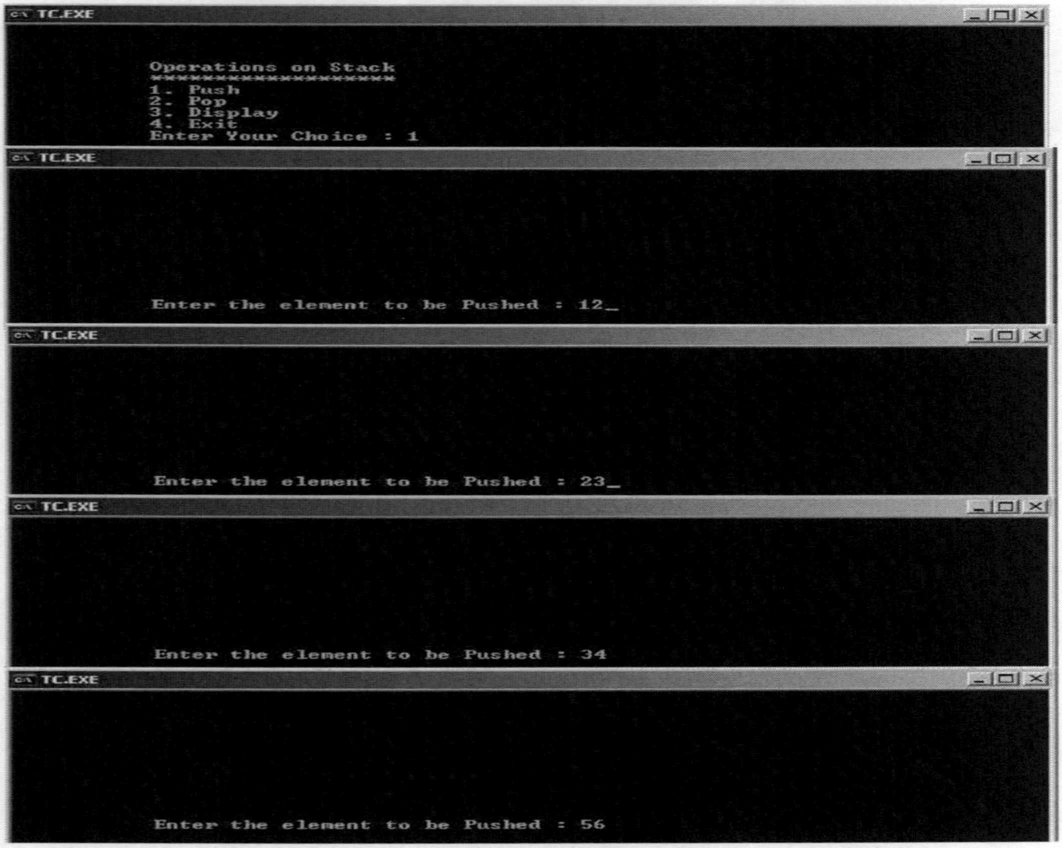

Figure 9.21 Four elements pushed into the stack

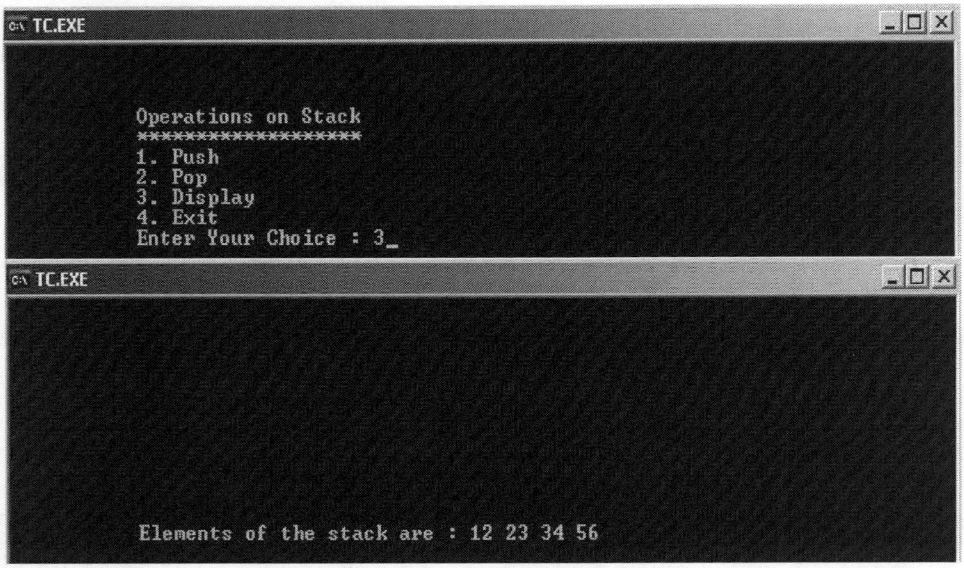

Figure 9.22 Displaying the elements in the stack

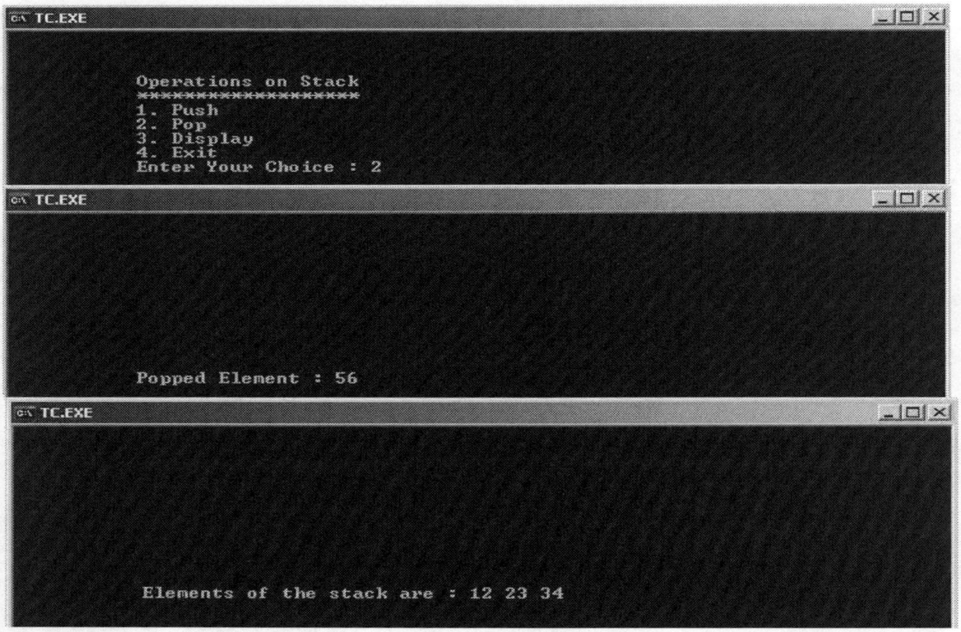

Figure 9.23 Popping one of the elements from the stack, displaying the remaining elements

Figure 9.24 Message indicating that the stack is empty (all the elements are popped out)

Figure 9.25 Message indicating that the stack is full (maximum size is reached)

The stack would be more efficient when implemented with the linked list, since, by now, you are familiar with the linked list. The following is the program for the same. I would appreciate if you can execute and authenticate the program. I have given a couple of screenshots in Figures 9.26 and 9.27 following the coding for a better understanding.

```
/* Stack Operations through Linked List */
#include<stdio.h>
#include<alloc.h>
```

```
#define maxsize 10
void push();
void pop();
void display();
struct node
{
int info;
struct node *link;
}*start=NULL, *new,*temp,*p;
typedef struct node N;
main()
{
int ch,a;
clrscr();
do
{
gotoxy(12,3);
printf("Stack Operations through Linked List");
gotoxy(12,4);
printf("************************************");
gotoxy(12,5);
printf("1.Push");
gotoxy(12,6);
printf("2.Pop");
gotoxy(12,7);
printf("3.Display");
gotoxy(12,8);
printf("4.Exit");
gotoxy(12,9);
printf("Enter your choice : ");
scanf("%d",&ch);
switch(ch)
```

```
{
case 1:
        push();
        break;
case 2:
        pop();
        break;
case 3:
        display();
        break;
case 4:
        clrscr();
        gotoxy(12,12);
        printf("Good Bye.....Have a nice day.....");
        sleep(2);
        break;
default:
        clrscr();
        gotoxy(12,12);
        printf("Invalid choice....Try again while the menu loads...");
        sleep(2);
        clrscr();
}
}
while(ch!=4);
}
void push()
{
new=(N*)malloc(sizeof(N));
clrscr();
gotoxy(12,3);
printf("Enter the item : ");
```

```
scanf("%d",&new->info);
new->link=NULL;
if(start==NULL)
start=new;
else
{
p=start;
while(p->link!=NULL)
p=p->link;
p->link=new;
}
}
void pop()
{
if(start==NULL)
{
clrscr();
gotoxy(12,12);
printf("Stack is empty");
}
else
if(start->link==NULL)
{
clrscr();
gotoxy(12,12);
printf("The deleted element is : %d",start->info);
free(start);
start=NULL;
}
else
{
p=start;
```

```
while(p->link!=NULL)
{
temp=p;
p=p->link;
}
clrscr();
gotoxy(12,12);
printf("The deleted element is : %d\n", p->info);
temp->link=NULL;
free(p);
}
}
void display()
{
if(start==NULL)
{
clrscr();
gotoxy(12,12);
printf("Stack is empty");
}
else
{
clrscr();
gotoxy(12,12);
printf("The elements are : ");
p=start;
while(p!=NULL)
{
printf("%d ",p->info);
p=p->link;
}}}
```

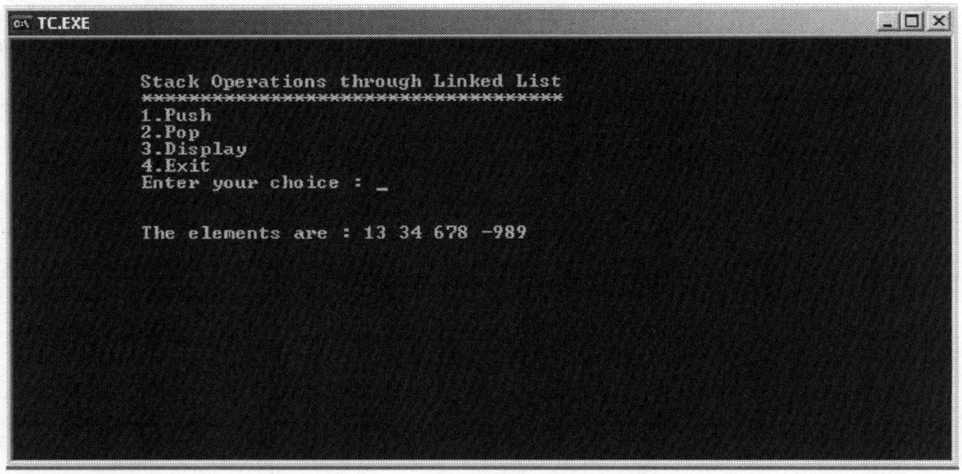

Figure 9.26 Four elements pushed into the stack

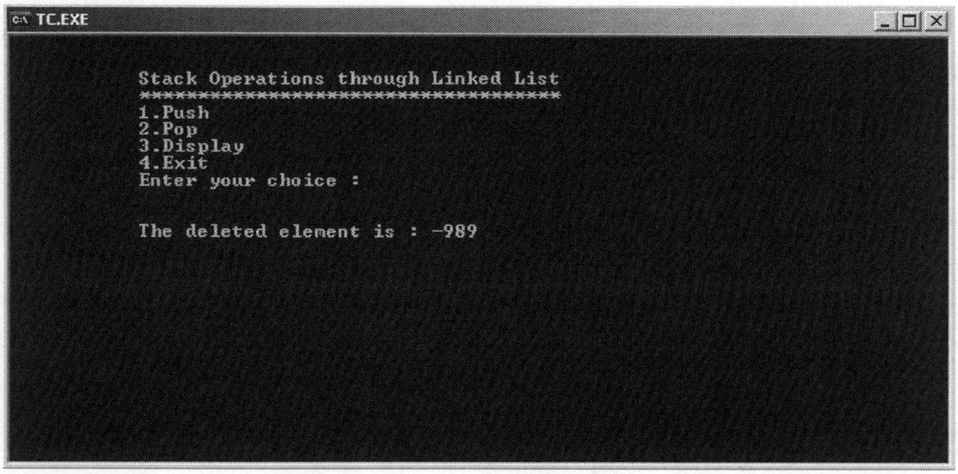

Figure 9.27 An element being popped out of the stack

9.5 QUEUE AND IMPLEMENTATION OF ITS OPERATIONS

A queue is a linear data structure in which data can be added to one end and retrieved from the other. Just like the queue of the real world, the data that goes first into the queue is the first one to be retrieved. That is why queues are sometimes called as *First-In-First-Out* data structures. Remember that we called the stack as *Last-In-First-Out* data structure and a queue is just opposite to that. In this regard, I would say, on a lighter note, that a queue is a fairer one, whereas a stack is not.

In case of queues, data is added to one end, known as *REAR* and retrieved from the other end known as *FRONT*.

Queue data structure provides services in computer science, transport and almost all our regular routines and in operations research (while we deal with the queuing theory). Queues also form an integral part of the circular lists, whose implementation, more or less matches with a queue, leaving a few minor differences.

The following program code gives you a fair idea of how a queue is implemented in C. The coding is followed by a few screenshots for an easier and thorough understanding of the concept. For a change, I have preferred using character inputs over the numerical ones into the queue. I would strongly recommend the readers to implement the idea using numerals too, as that will provide an overall advantage of understanding the concept.

```
/* Operations on Queue using Pointers */
#include <stdio.h>
#include <conio.h>
#include <stdlib.h>
struct list
        {
        char data;
        struct list *next;
        };
typedef struct list node;
void qinsert(node**,node**);
void qdelete(node**,node**);
void disp(node*);
main()
{
int opt;
node *front;
node *rear;
rear=front=NULL;
clrscr();
        do
```

```
        {
        gotoxy(12,3);
printf("Operations on a Queue");
        gotoxy(12,4);
        printf("********************");
        gotoxy(12,6);
        printf("1.Insert");
        gotoxy(12,8);
        printf("2.Delete");
        gotoxy(12,10);
        printf("3.Display the list");
        gotoxy(12,12);
        printf("4.Exit");
        gotoxy(12,14);
        printf("Enter the option : ");
        scanf("%d",&opt);
        switch(opt)
                {
                case 1:
                        qinsert(&front,&rear);
                        clrscr();
                        break;
                case 2:
                        qdelete(&front,&rear);
                        clrscr();
                        break;
                case 3:
                        disp(front);
                        getch();
                        clrscr();
                        break;
                case 4:
                        clrscr();
                        gotoxy(12,12);
```

```
        printf("Good bye...Have a nice day....");
                    sleep(2);
                    return;
                default:
                    clrscr();
                    gotoxy(12,12);
                    printf("Invalid Option....Please try again while the menu loads...");
                    sleep(2);
                }
        }while(opt!=4);
        getch();
}
void qinsert(node **front,node **rear)
        {
        node *newnode;    /* New node to be inserted */
        clrscr();
        newnode=(node*)malloc(sizeof(node));
        newnode->next=NULL;
        gotoxy(12,12);
        printf("Enter the character to push : ");
        fflush(stdin);
        scanf("%c",&(newnode->data));
        if(*front==NULL || *rear==NULL)
                {
                *front=newnode;
                *rear=newnode;
                }
        else
                {
        (*rear)->next=newnode;
        *rear=newnode;
                }
}
```

```
void qdelete(node **front,node **rear)
        {
        node *delnode;      /* Node to be deleted */
        if(*front==NULL || *rear==NULL)
                {
                clrscr();
                gotoxy(12,12);
                printf("Queue has become empty... Deletion not possible...");
                getch();
                }
        else
                {
                delnode=*front;
                (*front)=(*front)->next;
                free(delnode);
                clrscr();
                gotoxy(12,12);
                printf("Successfully deleted....");
                }
        }
void disp(node *f)
        {
        clrscr();
        gotoxy(12,12);
        if(f==NULL)
        {
        gotoxy(12,12);
        printf("The queue is empty");
        }
        else
        {
printf("The elements in the queue are : ");
        while(f!=NULL)
```

```
printf(" %c ",f->data);
            f=f->next; }    }       }
```

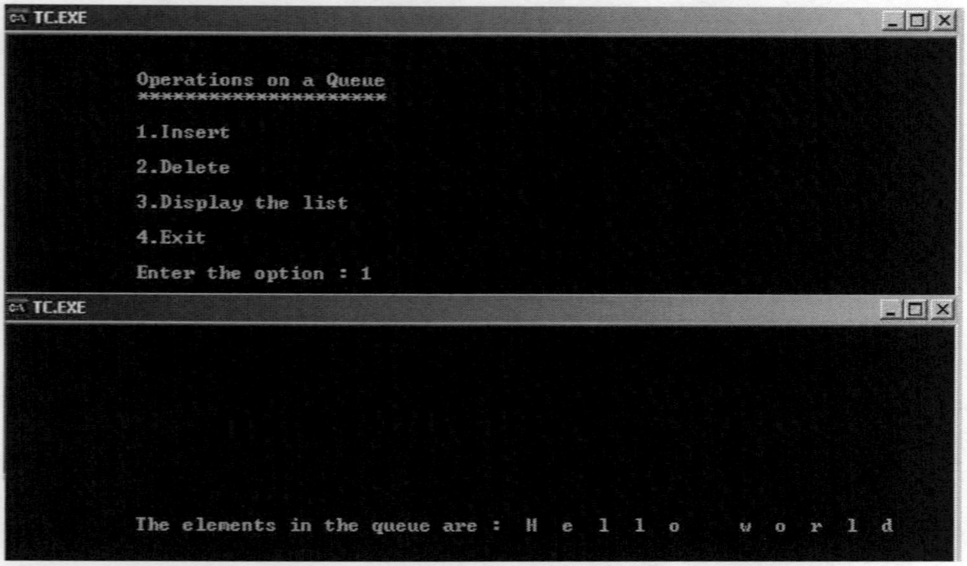

Figure 9.28 Character elements being inserted into the queue

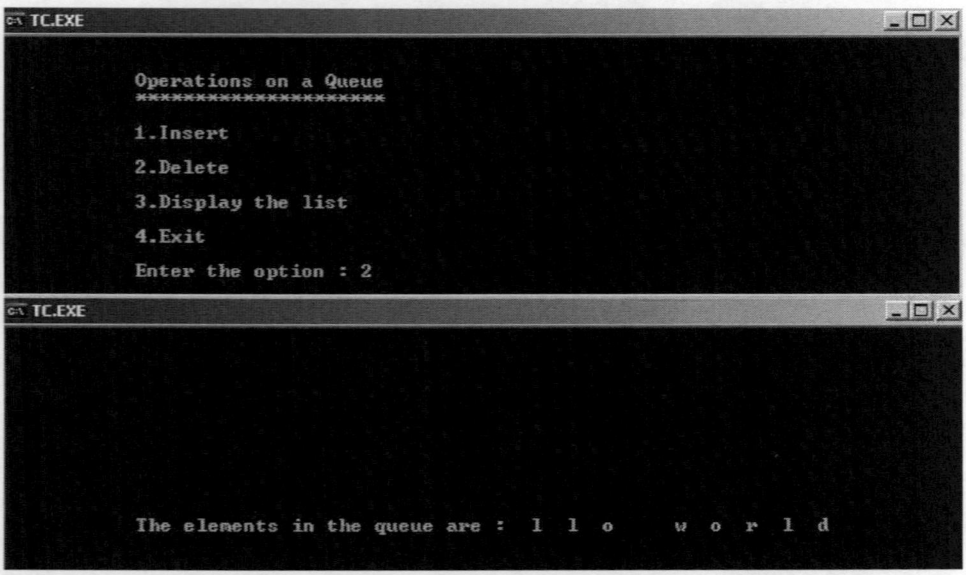

Figure 9.29 Two character elements being removed out of the queue from the ***FRONT***

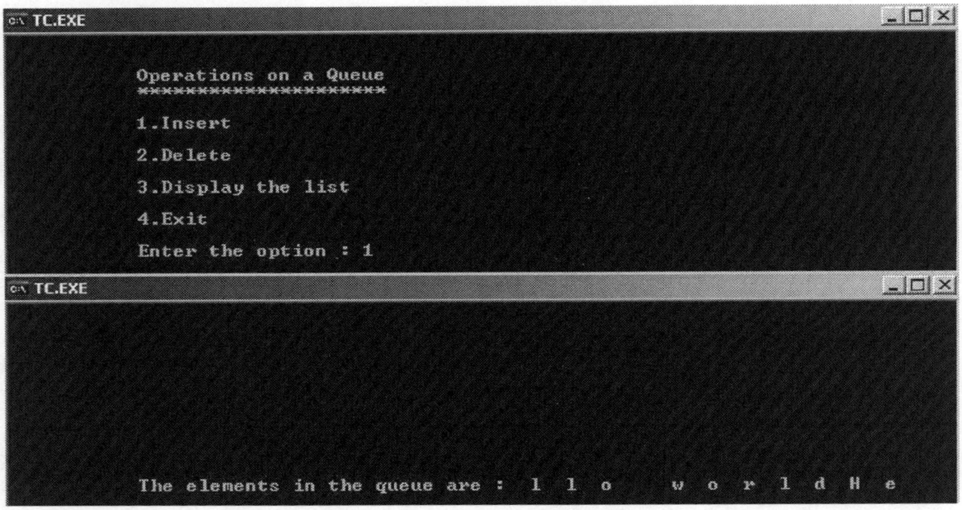

Figure 9.30 Further insertions of the elements done at the ***REAR*** end of the queue

The screenshots in the figures 9.28 through 930 show the execution of the program listed above. From the figures 9.29 and 9.30, we could easily conclude that any removal and insertion of the elements in the queue are done at the *FRONT* and *REAR* ends respectively.

As we go on deleting the elements from the *FRONT*, we will reach a stage when the queue will become empty and the message will be flashed instantly by the program. This is shown in the Figure 9.31.

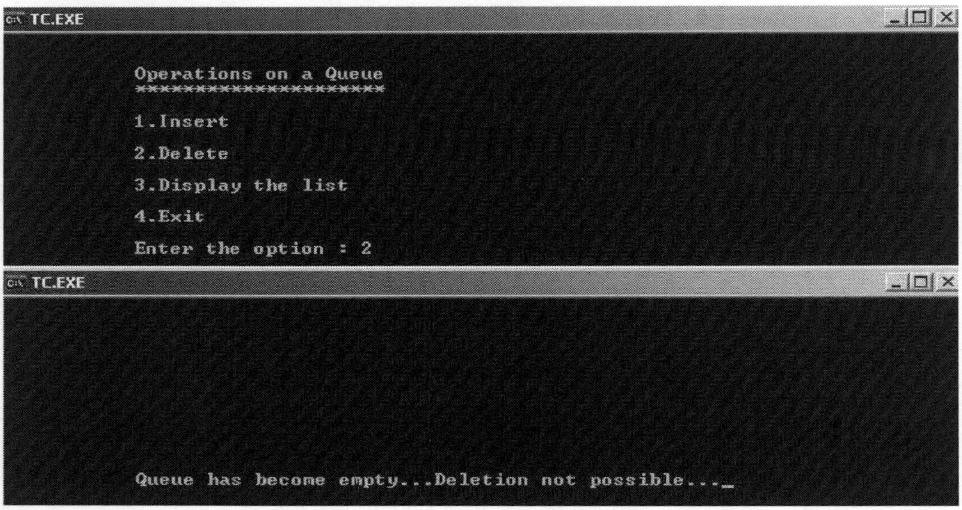

Figure 9.31 Status message after all the elements are removed from the ***FRONT*** end

I am sure that the readers of this chapter are feeling the wonder of C language in handling the real-life situations, and implementing pointers, would have become a child's play for you. I would appreciate if the same spirit is carried forward to the next aspect, which in my opinion is the ultimate application of the pointers, namely, binary trees and their implications.

9.6 INTRODUCTION TO BINARY TREES

All of us know what a tree is. Basically a tree will have roots and branches. Some trees like banyan have children too. The tree that we are going to discuss here too has children and technically we will use a different term to designate the children. We will not go deep into analyzing the tree and the way it works, as it will defeat the scope of this chapter, in which we are intended to discuss only the key aspects of the data structures available in C.

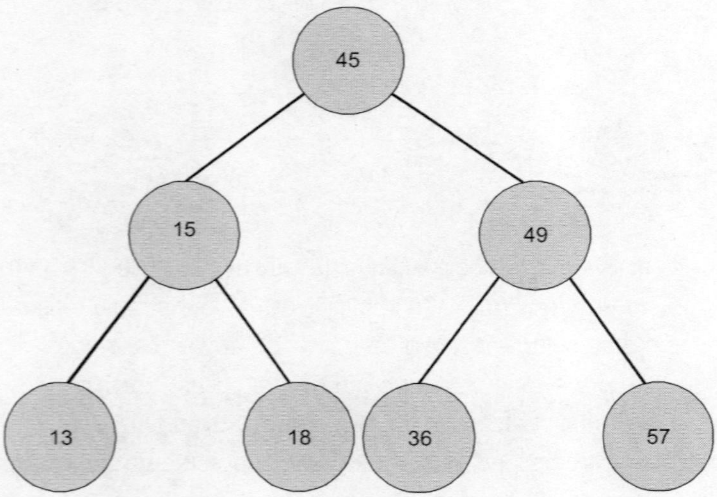

Figure 9.32 A typical balanced binary tree

The binary tree is the fundamental data structure used in computing science. It is very useful in storing the sorted data and retrieving the same rather rapidly. It comprises a parent node and links to up to two leaves called as child nodes. It is the relationship between the parent and child nodes and the way they are used that makes the binary tree such an efficient data structure.

Graphically, a binary tree can be represented as a tree drawn upside down. It starts with a root node and it will have two child nodes. Each child may have its own child nodes. The tree, ideally, would be structured into a perfectly balanced tree, with each node having the same number of child nodes to its left and to its right. This type of balanced tree would facilitate the fastest insertion and retrieval of data. A typical representation of the binary tree is shown in Figure 9.32.

We see from Figure 9.32 that the parent node starts in the value 45. The two child nodes on the left and right of that node have the values 15 and 49 respectively. The value of the left node is less than that of its parent and that of the right node is more than the value of the parent. Assuming that we remove the root node (having the value 45) and its right child nodes, then the node with the value 15 will become a new, smaller binary tree. Through recursive functions, we can easily insert and search for any value.

Varieties of operations are possible on a binary tree, of which searching the tree, inserting a leaf or a node, deleting a leaf or a node, traversing the tree (pre-order, in-order and post-order), sorting (heap sort or binary sort) are the significant ones.

When we think of the applications of a binary tree, they are innumerable. It has applications almost in all the disciplines associated with computing science, such as mathematics, statistics, bioinformatics, operations research, networking, graph theory and so on.

SUMMARY

The idea of this chapter was to introduce the basic data structures used in programming languages and a few of their implementations in C. We started with a very simple reasoning on the purpose and idea of learning the data structures. We then briefly browsed the different data structures in use. We then had an elaborate discussion on singly linked list and the way it is implemented through the C syntax. We had a peek at the doubly linked lists and circular lists. This was followed by elaborate discussions on stacks and queues and their implementations. I am confident that the readers of the book would feel the easiness in deploying the data structures in their applications and would largely appreciate the efficiency of these data structures. We concluded the chapter with a very brief peep into the binary trees. Binary tree is an ocean by itself and it has a variety of applications in the real world and quite often we end up using the nuances of the concept in one way or the other.

CRACK THE NUT

1. Consider the arithmetic expressions (expressions involving arithmetic operators). Think of a program that evaluates the same using stack.
2. Try to implement the circular lists.
3. Consider a polynomial and try to implement the same using a linked list.
4. Can you reverse a stack using recursion? Try writing a program for the same.
5. Given a stack, say S. Write a program to sort the stack (in ascending or descending order).
6. You are given with a singly-linked list. Find the n'th element in the list from the end of the list.
7. If you are given with two unsorted singly-linked lists, find an efficient way to sort and obtain a single list.
8. If the characters 'W', 'X', 'Y', 'Z' are placed in a queue (in that order) and then removed one at a time, in what order will they be removed?
9. To simulate people waiting in a line, which data structure would you prefer to use?
10. List out the areas in which data structures are applied extensively?

10

C EXCLUSIVES

10.1 STORAGE CLASSES

If you are learning about variables and functions and their usage, the learning process would be complete only when you understand the storage classes available in C. There are four storage classes in the language and they all invariably define the scope (visibility) and lifetime of variables and functions. The storage classes are represented by the keywords auto, register, static and extern. The following sections detail on these storage classes.

10.1.1 Auto Storage Class

So far we have written many programs and used many variables, you might be surprised that we have never used the keyword auto for designating a variable pertaining to that storage class. The reason is very simple. For all the local variables, this is the default storage class. So using the keyword ahead of the declaration is optional and considered to be redundant while coding the program. Consider the following program in which we have declared and initialized an integer variable 'a'.

```
main()
{
int a=10;
printf("%d",a);
}
```

The output of the program would be '10'. Now try changing the code by introducing the keyword auto in the declaration section and execute the program again. Can you guess what the output would be? Yes. You are right. We will get the same output as in the earlier case. The following is the changed coding for you to execute and check.

```
main()
{
auto int a=10;
printf("%d",a);
}
```

10.1.2 Register Storage Class

While we write programs using loops, we access the loop counters or subscripts in the case of an array rather very frequently than any other variable. In such cases, we cannot afford storing the variable that is accessed very frequently in the program, in the RAM. Instead, they would be stored in the registers (which are nothing but the memory devices), and accessed very easily while executing the program. There is one caution here for the readers. When you say register ahead of the declaration of variables, it does not necessarily mean that the variable shall be stored in a register. Depending on certain restrictions and the availability of the registers during the execution, the variable might get stored in the register. The following piece of code gives a fair idea of how to declare a variable of type register. The output of the program piece would be '10'.

```
main()
{
register int a=10;
printf("%d",a);
}
```

10.1.3 Static Storage Class

This is the default storage class for the global variables. Naturally it should be the default storage class, since we would anticipate the program to remember the value of the global variable throughout the execution of the program. This declaration serves the purpose. Again, like the other two storage classes described above, the keyword static need not be necessarily provided ahead of the declaration. The following piece of code would help you in understanding my statement more clearly. The variables 'a' and 'b', both are regarded as static, though the keyword is provided only for 'a'.

```
static int a=10;
int b=20;
```

```
main()
{
printf("%d\t%d",a,b);
}
```

Another interesting application of the static variables is in the recursive functions. Since, the static variables are capable of holding their values between functions and would continue to exist throughout the execution of the program, they are efficiently used during recursive calls. When a variable is declared as static within a function and is initialized with a value, it will not be reinitialized when the function is called any number of times and the variable is capable of holding its last known value when the function was invoked. The following piece of code would help you in getting this point in an easier way.

```
#include<stdio.h>
main()
{
clrscr();
call();
call();
call();
getch();
}
call()
{
static a=10;
a++;
printf("%d\t",a);
}
```

You can see that the function call() is invoked thrice from the function main(). What do you think the output of the program would be? Do you think 11 would be printed thrice? No. The output is 11 12 13. Surprised!!! The reason is quite simple. Please follow the explanation which would make things clear. The variable 'a' is declared static with an initial value of '10'. When the function call() is invoked first time from the function main(), the value of 'a' is incremented by '1' and has become '11'. When the function call() is invoked for the second time, the earlier value of 'a' (11) is retained since it is declared as static and hence incrementing it once again results in '12'.

When the function call() is invoked for the third time, the last value (12) is incremented by one and hence it becomes '13'. Hence we get the output as mentioned above.

We could see that the value of 'a' is retained between the function calls and it is not reinitialized each and every time when the function call() is invoked from the function main(). This aspect is very important in recursive functions, as the logic requires one or more variables and their values to be retained in between the function calls.

Another significant aspect of declaring a variable as static is that the variable is initialized by default with 'zero', which is not the case in other modes of declarations. In all the other modes, unless we declare the variable as static, the default value would be any 'garbage' value within the range of the data type to which the variable pertains. Declaring a variable as static is just another way to initialize the variable as '0'. May be, you can call it an implicit way of initializing the variable as '0'.

10.1.4 Extern Storage Class

This storage class is basically used to give a reference to a global variable or a function to all the program files, that is, the variable or the function is global, not just to functions defined and used in the same program, but also to all the program files available. This kind of storage class is frequently used in file access and their manipulations. A variable declared in one file can be used in another, just by declaring the variable as extern. If the variable is declared as extern and initialized with a value in one file, we cannot change its value when it is used in some other file. Remember that, we are allowed to use it in any number of files and we are not allowed to change its value while we use it. The following piece of code explains the way the extern variable works.

```
File-1: FIRST.C
#include"SECOND.C"
int a=10;
main()
{
call_extern();
}
File-2: SECOND.C
void call_extern();
extern int a;
void call_extern()
{
printf("%d",a);
}
```

When we compile the program, FIRST.C, we can get the output as '10' and as we change the value from the file FIRST.C, the same will be recognized by the file SECOND.C and hence correspondingly the output will vary. The role of extern variable is hence highly appreciated.

10.2 SCOPE AND LIFETIME OF VARIABLES

The scope and lifetime of a variable in C, unlike the other programming languages, is largely dependent upon the type of storage class that we have used while declaring the variables. The scope determines over which part or parts of the program, the variable would be recognized. The lifetime of a variable determines the period over which it would retain its value. The following table 10.1 gives the gist of the scope and lifetime of the variables pertaining to different storage classes. This is common for all the variables regardless of which data type they belong to.

Table 10.1 Scope and lifetime of variables of different storage classes

Storage class	Scope	Default Value and Lifetime
auto	These are declared at the start of a program's block such as in the curly braces. Memory is allocated automatically upon entry to a block and freed automatically upon exit from the block. The scope of automatic variables is local to the block in which they are declared, including any block nested within that block. For these reasons, they are also called local variables.	Default value will be **garbage** (within the limits of the data type to which the variable belongs), if they are not provided with any initial value. The value will remain throughout the execution of the program, but accessible only in the function in which it is defined.
register	Same as the *auto* variables and it is not necessary that the variable's value need to be stored in a register. It depends upon the restrictions posed by the hardware.	Default value will be **garbage** (within the limits of the data type to which the variable belongs), if they are not provided with any initial value. The value will remain throughout the execution of the program, but accessible only in the function in which it is defined.
static	Scope is throughout the execution of the program. The variable will be recognized in all the functions across the program and hence has a global scope.	Default value is '0' (unless we give some other value). The variable that is declared as *static* has the capacity of retaining its values between the function calls.
extern	Accessible from within any block and/or remains in existence for the entire execution of the program. Such variables are called global variables.	Default value is '0' (unless we give some other value). The value will remain throughout the execution of the program and accessible across different files.

10.3 COERCION AND FLOATING POINT ISSUES

I have heard many saying C language as a weakly typed language and to some extent the statement is true. In this section of the chapter we would analyse this aspect of C. More or less, this statement relates to the type conversion ideas that have been built-in with the language compiler.

When we start debating on type conversion, we may take two aspects, namely implicit and explicit. Explicit type conversions always happen with the knowledge of the programmers or the end-users. But the implicit type conversions happen without the programmer being aware of the same. Implicit type conversions come as a surprise to the programmers when they are not aware of the concept earlier. Let us first deal with implicit type conversions, which are completely taken care of by the compiler. As a programmer and end-user of a program, you have to have the knowledge of this implicit type conversion.

Implicit type conversion is technically termed as coercion. Again, we may classify the coercion based on whether the coercion is narrowing the result or widening it. Consider the following code fragment in C:

```
#include <stdio.h>
main()
{
float a, b;
int c;
clrscr();
scanf("%f%f",&a,&b);
c=a+b;
printf("%d",c);
getch();
}
```

When you execute the above program and give inputs for 'a' and 'b' as '2.3' and '4.3' respectively, the output that you anticipate must be '6.6'. But the output is just '6'. This is where coercion comes into picture. In this example, the result is being narrowed and the actual result is truncated instead of being straightforward.

The reasoning would be very simple when we go through the entire program very carefully. The real culprit in the above program is the declaration section, where we have declared the variables 'a' and 'b' as float, and 'c' as int. The values that we get for 'a' and 'b' are accepted and even stored as float. They are retrieved as float values during the addition process done using the statement $c = a + b$. But when the result is stored in 'c' which is an integer variable, the truncation happens, since an integer variable is not capable of storing a float result.

Now, you add the code **printf("%f",** *a + b*) in which you directly add the values of 'a' and 'b' (instead of storing the result in 'c' and then printing it) just before you print 'c' in the above program. Now check your output. Do you find any difference between the earlier output and the latest? Obviously, there should be some difference. You would be getting two outputs, one for the code printf("%f", a+b) and the other for the code printf("%d", c). The outputs will be '6.600000' and '6' respectively, for the same inputs for 'a' and 'b' as above. Obviously the former output is correct whereas the latter is not the anticipated one. I would not say that the latter output is incorrect. It is correct according to the type conversion ideas in C and incorrect according to the users like us. This is exactly called as coercion and in this case, the result, as I had mentioned earlier, is narrowed. Because of this kind of coercion, you lose the accuracy in result. This will have more impact in complex problems and applications where you tend to use more variables that hold float and associated values.

What do you think will be the moral of the entire story that we have discussed above? It is again going to be very simple to state. *"Do not try to store the result of operation executed among the variables of float and its associated data types, in a data type such as integer. If you do so, that would result in truncation of the result due to coercion".*

Let us take the other case, where coercion widens the type and the accuracy of the result is not lost at all. Consider the following code to elucidate the idea.

```
#include <stdio.h>
main()
{
float a,c;
int b;
clrscr();
scanf("%f%d",&a,&b);
c=a+b;
printf("%f",c);
getch();
}
```

When you execute the above coding by giving values '2.3' and '4' for 'a' and 'b' respectively, you would anticipate to get '6.3' and that is what you would exactly get too. There should be a big question in your mind now. When does the type conversion happen—before the addition or after that? Logically, the type conversion shall happen even before the values are added.

Please note that we have declared 'a' and 'c' as float type variables and 'b' as int. When we add the values of 'a' and 'b', internally 'b' is automatically promoted to float type and would be recognized as '4.000000' instead of just '4' and hence the actual addition will take place between '2.300000' and '4.000000' and this yields the result '6.300000'. This is where type widening happens and as you have just witnessed, the result of this type of coercion will never give rise to a loss in accuracy of the result and so could be appreciated.

All that we had discussed so far happen automatically without the knowledge of the programmers or the end-users. That is the reason why *coercion* was termed as implicit type conversion. The programmers too have an option of converting the types explicitly when they develop programs. Consider the following piece of coding.

```
#include <stdio.h>
main()
{
int a;
clrscr();
scanf("%d",&a);
printf("%f",(float)a);
getch();
}
```

Please follow the lines of code very carefully and observe that we had used a new piece of information inside the function printf(). What do you think the code (float)a would do and what would be the result of the code? This way of using a data type (within parentheses) followed by a variable (whose value we would be converting to the desired type) is called as implicit type conversion or technically, typecasting. The programmers find this useful in situations when they want the variable to be stored as one type and used as another type when they manipulate the same in calculations and finally display the results. In the above example, if we provide a value '4' for 'a', it will be stored as an integer and since we typecast the variable 'a' when we print the value of 'a', it will be printed as '4.000000'.

I would now put a query to you. What do you think will happen when you remove the typecast from the function printf() and use the format specifier as '%f'? That is, the function printf() is to be modified like the one below. Please execute the coding and try to get the result and check whether that is what you anticipated.

```
printf("%f",a);
```

Try the option where the format specifier used is '%d' and not '%f'. What would you get in such a case? Obviously these kinds of explorations will make your learning and understanding better. In some special cases (such as in the middle of a program, a variable of one type has to be treated as a variable pertaining to another type), as a programmer, we may need to use explicit type conversions and this typecasting comes in very handy and it would facilitate us in producing anticipated results.

The constant values used in a program are also influenced by "coercion". For example, try the following coding.

```
#include <stdio.h>
main()
{
float a=0.7;
clrscr();
printf("%d\t%d",sizeof(a),sizeof(0.7));
getch();
}
```

The answer will be '4' and '8' for the sizeof operators used in the function printf(). Variable 'a' is stored as float and hence occupies 4 bytes and the constant '0.7' is automatically recognized as double and hence occupies 8 bytes. Is it easy to restrict '0.7' to 4 bytes? Yes, it is really an easy task. Just change the existing printf() as the one below and execute the coding again.

```
printf("%d\t%d",sizeof(a),sizeof(0.7f));
```

Now the output will be 4 for both the sizeof operators. Just by putting an 'f' (uppercase or lowercase) after the float value, we could easily restrict the number of bytes occupied by the value.

Even though C is technically a weakly typed language, we cannot take it as completely weak. It is partially weak with respect to the data types and the explicit typecasting is considerably a useful aspect, as said, in special situations.

Floating point values and their usage in programming always have to be taken care of and to be given utmost attention. We had just seen the weak nature of the data type, especially the floating point and its associated types. If the programmers follow and thoroughly understand the way the floating point values behave, they can challenge any kind of outputs generated due to their existence.

Recall the illustration that I had already given in the Chapter 3, in the section 3.5 (Understanding and Debugging a C Program). The illustration implies just one of the issues with the usage of floating point values in a program. We will discuss more on debugging issues in Chapter 11.

10.4 THE C PREPROCESSOR (CPP) DIRECTIVES

I have often seen C programmers fondly using preprocessing techniques very efficiently. This may be due to the crispy nature of the preprocessors and the ease with which they work. To know more about the preprocessors in C, I would advise you to read the sections that follow.

The C preprocessor, as many of you think, is not the part of the compiler. In fact there is no relation between the compiler and the preprocessor. From the name itself, you may guess that preprocessor does the job beforehand. One more interesting fact about the C preprocessors is that they do not spot errors in your coding and do a job similar to that of a compiler. You may presume it to be a separate step in the compilation process. In most understandable terms, a C preprocessor is just a tool used in programming for substitution. C programmers fondly call the C preprocessors as CPP and we will also use the same terminology in our discussion from now on.

Invariably, all the preprocessor directives begin with the '#' symbol. Table 10.2 gives a list of few of the preprocessor directives used in C. Knowingly or unknowingly we have used '#include' directive in all our programs discussed so far. This directive is basically used to include the header files and hence the prototypes of the built-in functions that we had used in the program. Even if you have a look at the Chapter 9, where we have done programs on linked lists, stacks and queues we have used '#define'. Hence, don't you think that they are easy to use? If you learn the exact ways to use them, surely the CPP directives are going to be handy and versatile part of your programs.

Consider the following piece of code in which we have used '#define' directive to print an integer value in the function main().

```
#include<stdio.h>
#define VALUE 10
main()
{
int a;
a=VALUE;
clrscr();
printf("Value : %d", a);
getch();
}
```

In the example above, note the second line of coding containing the '#define' directive. Here '#define' is termed as a macro, *VALUE is termed as* macro template *and the value '10' is termed as* macro expansion.

When we compile and execute the above program, we will get the output '10'. When you analyse the way the program gets compiled and executed, it will be a bit different from the normal programs, in which we would not have used a CPP directive. Here, before the compilation, the substitution process takes place. The preprocessor would browse through the entire coding for the occurrence of VALUE and substitute the same with the value '10'. When the compiler executes the printf() statement inside the function main(), then the value that is substituted for VALUE in the line 'a=VALUE' is recognized by the compiler and hence could print a's value, which is nothing but that of VALUE's.

Table 10.2 Few preprocessor directives in C

Preprocessor Directive	Used for
#include	Including header files, which are the standard libraries in C
#define	Defining a macro
#undef	Removes a macro that has been defined earlier
#ifdef	Checking if the macro is defined
#ifndef	Checking if the macro is not defined
#if	Testing if a compile time condition is true
#else	This is the alternative for #if and executes the statements when the #if returns a false
#elif	The combination for #else and #if
#endif	Marks the termination of the preprocessor conditional
#	Null directive and has no effect
##	Used for forming strings out of the given arguments

Consider the following piece of coding as another example of using a '#define' directive.

```
#include<stdio.h>
#define SQUARE(X) (X*X)
main()
{
int a;
```

```
printf("\nEnter the value of a : ");
scanf("%d",&a);
printf("Square = %d",SQUARE(a));
getch();
}
```

Observe the way we have used the '#define'. It is quite different from the way we had used that earlier. We are in fact, trying to use that as a function. Is that possible with a macro? Yes. Of course it is possible and that is the beauty of a macro. A macro, being just a substitution tool, is used flexibly in programs, just like a function call. So, what exactly happens in the above coding?

When the CPP looks for the availability of a macro template in the program, it comes across the line of code, 'printf("Square=%d",SQUARE(a))'. Immediately it substitutes the expression 'SQUARE(a)' with the equivalent expression (a*a), since the macro expansion for the given template is (X*X) and in terms of 'a' it has to be (a*a). Once the substitution completes, the compilation is done and subsequently the program is also executed and hence we get a result, which is the square of the integer that we give through the function scanf(). These types of macros are called as parameterized macros, which work very similar to the functions. We will later in this chapter see how exactly the macros differ from functions and which is more efficient. I am sure there is no difficulty in understanding the way the macro works here.

Now let us take another CPP directive from the list and know how that works. Consider the usage of '#undef'. Typically this directive is always paired with the '#define' directive. If we want to remove a particular macro that we would have earlier defined with '#define', then we would have to use '#undef' directive to achieve that. Consider the following coding to understand it better.

```
#include<stdio.h>
#define HELLO 20
main()
{
int a;
clrscr();
a=HELLO;
printf("Value = %d",a);
#undef HELLO
#define HELLO 10
a=HELLO;
```

```
printf("\nValue = %d",a);
getch();
}
```

When you execute the above coding, you would get the outputs '20' and '10'. The reason is again interesting and simple. When we first make use of the directive '#define' we had used a macro template HELLO and we had given the expansion for the same as '20'. As you go down the coding, we had undefined the macro template HELLO and immediately we had defined the same with a new value '10'. In both the cases the value is assigned to 'a' and printed. Hence when the first printf() is executed, we get the output as '20' and for the second one, we get the output '10'. Another interesting point that you may note here is that we had used the CPP directives inside the function main() just like any other C statement. It stands valid to use the CPP directives in this manner.

Quite surprisingly, you can use the '#undef' directive for a macro template, which has not been defined earlier anywhere in the program. This only ensures that the definition, if it had been made already, will be removed and other than that this will not serve any purpose. Also, note that the '#undef' directive will not perform a macro replacement.

Let us conclude our discussion on the CPP with learning one more directive in Table 10.2, namely #ifndef and #endif (both come as a group and not alone). Consider the following piece of coding.

```
#include<stdio.h>
#ifndef A
#define A 20
#endif
main()
{
int x;
clrscr();
x=A;
printf("%d",x);
getch();
}
```

We see that we first check whether the macro template 'A' is defined using the '#ifndef' directive. This simply means that we are checking whether the template is not available already.

If this checking returns a TRUE, then we go for defining the template 'A' and providing it an expansion '20'. The output for this program would be '20'. If the macro template had already been defined with a different expansion, then that would be used and not '20'. I would advise you to test the same by including the line '#define A 10' just before the #ifndef directive in the above coding and see what happens to the result. The output in such a case would be '10' and not '20', since the condition '#ifndef' fails.

The '#ifndef' and '#endif' can be used within the function main() too, just like the way we used the other directives '#define' and '#undef'. I leave it to the readers to explore the other CPP directives that are left from this discussion. Please note that these are not the only ways that one can use the CPP directives and in reality, they are used more efficiently in developing meaningful applications.

When you are using the CPP directive '#define' with parameters you should be careful and should not confuse them with the functions. The major difference between the parameterized macros and the functions is that the former is a substitution and the latter is not. The functions are invoked and not substituted. Table 10.3 gives the major differences between these two and I would advise the readers of this book to read and carry them forward and try to use macros more efficiently and to not use them for replacing functions. If you do so, you will be messing up the entire thing

Table 10.3 Major differences between macros and functions

Preprocessor directive - #define	Functions
The macro expansions are looked for in the program and are just substituted in the code even before the compilation starts. If there are errors, they may possibly not be notified by the compilers. In simpler terms, macros just work like a 'find' and 'replace' in your text editors.	These are invoked and are not mere substitutions, and possibility of occurrence of errors are checked during compilation. If there are syntax errors, that has to be fixed before executing the function.
Macros do not occupy memory when they are defined and used. They are simply used like constants.	Functions do occupy memory and they are allocated with specific addresses in the memory. We can pass and return these addresses to and from the other functions in the program.
A macro would return a 'code' and not a value.	A function would return a 'value' and not a code.
A macro is substituted and not invoked.	A function is invoked and not substituted.
A macro substitution does not need additional overheads such as maintaining the stack positions, etc.	When a function is called, the stack positions and the status have to be maintained.

and finally defeat the very purpose of modular programming. Be very firm in using a function where it proves to be more effective and efficient and do not go for a macro there. Also, use a macro where its usage is effective and handy and avoid using functions there. The point is that you cannot skip either of them just because you find one of them easier and the other one a bit difficult.

10.5 DYNAMIC MEMORY ALLOCATION

So far in all our programming examples, we have been following what is called as "Static Memory Allocation". But we may come across situations where we have to deal with data that is dynamic in nature. That is, the number of data items that has been supplied to the program changes dynamically according to the requirement. Consider our earlier programs that we have done with arrays. We have been declaring the arrays of a static size, say, of 100. When we used such arrays, we probably have not used all the 100 cells of the array and the number of elements that we have been using to store in the array was always less than that of its size (regardless of the type of the array that we had used). But, whether we are using the entire array or not, the memory for the array is allocated for the size that we had provided in the declaration section.

For example, if we had declared an integer array such as the one below:

```
int a[100];
```

Here, the array is capable of holding 100 elements of type integer and hence we conclude that the size of the array is '100'. While executing the program, 200 bytes of memory will be allocated for the array declared above (since one integer takes 2 bytes in the memory). But while we use the array for storing the elements, we may go for only 10 or 20 elements (as said, please go through the programs on integer and character arrays that have been done earlier). In case, we are using only 10 elements out of 100, then the remaining 90 cells are unused and an equivalent memory of 180 bytes are not used. Hence, as programmers, we are wasting the memory space. Now consider the problem from the other point of view. What if we need to enter additionally 10 elements beyond the declared size of the array? This is also not possible since we have been allocated only 200 bytes of memory and hence we cannot exceed even by one byte when we give inputs.

Thus, we are having problems at both the extremes, when we want to use less number of elements or when we want to use more elements, of the declared size of 100. This situation is tackled by C through its dynamic memory allocation. This is one way we optimize the usage of memory and manage the memory utilization.

In C, there are four library or built-in functions that come in handy during dynamic memory allocation. They are summarized in Table 10.4 and we will do a brief study on them and analyze how efficiently they could be used in programming, especially when we are in a situation where we either have more or less data to manipulate. Remember to include the header file stdlib.h or alloc.h to use the dynamic memory allocation functions.

Table 10.4 Library or built-in functions in dynamic memory allocation

Function Name	What it does?
malloc()	Allocates memory chunk requests from the program in terms of bytes and returns a pointer to the first byte of the allocated memory chunk
calloc()	Allocates memory chunk requests from the program in terms of bytes, initializes them to zero and returns a pointer to the first byte of the allocated memory chunk
free()	Releases the space allocated by *malloc()* or *calloc()*
realloc()	Reallocates and modifies the space allocated earlier and returns a pointer to the first byte of the newly allocated memory chunk

10.5.1 malloc()

You can allocate a block of memory using this function. After allocating the block of memory, the function malloc() would return a pointer to void (remember that void is one of the primitive data types in C) and you can assign the same to any type of pointer based on your requirements. The following is the general method to use the function malloc().

```
ptr=(cast-type*)malloc(number of elements*byte-size); (ptr – pointer variable)
```

As a specific example, I can give you such as the one below:

```
x=(int*)mallco(n*sizeof(int)); (n – number of elements of type integer)
```

When the above line of coding is used in the program, we can see that the memory allocated will be exactly according to the number of data items of type integer that we would use in the program. Thus we are able to achieve the same idea, which we did using the arrays earlier. The following program code would help you in further understanding the concept.

```
#include<stdio.h>
#include<conio.h>
#include<stdlib.h>
void main()
{
int no, *pt,i;
clrscr();
printf("Enter no of Students : "); /* Getting the number of data items that has to be stored */
scanf("%d",&no);
pt=(int*)malloc(no*2); /* Dynamic Memory Allocation is done here */
```

```
/* Checking whether the memory is allocated or not */
if(pt== NULL)
{
printf("\n\nMemory allocation failed!");
getch();
exit(0);
}
/* Getting the input data items */
printf("\nEnter roll no of students : ");
for (i=0;i<no;i++)
scanf("%d",(pt+i));
/* Displaying data items */
printf("\Entered roll nos are : ");
for (i=0;i<no;i++)
printf("%d\t",*(pt+i));
getch();
}
```

The output of the above program is shown in Figure 10.1. If you observe the program carefully, immediately after getting the number of data items to be stored, we check for the availability of memory for allocation and if not available, we flash the message that sufficient memory is not available. If memory is available, then everything goes fine.

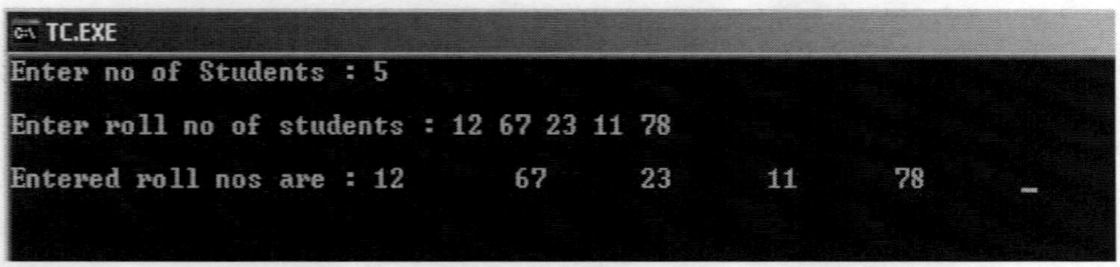

Figure 10.1 Output of the program using malloc()

Also, note the line of the coding pt=(int*)malloc(no*2). Can you guess why we have used the multiplication no*2 inside the malloc()? The reason is that we are trying to input integers as data items and each integer will take '2' bytes in memory and we have to reserve memory for as many integers as would be entered (in this case the number of integers that we would enter is stored in the variable 'no'). In case you forget the number of bytes occupied by a particular data type, you need not worry and C has provided the alternative for you. Instead of the existing code, you may use the code pt=(int*)malloc(no*sizeof(int)) , which will produce the same effect as the earlier code. Also note that the memory chunk allocated by malloc() is typecast into integer data type since the function malloc() returns void by default.

10.5.2 calloc()

This is another function that does the same job as the function malloc(). Even though the tasks of both the functions are one and the same, calloc() differs from malloc() in that it initializes the allocated memory chunk to zero apart from allocating it. Also, this accepts two arguments and malloc() requires only one. The following is the general syntax of the function calloc().

```
ptr=(cast-type*)calloc(number of elements,byte-size); (ptr – pointer variable)
```

Follow the same example given in the section above for understanding calloc(). In the program coding, replace the line pt=(int*)malloc(no*2) with the line pt=(int*)calloc(no,2) . You will get the same results. Also try using the line of code pt=(int*)calloc(no,sizeof(int)) and again there would not be any change in the output.

10.5.3 free()

When we use variables in a program, their memory is managed by the storage classes that we had already explored in the very beginning of this chapter. They are compile-time variables and hence easily managed. But when we dynamically allocate the memory for the variables, we have to take care that the memory is released after the usage. This is achieved through the function free(). Basically we do this since the memory already occupied, if released properly, could be used in the future for other purposes. In the program coding listed above, we have not released the memory allocated for the variable *pt through the function calloc(). This can be done by using the code free(pt) after printing the data items that you had obtained is displayed. It is always wise to release the memory used, using this function.

10.5.4 realloc()

The function realloc() changes the size of the block of memory that was previously allocated with the function malloc() or calloc(). The following is the prototype of the function realloc().

```
void *realloc(void *block, size_t size);
```

After the reallocation, a pointer is returned, which might be different from the original address that was returned with the functions malloc() or calloc(). The following program would be able to explain the way the function realloc() works.

```
#include <stdio.h>
#include <stdlib.h>
#include <string.h>
main()
{
char word[80], *message;
clrscr();
puts("Enter a line of text : ");
gets(word); /* Getting the first string */
/* Allocating memory for the string using malloc() */
message = (char*)malloc((strlen(word)+1)*sizeof(char));
strcpy(message, word);
printf("\nThe first string is : ");
puts(message); /* Displaying the first string */
puts("\nEnter another line of text : ");
gets(word); /* Getting the second string */
/* Reallocating the memory based on the length of the new string obtained by appending the second string
with first */
message = realloc(message,(strlen(message) + strlen(word)+1));
strcat(message,word);
printf("\nThe new string after appending the second with the first is : ");
puts(message); /* Displaying the new string */
getch();
}
```

String inputs are used as the data items for a better understanding of the function realloc(). See that the memory is allocated using the usual function malloc() according to the length of the string. The length of the string is known through the function strlen(). You may also see that '1' is added with the length of the string to facilitate the appending of NULL ('\0') character to the string obtained.

Before the second data item is got, the memory with new chunk of memory, which is nothing but the existing memory plus the memory that might be required for my new data item (with one byte extra for storing the NULL character), is reallocated Figure 10.2 shows the output of the program listed above.

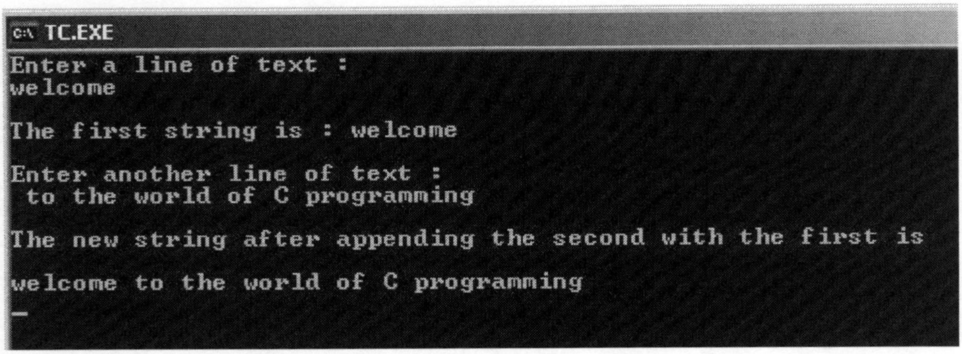

Figure 10.2 Output of the program using realloc()

SUMMARY

As the title of the chapter suggests, we have covered most of the miscellaneous topics in C, which were possibly not dealt with in any of the chapters earlier. As most us know that C is a sea by itself and more concepts, techniques and ideas are still being explored, I have wrapped up with a few important ideas, which I felt, are essential for the students to learn.

This chapter kicked off with a discussion on the storage classes. The four storage classes were detailed and the ideas explained were supported with program fragments. The significance of using static variables in functions was thoroughly discussed. This was followed by a very brief and significant discussion on the scope and lifetime of variables used in C, which was again based on the storage classes.

The concept of coercion and a few floating point issues followed in which we elaborately analyzed the weak typing in the language. The merits and demerits of weak typing were discussed and it was concluded that C's weak typing was not completely weak and it is just partial. The same was justified with a few examples.

The C preprocessor directives, which are considered to be the versatile ideas in the language was also discussed with reference to a few important ones. Most importantly, I have tried to list out the major differences between the functions and macros (which are preprocessor directives) towards the end, which, I am sure, would help the programmers (especially those who feel comfortable in working with macros) to have a fair idea of when to write functions and when to write macros. I have also concluded the section by saying that you cannot skip both and you are not supposed to use one in place of another, which would surely spoil the program or the application that you develop.

Towards the end of the chapter, I have given a lucid explanation on dynamic memory allocation and the importance of the same from the point of view of memory management. The functions such as malloc(), calloc(), free() and realloc() were thoroughly discussed with necessary programming examples. I can confidently say that dynamic memory allocation in C is a small step towards the memory management that is done at higher levels in operating systems.

I hope the flow of ideas in this chapter are surely helpful to the readers in carrying them forward without any hindrances. While the fact is that this is not the end of learning C, equally true is that you are now close to your target since you have compiled more concepts till this chapter. Now, please go and crack the nut!!!

CRACK THE NUT

Identify the outputs for the following programs:

1.
```
main()
{
extern int i;
i=20;
printf("%d",i);
}
```

2.
```
#define square(x) x*x
main()
{
int i;
i = 64/square(4);
printf("%d",i);
}
```

3.
```
#include <stdio.h>
#define a 10
main()
{
#define a 50
printf("%d",a);
}
```

4.
```
#define clrscr() 100
main()
{
clrscr();
printf("%d\n",clrscr());
}
```

5.
```
#define f(g,g2) g##g2
main()
{
int var12=100;
printf("%d",f(var,12));
}
```

6.
```
#define prod(a,b) a*b
main()
{
int x=3,y=4;
printf("%d",prod(x+2,y-1));
}
```

7.
```
#ifdef something
int some=0;
#endif
```

```
main()
{
int thing = 0;
printf("%d %d\n", some ,thing);
}
```

8.
```
#if something == 0
int some=0;
#endif
main()
{
int thing = 0;
printf("%d %d\n", some ,thing);
}
```

9.
```
#define DIM( array, type) sizeof(array)/sizeof(type)
main()
{
int arr[10];
printf("The dimension of the array is %d", DIM(arr, int));
}
```

10.
```
main()
{
static int var = 5;
printf("%d ",var—);
if(var)
main();
}
```

11.
```
#include<stdio.h>
#include<stdlib.h>
int main()
{
int *p;
p = (int *)malloc(256 * 256);
if(p == NULL)
printf("Allocation failed");
return 0;
}
```

12.
```
#include<stdio.h>
#include<stdlib.h>
int main()
{
int *p;
p = (int *)malloc(20); /* Assume p has address of 1314 */
free(p);
printf("%u", p);
return 0;
}
```

13.
```
#include<stdio.h>
#include<stdlib.h>
int main()
{
char *ptr;
*ptr = (char)malloc(30);
strcpy(ptr, "RAM");
printf("%s", ptr);
```

```
free(ptr);
return 0;
}
```

14.
```
#include<stdio.h>
#include<stdlib.h>
int main()
{
int *a[3];
a = (int*) malloc(sizeof(int)*3);
free(a);
return 0;
}
```

15.
```
#include<stdio.h>
#define SWAP(a, b, c) (c t; t=a, a=b, b=t)
int main()
{
int x=10, y=20;
SWAP(x, y, int);
printf("%d %d\n", x, y);
return 0;
}
```

16.
```
#include<stdio.h>
#define SQR(x) (x*x)
int main()
{
int a, b=3;
a = SQR(b+2);
printf("%d\n", a);
```

```
return 0;
}
```

17.
```
#include<stdio.h>
#define SQUARE(x) x*x
int main()
{
float s=10, u=30, t=2, a;
a = 2*(s-u*t)/SQUARE(t);
printf("Result=%f", a);
return 0;
}
```

18.
```
#include<stdio.h>
#define CUBE(x) (x*x*x)
int main()
{
int a, b=3;
a = CUBE(b++);
printf("%d, %d\n", a, b);
return 0;
}
```

19.
```
#include<stdio.h>
#define SWAP(a, b) int t; t=a, a=b, b=t;
int main()
{
int a=10, b=12;
SWAP(a, b);
printf("a = %d, b = %d\n", a, b);
return 0;
}
```

20. What will be the output of the program?

```
#include<stdio.h>
#define FUN(i, j) i##j
int main()
{
int va1=10;
int va12=20;
printf(«%d\n», FUN(va1, 2));
return 0;
}
```

21.
```
#include<stdio.h>
#define MAN(x, y) ((x)>(y)) ? (x):(y);
int main()
{
int i=10, j=5, k=0;
k = MAN(++i, j++);
printf("%d, %d, %d\n", i, j, k);
return 0;
}
```

22.
```
#include<stdio.h>
#define MIN(x, y) (x<y)? x : y;
int main()
{
int x=3, y=4, z;
z = MIN(x+y/2, y-1);
if(z > 0)
printf(«%d\n», z);
return 0;
}
```

23.
```
#include<stdio.h>
#define MESS junk
int main()
{
printf("MESS\n");
return 0;
}
```

24.
```
#include<stdio.h>
#define MAX(a, b) (a > b ? a : b)
int main()
{
int x;
x = MAX(3+2, 2+7);
printf("%d\n", x);
return 0;
}
```

25.
```
#include<stdio.h>
#define PRINT(i) printf("%d,",i)
int main()
{
int x=2, y=3, z=4;
PRINT(x);
PRINT(y);
PRINT(z);
return 0;
}
```

26.
```
#include<stdio.h>
#define MAX(a, b, c) (a>b ? a>c ? a : c: b>c ? b : c)
```

```
int main()
{
int x;
x = MAX(3+2, 2+7, 3+7);
printf(«%d\n», x);
return 0;
}
```

27.
```
#include<stdio.h>
int main()
{
float f=43.20;
printf("%e, ", f);
printf("%f, ", f);
printf("%g", f);
return 0;
}
```

28.
```
#include<stdio.h>
    int main()
    {
    float fval=7.29;
    printf("%d\n", (int)fval);
    return 0;
    }
```

29.
```
#include<stdio.h>
    int main()
    {
    float *p;
    printf("%d\n", sizeof(p));
```

```
return 0;
}
```

30.
```
#include<stdio.h>
#include<math.h>
int main()
{
printf("%d, %d, %d\n", sizeof(3.14f), sizeof(3.14), sizeof(3.14l));
return 0;
}
```

11

ERRORS, BUGS AND DEBUGGING

11.1 AM I ALLOWED TO MAKE ERRORS?

Any beginner in any programming language would be asking "Am I allowed to make errors while I program?" I always tell my students, "The best way to learn a programming language is by making errors". After all, we learn so many things in life only by making errors and what if we make errors when we write a program?

The next obvious question by any programmer would be "Do I have any limit in making errors, that is, can I make errors even if I claim to be an expert in that programming language?" Again the answer would be "Yes". There is no limit for the errors you make, provided you learn every time you make one. But make sure that every time you make an error, it is a new one and not the one already made. This will help you in learning newer things about the language.

The first thing that I would tell my students when I start teaching C language to them would be to have a record of all their errors whenever they make one. This would help them in understanding the language in a better way and they would invariably become mobile compilers/interpreters in fixing up the errors rather easily. It would be child's play for the programmers to fix up the errors, who would know when, how and where the error would probably creep up in a program.

This chapter deals with all these essential things that a programmer, particularly in C language, is supposed to know and preserve. Please go ahead and I wish you all to become perfect "debuggers" by yourself after getting to know how to do and how not to.

I would also request you to continue making newer errors and keep your error database growing, as C is a God's Programming Language and to explore the God, you have to make errors without hesitating and without knowing that you are making one.

11.2 WHY ERRORS/BUGS GET INTO A PROGRAM?

When you could understand the reason behind the occurrence of errors, probably the rate of making errors while you write a program would come down. What do you think is the reason for getting errors while you program? What kind of errors do you tend to make?

In my understanding, there are two possible errors that anyone, including me, can make while editing a program. One section of errors is the logical errors and the other one is the syntax errors. Syntax errors are again known by the names, compiler errors and linker errors. While the syntax errors are noticed by the compiler/linker and flashed to you during compilation/linking stage, the logical errors are left unnoticed. It is sometimes dangerous to commit logical errors, as you may not be able to locate the errors easily and you probably would be wasting a lot of time in fixing it.

While the syntax errors arise due to misspelling of keywords or function names, skipping the prototyping, skipping the headers and so on, the logical errors arise due to wrong thinking. I would rather advise those who begin programming in C, to think correctly, so that they can act accordingly. I used to frequently tell my students to take a pen and paper and sketch out the program and its output (especially when they work with concepts such as arrays, pointers, structures, etc.) before they actually edit the program and compile that. Sketching out the program and tracing the output would give an elaborate picture of the program, since you would have given deep thought in tracing out the program and you would have eradicated all possibilities of getting an error by the time you open your editor to key in the program and compile it.

11.3 STRUCTURED PROGRAMMING DOES IT ALL

There is a famous saying "Plan your work and work your plan". Writing a program is like building a house. We never build the house without a proper plan and you would probably know the outcome of a house that is built without a proper plan. You have to make out a good blueprint of the program and plan it in advance. Another important aspect of a planned program is that it tends to have fewer bugs compared to an unplanned program. Also, as I had mentioned, to get a good structure for your program (just like for your house), it is always good to make out a plan.

When a program is structured, we normally break the program down into different sub-programs (divide and rule, as it is easy to say). This would facilitate the compiler to check the sub-programs separately and don't you think this has a great significance when you debug the program?. As you finish a sub-program, you have the opportunity to debug the program. Remember that when you do not debug a program from the beginning till the end, it is always going to be a Herculean task for you when you do it after the completion of the program. In such a case, you would sit and scroll through the entire program for the bugs, whereas, the task would be simpler in case you go for implementing modular programming and debugging the individual modules as and when you finish coding them.

Another interesting thing that you must note here is that the sub-programs could be used in any of the programs that you may later write and since you have already debugged the sub-program, you have the advantage of including a coding that is already debugged and hence save a lot of time.

11.4 COMPILER ERRORS

Every programming language has something significant called as syntax rules. They are the grammar of the language. When you learn English, you tend to follow the grammar of the language and any misuse of the language or word is not allowed while you learn the language. Similarly the syntax rules are the backbone of any programming language and they define the language as a whole.

For example, consider that we are driving a two-wheeler or any such vehicle on the road. Do we have certain rules to follow when we drive? The most important is the driving license, which is similar to knowing the methodology of using the language syntax. The other things that we must essentially know while using a vehicle on the road are recognizing the traffic symbols, obeying the traffic signals and the lane systems. These things are similar to planning out a program and writing a structured program. Given the many rules and regulations to follow on the road when you are using a vehicle, who is there to enforce them? Yes, you are right. We have the traffic cops on the road who ensure that these rules are properly followed by you and me. A compiler is similar to a traffic cop. Compiler programs are designed in such a way that they enforce the syntax rules and hence you get a syntax error when you deviate from the normal syntax that has to be followed. As the name suggests, the compiler will compile the errors and consolidate them to you, unlike the interpreters, which does line by line interpretation and would probably not allow you to proceed unless you fix up the error in a particular line of a program. This methodology of the compilers, though seems to be good, sometimes leads to confusing the programmers. For example, if a compiler finds an error in a program in the line number 15, it may report the couple of lines that follow the actual line where the error has occurred, also as error-prone. So it is always good to know how the compilers report the errors and make yourself tuned up to that.

Make it a point that the first error that the compiler would report to you is the true error and once you fix that error, the other errors that followed the first error might vanish. Consider the following program in order to exemplify this behaviour of the compiler.

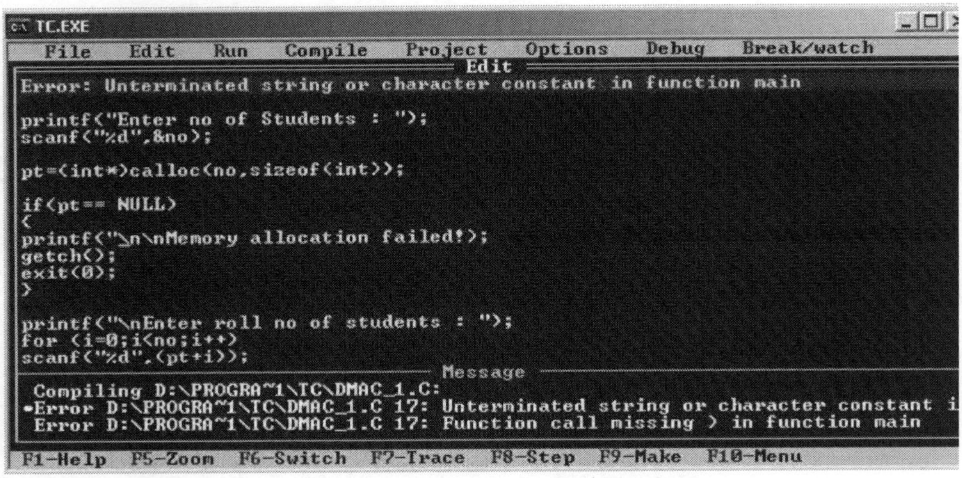

Figure 11.1 Compiler-generated errors

You can see from Figure 11.1 that two errors are generated in the program—one saying "Unterminated string or character constant in function main" (the line is indicated with a dark single-headed arrow) and the other saying "Function call missing) in function main". But the real error is that we have missed out typing the closing quotes after the string and this is what the first error has indicated, and once we fix this error, the second one would naturally disappear. Obviously, if you have a closer look at the second error, it does not sound logical, since we have used a closing parenthesis to indicate that the function is closed. The figure 11.2 shows that both the errors have disappeared on clearing the first error.

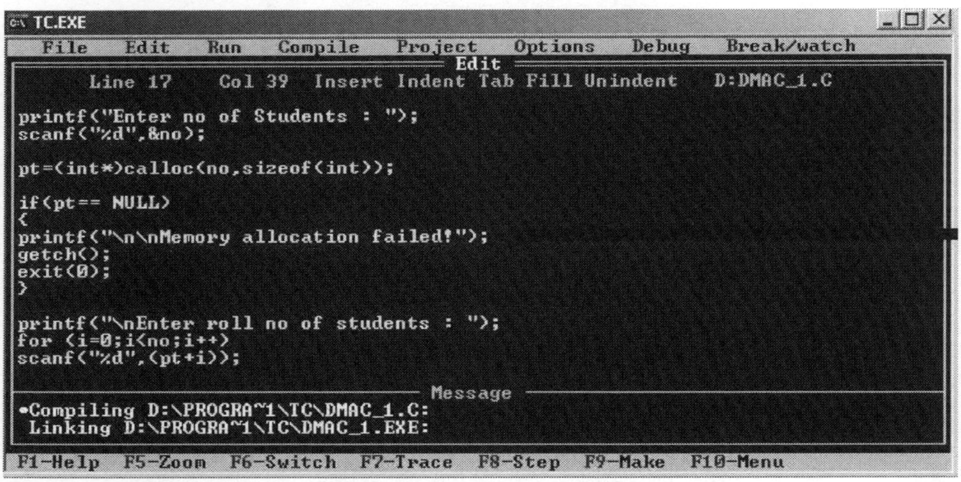

Figure 11.2 Both errors (indicated in the figure 11.1) disappear as we clear the first error

Compiler is just a software that is created by the language designers and hence are subject to generate errors that need not be completely accepted. In simple words, the errors generated by the compiler may be wrong in many cases. As a programmer, you shall not assume or take for granted that the line number that is indicated by the compiler is indeed having an error. Basically a compiler scans the source code sequentially from the first line to the last line and when it does so, it may not detect any error until many lines below where the actual error is. It may indicate a line in the source code assuming that something wrong is out there. We can take one thing for sure that when an error is indicated by the compiler, the error is actually available in any of the lines above the indicated line and not below. I have illustrated this using Figure 11.3, where the error is indicated by the compiler to be in line 13, whereas the actual error is in line 11. You will notice that the line number where the cursor has positioned itself is 11 (indicated with a dark ellipse and an arrow) and that is the actual point in the program where we have the error. When you notice down the window, we see that the compiler has indicated that line 13 has the error (marked with a dark ellipse in the figure). Rationally, the compiler would be doing a similar kind of scan throughout the program, regardless of the number of lines in the program.

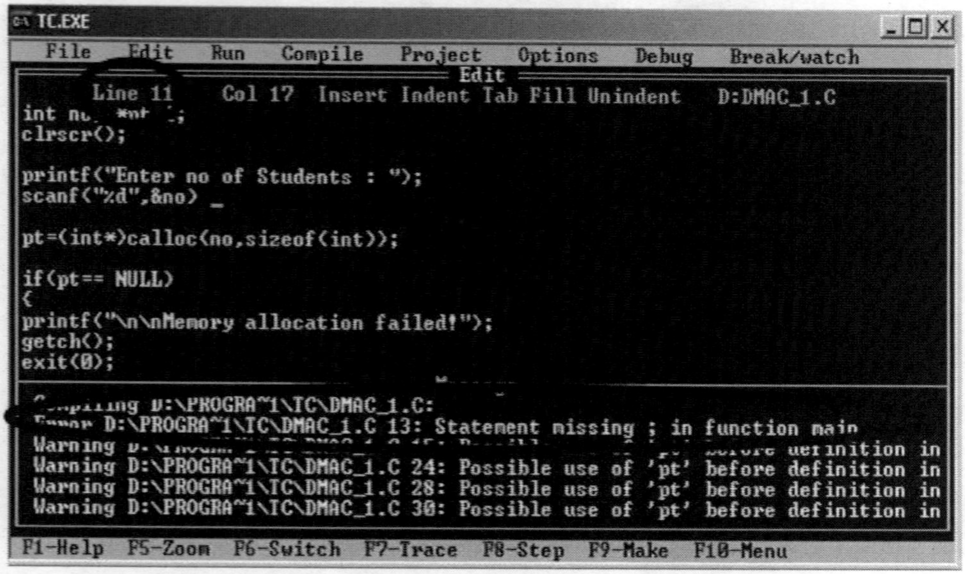

Figure 11.3 Way the compiler scans the program for errors

In Figure 11.3, you would also notice that warnings are generated along with the error. Warnings cannot be actually termed as syntax errors. They may arise due to logical errors that one may commit in a program. The compiler generates a warning just to let the users know that the particular line under scanner is suspicious, though legal. Consider the Figure 11.4, where one warning is generated, though the program is correct and successfully compiled.

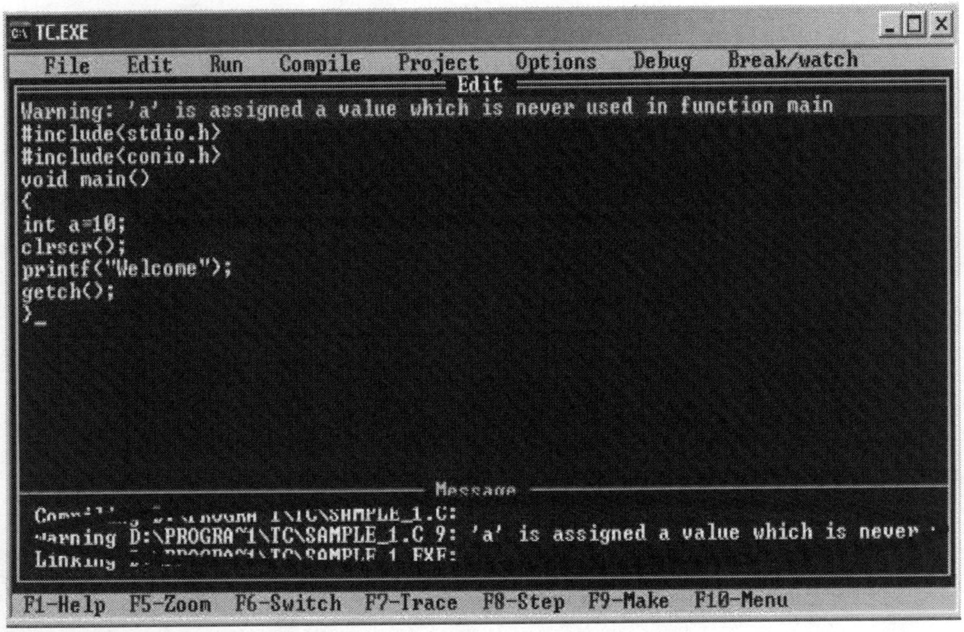

Figure 11.4 Warning is generated, though the compilation is successful

The obvious reason for getting such a warning is that, we have declared and defined a variable 'a' (with an initial value '10') and we never have used the same anywhere in the program, except at the point of its declaration. So, the compiler is a bit worried about you declaring a variable and assigning a value to it, but it has never been used and hence flashes a warning to you.

To conclude this discussion, I would like to caution you on one common mindset programmers of any language have. Just because your program has no errors or warnings and has successfully compiled, it does not necessarily mean that your program is absolutely fine. It is only an indication that your program is free from all syntax errors and still it may have logical errors in it, which has to be taken care of by the programmers themselves. Beware that your job is not to just develop syntax-free programs, but sensible ones.

11.5 LINKER AND RUNTIME ERRORS

Unlike the compilers, linkers are brilliant and generate errors that are always meaningful and to the point. They neither are ambiguous nor allow the programmers to go ambiguous. Linker is software that links object files (compiled machine code obtained from the source code) and the built-in libraries (normally the headers that we include in our program). As a result of linking we would be getting the executable form of the source file and the program is now ready for further execution.

I would like to put forward one more significant point here. Most of the programmers, who work with compiled languages (such as C) assume that the job ends with a compiler and it is enough to get rid of compiler errors. But, it is the linker that plays the key role in producing the executable file from the source code. Hence, along with the compiler errors, try to fix up the linker errors too.

The methodology used by the linker is just the matching of things such as functions, variables, etc. What do you think is the most common linker error that you would get while you write and compile programs? Obviously it would be the mistake that you do while spelling the names of functions or using wrong symbols. The linker always matches with the built-in libraries as said earlier and when it does not find a match, it would flash error. Figure 11.5 shows one such error. Please note that, when you get a linker error, you would not be able to get the cursor back in the editing screen just by pressing enter key. You have to use the keystroke F6 to get the cursor active in the editor.

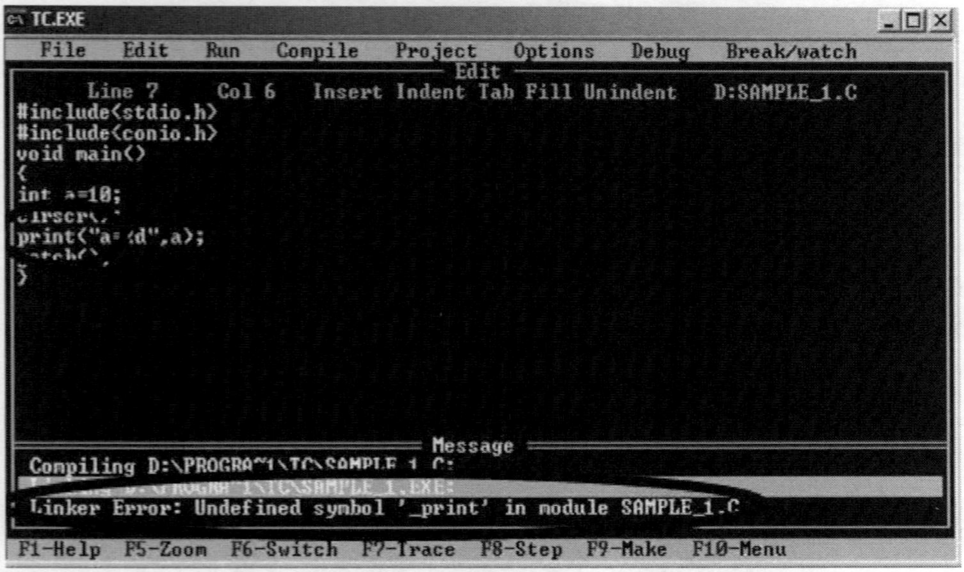

Figure 11.5 Linker error due to misspelling the function *printf()*

You can see that the function printf() is spelt wrongly by the programmer as *print()*. Due to this, the compiler could not find a match either in the headers or in the form of user-defined function.

On the other hand, we have runtime errors, which happen during execution of a program. Notice the program in the figure 11.6 and the error generated when we execute the program.

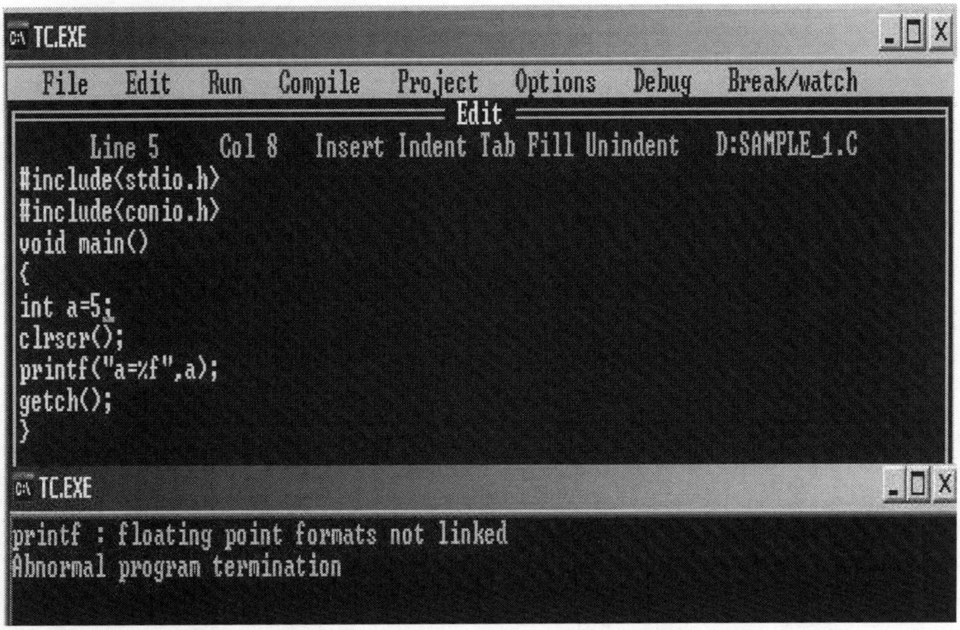

Figure 11.6 Runtime error due to improper usage of floating point value

The above error occurred since we had tried to print an integer value in the floating point format, since it is allowed in this manner (we may do explicit typecasting to print, but not like this). Errors of these types are recognized only during the execution of the program and hence categorized under the runtime errors. It is the sole responsibility of the programmers to take care of such errors and ensure that these errors do not occur very frequently. Another runtime error occurs when there is an improper use of recursion. Normally when recursive functions are defined, we are expected to give a condition inside the recursive function to terminate the function call and get the control back to the caller (the function main() in most of the cases). When we skip/forget to provide a condition inside the recursive function, it would result in stack overflow and would result in runtime error/ abnormal program termination. Figure 11.7 illustrates a simple program where we have used recursion (the function main() called recursively) and since we have not provided any condition inside the function main(), this would give rise to abnormal program termination.

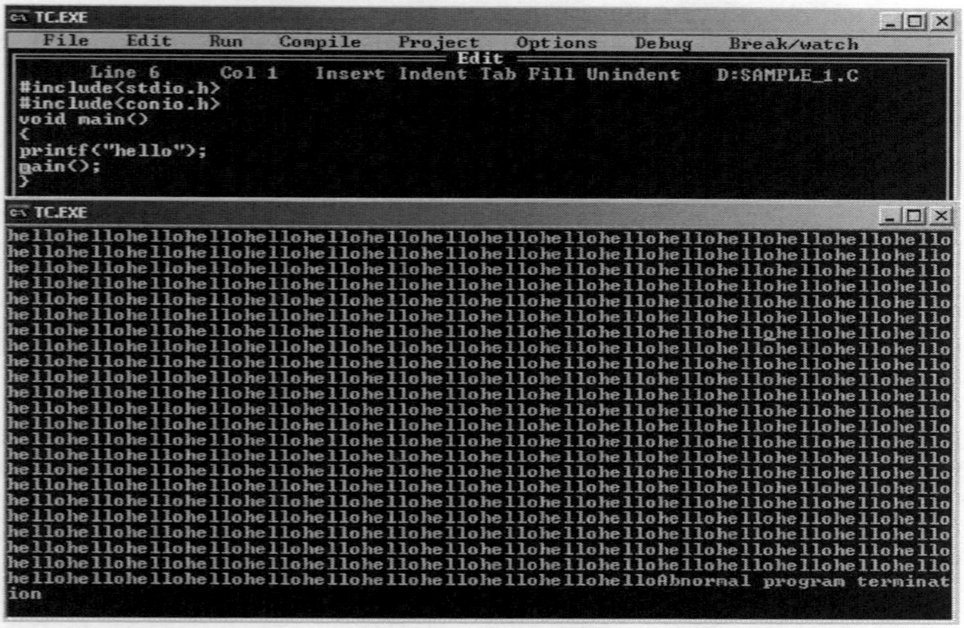

Figure 11.7 Runtime error due to improper definition of recursive function

When you make minor alterations in the program in Figure 11.7, you may get rid of this issue. Figure 11.8 shows the modified program with the output, where we do not encounter a runtime error.

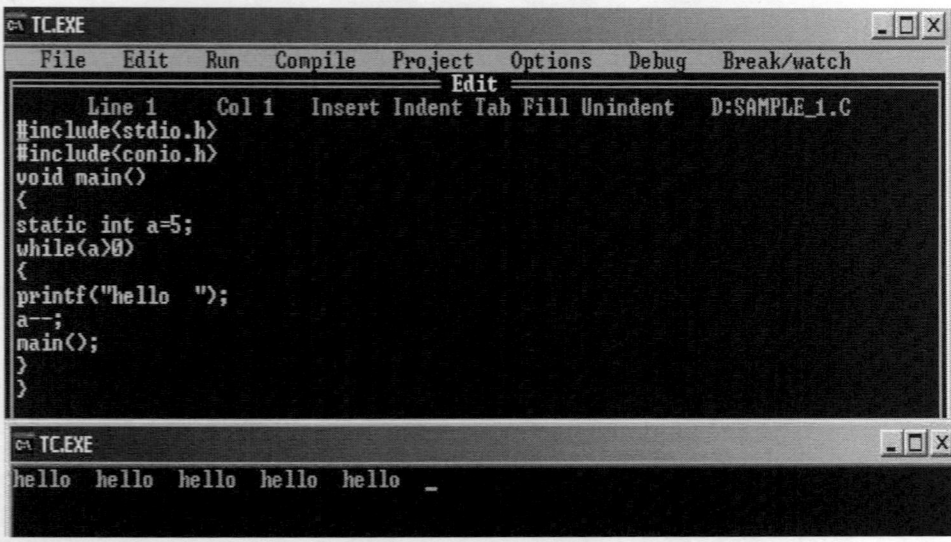

Figure 11.8 Runtime error fixed by providing a condition

We see that in the program of the Figure 11.8, a *while* loop is provided and a condition is tested and based on the return value of the condition, the loop statements are executed recursively. Thus, we avoid the occurrence of the runtime errors.

11.6 WAYS TO ENCOUNTER ERRORS

There are umpteen ways to encounter errors. One easy way is use a *printf()* statement wherever you feel that there is a possibility of occurrence of a logical error. This would facilitate you to have a thorough look into the entire program and obviously once you get the values or messages printed out, you will easily trace the errors and fix them up. To me, *printf()* is a good and a primary source of debugger. So use them liberally in your programs.

I generally suggest another easier method for the students when I teach them C and it is the method of discussion and explanation to others by going through the entire program and in the process of explaining the program to your fellow students or readers, you may land up in finding bugs, which you would easily fix up.

When you debug the programs for errors, do not just look for the errors and try to remove them. Discover the source of the errors, so that you would be making yourself much stronger while compiling complex programs. For this, you must be antagonistic and crave for learning about the errors. Always write down the program. Do not just start keying in the program and try to find out the errors and fix them up.

When I give programs to debug to my post graduate students, I have seen a few of them mindlessly changing the code and manipulating it to get the anticipated results. As I had mentioned earlier, do not just try to get a successful compilation for your programs and try to get as many errors and keep a database of errors. Take it as a rule of thumb that "A programmer who compiles a program successfully at the first go may be a good programmer but surely not an efficient programmer". The reason is very simple. The programmer would never look into the program for the wonderful concepts that may be hidden in that program since the program was compiled successfully. So, the magnitude of learning in this case would be less.

Another common point where the students or the programmers would make errors while writing a program is their interpretation of the input data. They tend to assume invalid options about the input data and hence possibly get errors due to that.

Programmers must patiently test each and every feature of the program they have written to identify the errors. They should always be sceptical about their programs and refine it to the stage where they would finally get convinced.

11.7 DEBUGGING TOOLS

11.7.1 Using *assert* macro

Utilizing the macro assert is the best and easy resource for debugging C programs. This macro is available in *assert.h* and you have to include this in your program just like any other header files. The following is the prototype of the macro *assert*.

```
void assert(int test);
```

If the expression evaluates to true, the program will not be terminated and would proceed. If the expression evaluates to false, then it would result in the termination of the program and an error message would be prompted in such a case.

Consider the following program listing in which I have used assert macro, just like an *if* statement.

```
#include<stdio.h>
#include<conio.h>
#include<assert.h>
void main()
{
int a;
clrscr();
printf("\nEnter value for a : ");
scanf("%d",&a);
assert(a!=12);
printf("Hello %d",a);
getch();
}
```

In this program, I am trying to check the value of 'a' that I get from the keyboard, using the assert macro and if the macro evaluates to false (which is to say that I had entered '12' for 'a'), then the program gets aborted and I get the following error message:

```
Enter value for a : 12
Assertion failed: a!=12, file SAMPLE_1.C, line 11
Abnormal program termination
```

This macro would facilitate the program to work correctly as long as it remains true. But for an advanced level of error checking and handling, this macro cannot be used. In other words, for complex and commercial applications, the *assert* macro would not be that much useful. This possibly finds more usage at the level of students, when they begin to explore the concept of exception handling.

11.7.2 Tracing the Program

Tracing the program is another good method that I would suggest for fixing up your errors and getting down the rate of errors that occur in your program. By tracing the entire program, you probably understand the flow of control and it would then be easy for you to locate the bugs and remove them. Tracing is possible with the *F7* key in C programming. Remember that you can trace only those programs that are free from syntax errors. Once you give the keystroke *F7*, then a light green rectangular bar will appear on the editor and as you keep pressing the key with necessary pauses, the bar will scroll through the program indicating the flow of control and finally would disappear after executing the closing curly brace. Figure 11.9 shows the combined screenshots when a simple program is traced and see that the control is indicated as a rectangular bar when we trace the program. Please see that I have shown only a part of tracing and you can try out the remaining. Remember, you need to give the *enter* keystroke (usual way of letting the compiler know that you have given an input to the program) after giving the input for 'a' and not *F7* when you trace the program.

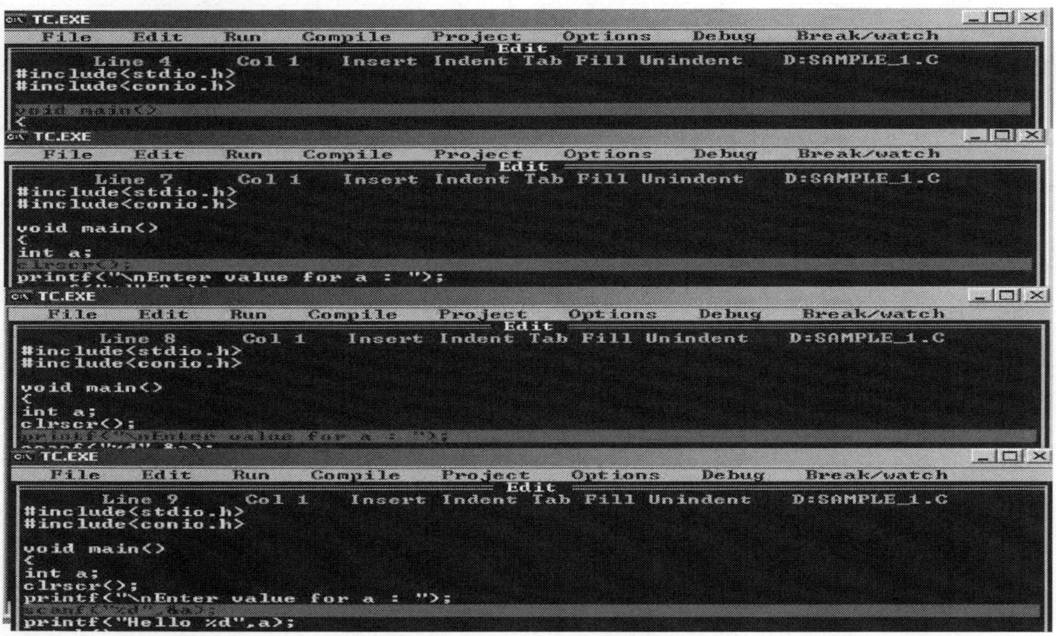

Figure 11.9 Tracing a program using F7 key

11.7.3 Adding Watches

While you trace the program for bugs, you can watch the progress and status of variables that you have used in your program and get to know what is happening really throughout the program. Consider the following program and let us add a watch (you can add a *watch* for one or more variables through the *break/watch* option in the menu or using the hotkey combination *Ctrl+F7*).

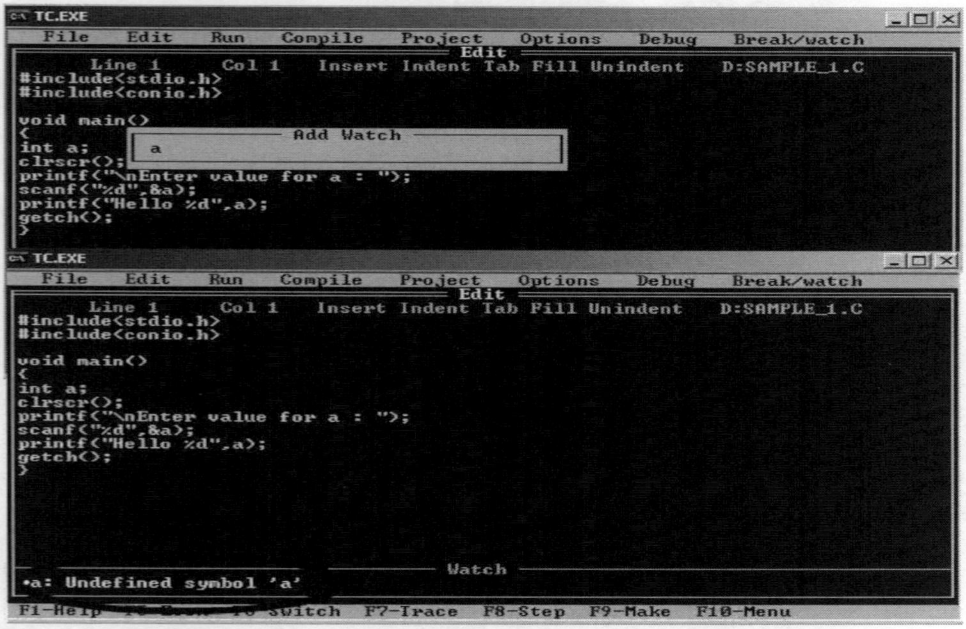

Figure 11.10 Adding a *watch* before tracing

```
#include<stdio.h>
#include<conio.h>
void main()
{
int a;
clrscr();
printf("\nEnter value for a : ");
scanf("%d",&a);
printf("Hello %d",a);
getch();
}
```

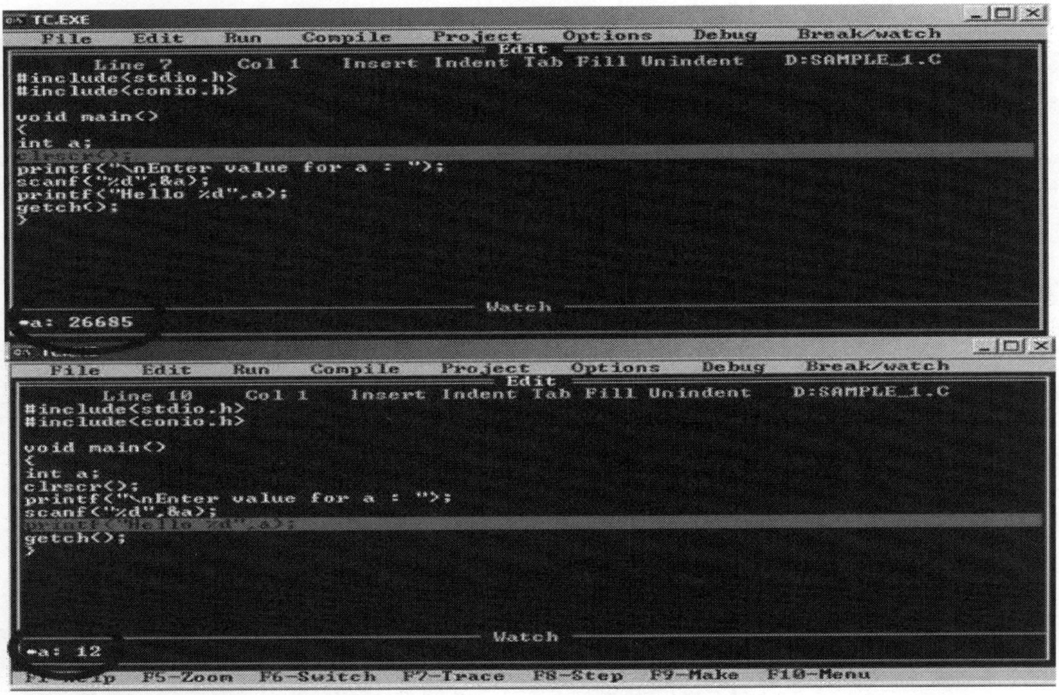

Figure 11.11 Value of the variable 'a' when watched

Figure 11.10 shows the screenshot when you open the watch window and a variable is added for watching its progress. In our example, we had used only one variable in the program, which is 'a'. Immediately after you add a watch, you may notice that the watch area down the screen becomes active and flashes a message *a : Undefined symbol 'a'*. The reason is that you have not started tracing the program. The variable 'a' will be declared and allocated with memory only when the particular line of the program is executed and from then on the variable will be **watched** for its initial value (usually garbage) and progress after that.

Figure 11.11 shows the initial value of the variable 'a' (26685), which is watched and as we progress the value changes to the one that we give from the keyboard (12).

11.8 COMMON ERRORS THAT YOU MAKE

As I had mentioned time and again, getting errors always makes you learn. So do not bother when you get errors in your program and be happy that you have got one. Errors are the ones that bless you in disguise and you master the language by removing the errors and bugs in the program. I have figured out some of the common bugs that a programmer, whether a beginner or not in C, would possibly commit while writing simple to complex programs.

11.8.1 Using = instead of ==

This is one of the common errors that can be made, not only by the beginners, but also experienced programmers. If you use, '=' or '==', you are syntactically correct and you won't be notified that you have an error. Consider for example, the following code.

```
#include<stdio.h>
#include<conio.h>
void main()
{
int a=10;
clrscr();
if(a=20) <-----------------
printf("\The value of a is %d", a);
else
printf("\nSorry I cannot print the value of a");
getch();
}
```

What do you think will be the output of this program? Observe the line indicated by the arrow mark in the program. What actually are we intending to do in that line? Actually we are trying to check whether the value of 'a' (which is initialized with '10') is equal to '20' or not. This means we are supposed to use the equality operator '=='. But in the process of typing we have used '=' in the place of '==' and syntax allows us to do so and so there is no chance that you would get an error. Some versions of C compilers issue a warning message saying "Possibly incorrect assignment statement in function main()". But you cannot expect the same warning from the version that you might be using.

So the output which you anticipate would be "Sorry I cannot print the value of a" since the condition is supposed to evaluate to false. But the output that you would get on execution would be "The value of a is 20".

Can you trace out why such an output is generated? The culprit is '='. When that particular line (indicated with arrow) is executed, assignment operation is done instead of comparison and hence the variable 'a' that was initialized with '10' is assigned a new value '20' and the condition is reduced to if(20). This type of representing a condition is called as implicit checking in C and this is equivalent to if(20!=0). Obviously the condition would return a *true* and hence the corresponding statement is executed.

Note that just one typing error has so many meanings internally and produces an output that you would never expect. When you correct the condition inside *if* by saying if(a==20), it would evaluate to *false* and you would get the output correspondingly.

11.8.2 Missing out Break Statement Inside Switch

Another common error that one would make while writing programs is missing out *break* statement when using *switch-case* block. Consider the following program code.

```
#include<stdio.h>
#include<conio.h>
void main()
{
int opt;
clrscr();
printf("Enter the option : ");
scanf("%d",&opt);
switch(opt)
    {
    case 1:
        printf("\n\nHello");
    case 2:
        printf("\n\nHow are you?");
        break;
    case 3:
        printf("\n\nAre you there?");
        break;
    default:
        printf("\n\nGood bye");
    }
getch();
}
```

When you execute the above program by giving '1' for the option, then you would anticipate only the 'case 1' to get executed and "Hello" to be printed. But much against your expectations, the compiler would execute 'case 2' also and along with "Hello" you would get "How are you?"

This is because we have not given a break after the 'case 1' and hence the compiler would consider the cases '1' and '2' as one case and would stop the execution upon executing the two cases. If you forget to give *break* to the second case and give option '1' then you would get the outputs corresponding to the cases '1', '2' and '3'.

Again these kinds of errors are not notified to you and it is your responsibility to ensure that these kinds of errors are avoided by giving more care. As an exercise, try giving a *break* to the first *case* and skip the *break* from the second one and see what happens. This kind of explorations will only improve your debugging skills.

11.8.3 Knowing the Array Bounds

When you use arrays and subscripts for them, you must be carefully giving the start and end values to the subscripts, since C does not support bounds checking. This means, there is no check done by C whether the array subscript exceeds the array size. This is a snag in the language and was corrected only later in Java. So, the programmers have to take the responsibility again in such a case and see to it that the array subscript does not exceed the array size.

Another important aspect regarding the arrays in C is that the array index starts from '0' and most of the users forget this idea while working with arrays and start counting from '1' and this may lead to various types of unpredictable errors.

11.8.4 Supplying Semicolons at the End of Loops and Conditions

'Semicolons' are the biggest ever culprits, which may give unwanted results. Invariably, I have seen most of the C programmers take it for granted and put a semicolon at the end of each and every statement. Probably, they mistake the rule regarding the semicolon. They should understand that semicolon has to be used only for statements that are complete in all aspects.

When loops and conditions are used in programs, they become complete only when their blocks are complete and not at the point when they check the conditions. Consider the following program.

```
#include<stdio.h>
#include<conio.h>
void main()
{
int a=10;
```

```
clrscr();
while(a!=10);
printf("Welcome");
printf(" to C programming");
getch();
}
```

When the program is executed, we will get the output "Welcome to C programming". When you remove the semicolon after the while statement (indicated inside the circle), you will get the output " to C programming". The reasoning is quite simple. When a semicolon is put after the *while* statement, the C compiler interprets the statement as the one below.

```
while(a!=10)
;
```

Thus the compiler assumes a NULL statement after the *while* and hence on getting the return value *true*, it executes the NULL statement, which does not have any effect. Following the NULL statement, we have two statements, which are not part of the *while* loop, but that of the program. So they tend to execute whether or not the condition inside *while* is *true* or *false*. This is the reason why we get an output "Welcome to C programming" even when the condition inside *while* is *true* in this example (see that we have initialized 'a' as '10' and the condition 10!=10 is absolutely *true*. When the condition is *true*, the programmer would not expect the statement following *while* statement to execute, but, a semicolon is provided by the programmer and this would execute both the statements following the *while* statement, for the reasons just mentioned.

Try executing the program again without a semicolon and see the difference. Even using a semicolon at the end of the function *main()* would create similar problems.

11.8.5 Importance of using NULL Characters at the End of the Strings

Another important, yet, most forgotten aspect when a code is written using character arrays in C is not properly appending the array with a NULL character (\0). If this is not appended to the character array or a string, the C compiler would be ambiguous while processing the data related with the array and it may continue to process the array even after the actual string has terminated. Consider the following code.

```
#include<stdio.h>
#include<conio.h>
```

```
void main()
{
char a[100];
int i;
clrscr();
printf("Enter the string : ");
for(i=0;(a[i]=getchar())!='$';i++);
printf("\n\nThe string is : %s",a);
getch();
}
```

When you execute the above code, there would not be any problem till the point where you get the string. When you display the string using the *printf()* function, it will display the string with some additional characters, since after reading the string, we have not appended the NULL character (\0) to it. It will produce an output as shown in the figure 11.12.

Figure 11.12 String displayed with garbage characters, as NULL is not appended after reading the string

In the above program, include the coding a[i]='\0' immediately after the *for* loop. This will remove the bug and you would get an output where no garbage characters would be accompanying the output.

11.8.6 Ignoring Space for NULL Character During Dynamic Memory Allocation

When you dynamically declare a character array, remember to provide a separate position in memory for the NULL character (\0). Under normal conditions, you tend to forget to do this. Consider the following coding.

```
#include<stdio.h>
#include<conio.h>
#include<stdlib.h>
void main()
{
char *first,*second;
clrscr();
first=(char*)malloc(sizeof(char));
second=(char*)malloc(strlen(*first));  <----------
printf("\nEmter the first string : ");
scanf("%s",first);
fflush(stdin);
printf("\nThe first string is : %s",first);
strcpy(second,first);
printf("\n\nThe copied string is : %s",second);
getch();
}
```

Observe the line indicated by the arrow. There we are trying to dynamically initialize the second character array based on the length of the first one and we have not included one space for the NULL character. So the second string would be accommodative for all the characters of the first string and it would not have space for the NULL character which is supposed to be appended at the end. When you replace the line indicated above with the code **second=(char*)malloc(strlen(*first)+1);** the bug would be removed and you will have no problems now with regard to the second array.

What I have listed in this section is just a few logical errors that any beginner or an intermediary level programmer in C could make. There are many such errors from the basic level to the complicated level that one can make while writing programs in C and I would leave it to the readers of this book to explore them since I believe by now they would have become experts in finding out and fixing up the errors.

SUMMARY

Theoretically we are concluding our discussion with this chapter and we would move to doing more exercises in the next one. I am now confident that, as readers of this book and as programmers, you would now feel easier to communicate with the C language, since you now know how to induce errors/bugs in a program and clear them as well.

As I had mentioned, learning by making errors is the best way of learning anything in life and programming is not an exception to this idea. So, be ready to make as many errors as possible in your programs and I assure you that you would be branded as the best programmers, since you only would know how to fix them.

We started off the chapter with a very simple question "Am I allowed to make errors?" and I have given the answer very clearly in the paragraph above. Following the justification that I had given for my answer, I could explain you the reason for you to get the bugs in your programs and what your reaction has to be when you get one.

Following that I had elaborated on how important it is to have a modularized and structured programming. I had illustrated the compiler and linker generated and runtime errors and how careful you should be when you handle these types of errors. I had also indicated that by using the function *printf()* very liberally in your programs, you may encounter the errors and fix them very easily.

I had listed out a few debugging tools along with examples, which, I am sure would come in handy when you debug your programs. I had concluded the chapter with a list of common errors that you may commit and possible ways of taking care of them and not allowing them in your programs.

CRACK THE NUT

Identify the outputs for the following programs:

1.
```
#include<stdio.h>
void main()
{
int a = 320;
char *ptr;
ptr =( char *)&a;
printf("%d ",*ptr);
getch();
    }
```

2.
```
#include<stdio.h>
#include<string.h>
void main()
{
char *ptr1 = NULL;
char *ptr2 = 0;
strcpy(ptr1," c");
strcpy(ptr2,"questions");
printf("\n%s %s",ptr1,ptr2);
getch();
    }
```

3.
```
#include<stdio.h>
void main()
{
char arr[10];
arr = "world";
```

```
    printf("%s",arr);
    getch();
    }
4.  void main()
    {
    int i=5,j;
    j=++i+++i+++i;
    printf("%d %d",i,j);
    }
5.  void main()
    {
    int i=1;
    i=2+2*i++;
    printf("%d",i);
    }
6.  void main()
    {
    int a=2,b=7,c=10;
    c=a==b;
    printf("%d",c);
    }
7.  void main()
    {
    int x;
    x=10,20,30;
    printf("%d",x);
    }
```

8. void main()

```
{
int a=0,b=10;
if(a=0)
{
printf("true");
}
Else
{
printf("false");
}
}
```

9. void main()

```
{
int a;
a=015 + 0x71 +5;
printf("%d",a);
}
```

10. void main()

```
{
printf("%d %d %d",sizeof(3.14),sizeof(3.14f),sizeof(3.14L));
}
```

11. void main()

```
{
int x=100,y=20,z=5;printf(«%d %d %d»);
}
```

12.
```
void main()
{
int a=2;a=a++ + ~++a;
printf("%d",a);
}
```
13.
```
void main()
{
int a;
a=sizeof(!5.6);printf("%d",a);
}
```
14.
```
void main()
{
float a;
(int)a= 45;
printf("%d",a);
}
```
15.
```
void main()
{
int i=5;
int a=++i + ++i + ++i;
printf("%d",a);
}
```
16.
```
extern int x;
void main()
{
clrscr();
do{
do{
printf("%o",x);
```

```
}
while(!-2);
}
while(0);
getch();
}
int x=8;
```

17.
```
void main(){
int x=011,i;
clrscr();
for(i=0;i<x;i+=3){
printf("Start ");
continue;
printf("End");
    }
getch();
}
```

18.
```
void main(){
int i,j;
i=j=2,3;
clrscr();
while(—i&&j++)
printf("%d %d",i,j);
getch();
}
```

19.
```
void main(){
static int i;
clrscr();
for(++i;++i;++i) {
```

```
printf("%d ",i);
if(i==4) break;
}
getch();
}
```

20. void main(){

```
int i=1;
clrscr();
for(i=0;i=-1;i=1) {
printf("%d ",i);
if(i!=1) break;
    }
getch();
}
```

21. void main(){

```
int i=1;
clrscr();
for(i=0;i=-1;i=1) {
printf("%d ",i);
      if(i!=1) break;
}
getch();
}
```

22. void main(){

```
char c=125;
clrscr();
do
printf("%d ",c);
while(c++);
```

```
getch();
}
```

23.
```
void main(){
int x=123;
int i={
printf("c" "++")
};
for(x=0;x<=i;x++){
printf("%x ",x);
}
getch();
}
```

24.
```
#include<stdio.h>void main()
{
int a= sizeof(signed) +sizeof(unsigned);
int b=sizeof(const)+sizeof(volatile);
printf("%d",a+++b);
}
```

25.
```
#include<stdio.h>
void main()
{
volatile int a=11;
printf("%d",a);
}
```

SELF-LEARNING EXERCISES

The exercises given here cover all the chapters that are covered under C (excluding the first chapter since it deals with the introductory ideas in computing science and number systems). I have tried to cover the maximum ideas in these exercises. I wish the readers of this book to attempt these very sincerely and hence enhance their skills.

Basic Ideas (using data types, conditionals, loops, case)

1. Input two numbers and work out their sum, average and sum of the squares of the numbers.
2. Write a program to read a "float" representing a number of degrees Celsius, and print as a "float" the equivalent temperature in degrees Fahrenheit.
3. Use the *if-else* conditional structure to check the odd or even nature of a number.
4. Write a program to print several lines (such as your name, address, etc) (Hint: You may use either several printf instructions, each with a newline character in it, or one printf with several newlines in the string).
5. Write a program to find whether a given integer is positive or negative.
6. Use the if-else conditional structure to check whether a number *'m'* is a multiple of another number *'n'*.
7. Use if-else-if conditional structure to classify the operators given as arithmetic, logical, *etc.*
8. Use the case control structure to implement the above logic in 7

9. Assume that you input a floating point number in centimetres, print out the equivalent number of feet (integer) and inches, with the inches given to an accuracy of one decimal place (Hint: Assume 2.54 centimetres per inch, and 12 inches per foot)
10. Write a program fragment to find the reversed form of a four-digit integer using while loop
11. Write a program fragment to find the factorial of an integer using for loop
12. Write a program fragment to find the sum of the digits of a four-digit integer using a do-while loop (display the result in single digit, i.e., the sum of the digits of the number 1234 shall be displayed as '1' and not as '10')
13. Given as input an integer number of seconds, print as output the equivalent time in hours, minutes and seconds
14. Write program fragments to find out the area, volume, surface area and perimeter of different two- and three-dimensional figures
15. Write a program to print the Fibonacci series consisting of 'n' terms (A Fibonacci series has the first two terms as 0 and 1 and the remaining terms are obtained with the addition of the preceding two terms)
16. Write a program to find the Armstrong numbers between 1 and 1000 and print them (An Armstrong number is that which is the sum of the cube of digits of a given number, e.g. $153=1^3+5^3+3^3$)
17. Write a program to print all the prime numbers in a given range of values (get the lower and upper limits from the keyboard)
18. Write a program to find the sum of first 'n' natural numbers (get the value of 'n' from the keyboard)
19. Write a program to perform the basic arithmetic operations using switch-case structure
20. Write a program to print a Pascal's triangle
21. Write a program to print a Floyd's triangle
22. Write a program to find whether the given number is a palindrome or not
23. Write a program to print all the multiples of a number 'n' in a given range of values
24. Write a program to calculate the simple interest
25. Write a program to calculate the compound interest
26. Read a positive integer and perform the following: If the number is even, halve it; if it's odd, multiply by 3 and add 1. Repeat this process until the value is 1, printing out each value. Also print out how many of these operations you performed. If the input is less than 1, print the input as 'invalid'.

27. Write a program to display the size of all the basic data types using the sizeof operator.
28. Write a program to read a character from the keyboard and print its ASCII value.
29. Write a program to display all the character used in C along with their ASCII values.
30. Write a program to find all the possible roots of a quadratic equation.
31. Write a program to swap two values without using a temporary variable.
32. Assume that we enter numbers from 1 to n, not in any particular logical order. We enter the numbers randomly. In doing so, we enter only n-1 numbers. After entering the last number from the keyboard, the program that you write shall flash with the number that you had missed to enter.
33. Write a program to implement a menu driven programming with the help of switch-case structure and do-while loop (choose any 10 options and write the program).
34. Write a program to print only the uppercase and lowercase alphabets on the screen (use continue statement for the program).
35. Write a program to generate prime numbers up to the given value 'n' (for example, if the 'n' is 10, then the output shall be 2,3,5,7).
36. Write a program to generate the first 'n' prime numbers (for example, if 'n' value is 10, then the output shall be 2,3,5,7,11,13,17,19,23,29).

Integer Arrays

37. Write a program to read an array of 'n' integers and count the number of positive, negative, odd and even numbers and zeroes as well.
38. Write a program to read an array of 'n' integers and find out the number of positive and negative numbers in it.
39. Write a program to read an array of 'n' integers and find out the number of odd and even numbers and zeroes in it.
40. Write a program to read an array of 'n' integers and find the sum of them.
41. Write a program to read in numbers until the number –1 is encountered. The sum of all number read until this point should be printed out.
42. Write a program to read in 10 numbers and compute the average, maximum and minimum values.
43. Write a program to read an array of 'n' integers and find the sum of the odd and even numbers separately.
44. Write a program to convert the input decimal number to binary (use arrays for storing the binary result).

45. Write a program to convert the input binary to decimal (use arrays for storing the input decimal number).
46. Write a program to read an array of 'n' integers and find the sum of positive and negative numbers separately.
47. Write a program to read an array of 'n' integers and find the biggest among them.
48. Write a program to read an array of 'n' integers and find the smallest among them.
49. Write a program to read an array of 'n' integers and find the second biggest among them.
50. Write a program to read an array of 'n' integers and find the second smallest among them.
51. Write a program to read an array of 'n' integers and swap the biggest and smallest among them.
52. Write a program to read an array of 'n' integers and swap the second biggest and second smallest among them.
53. Write a program to read an array of 'n' integers and sort them in ascending order.
54. Write a program to read an array of 'n' integers and sort them in descending order.
55. Write a program to read an array of 'n' integers and sort the first half of the array in ascending order and second half in descending order.
56. Write a program to read an array of 'n' integers and check whether a given integer exists in the array or not.
57. Write a program to read an array of 'n' integers and check the number of occurrences of a given integer in the array.
58. Write a program to read an array of 'n' values and find the average of them and print the same.
59. Write a program to read an array of 'n' values and find out the arithmetic mean, standard deviation and variance.
60. Write a program to find the median of the given 'n' values.
61. Write a program, where you read two arrays 'a' and 'b' and sort them separately and merge the contents in sorted order. Check out for any duplicate entries and try to avoid them while you manipulate the arrays.
62. Write a program to read a matrix of order mXn and find the inverse of the same.
63. Write a program to read a matrix of order mXn and find the determinant value of the same.
64. Write a program to read two matrices of order mXn and find the sum of them (check the compatibility for adding them).
65. Write a program to read two matrices of order mXn and find the difference between them (check the compatibility for adding them).

66. Write a program to read two matrices of order mXn and find the product of them (check the compatibility for multiplying them).
67. Write a program to read a matrix of order mXn and find the transpose of the same.
68. Write a program to read a matrix of order mXn and find whether the upper triangle and lower triangles are equal.

Character Arrays

69. Write a program to read a string and convert all the uppercase alphabets in it to lowercase.
70. Write a program to read a string and convert all the lowercase alphabets in it to uppercase.
71. Write a program to read a string and find out the abbreviation of the same.
72. Write a program to read a string and check the number of occurrences of a particular character.
73. Write a program to read a string and encrypt the same.
74. Write a program to read a string and decrypt the same.
75. Write a program to read a string and count the number of uppercase, lowercase alphabets, digits and special characters in it.
76. Write a program to read a string and count the number of words, lines and characters in it.
77. Write a program to read a string and count the number of sentences in it.
78. Write a program to count the number of occurrence of vowels in a given text and indicate the number of times each vowel occurring in the test in terms of percentage.
79. Write a program to read a string and check whether it is a palindrome or not (a palindrome is one which reads the same way when it is read backwards).
80. Write a program to read a string and replace a comma (wherever appears) with a semicolon.
81. Write a program to read a string and copy it to another array.
82. Write a program to read a string and find the length of it.
83. Write a program to read a string and extract the first 'n' characters from it and store it in another array.
84. Write a program to read a string and extract the last 'n' characters from it and store it in another array.
85. Write a program to read a string and extract the middle 'n' characters from it and store it in another array.
86. Write a program to read a string, which is an assignment statement in Pascal and convert it to a FORTRAN assignment statement.

87. Write a program to read a string, which is a FORTRAN assignment statement and convert it to a Pascal assignment statement.
88. Write a program to read a string and ignore any extra spaces in between the words.
89. Write a program to read a string and check the number of palindromes in it.
90. Write a program to read a text and check the number of occurrences of a particular word in it.
91. Read an integer value. Assume it is the number of a month of the year; print out the name of that month.
92. Write a program to read a string and skip a particular word from it and copy the remaining part to another array.
93. Write a program to read a string and replace an existing pattern with another one.
94. Write a program to read two strings and concatenate them.
95. Write a program to read two strings and compare them.
96. Write a program to read a word and arrange the alphabets in that in alphabetical order.
97. Write a program to read a nucleotide or protein sequence and print the sequence in uppercase letters and also find the length of it.
98. Write a program to compute the percentage of AT and GC content in a given DNA sequence.
99. Write a program to arrange the given 'n' names in alphabetical order.
100. Write a program to compare the given string with the existing list of strings and store the positions where they occur in an integer array.

Structures and Unions

101. Write a program to process the students' records by getting the inputs for marks of five subjects and find out the total, average and grade.
102. Write a program to do the basic arithmetic operations with two given complex numbers.
103. Write a program to create a database of books and authors with other details such as publisher's name, year of publication, number of pages and price of the book.
104. Write a program to create a small database of a library.
105. Write a program to create a database of a cricket team.
106. Write a program to create a database of employees in a particular industry.
107. Repeat the programs from 101 through 106 using unions.

Pointers

108. Rewrite all the programs that you have done so far with pointers (this include the programs that you would have done using loops, conditions, numerical and character arrays, structures, unions, etc.). Do all the programs with the memory pictures for better understanding.
109. Write a C program to read through an array of any type using pointers. Write a C program to scan through this array to find a particular value.
110. Figure out the memory map for the singly linked list program.
111. Write C coding to implement the doubly linked list. Do go through the basics of a doubly linked lists before you start writing a program.

Functions

112. Rewrite the programs 1 to 36 using functions.
113. Write functions to find the nCr and nPr.
114. Write a function to find out the GCD of two numbers.
115. Write a recursive function to generate the first 'n' terms of Fibonacci series.
116. Write a recursive function to find the minimum and maximum of numbers.
117. Write a function to swap two numbers without using a third variable.
118. Write a program to find the power of one number raised to another.
119. Write a program to concatenate two strings using functions and pointers.
120. Write a function to search for a given string and if it is available, replace it with another string, wherever it appears in the input string.
121. Write a function to print all the single digit numbers in words.
122. Write a function to print all the two digit numbers in words.
123. Write a function to print all the three digit number in words.
124. Write a function to find the length of the given string using pointers.
125. Write a function that finds out the number of palindromes in the given text.
126. Write a function that checks whether the given string is a substring of another using pointers.
127. Write a function to convert the lowercase array to uppercase array using pointers.
128. Write a function to find out the number of words, lines and characters in the given text using pointers.
129. Write a function to find out the number of uppercase and lowercase alphabets, digits and special characters in the given text using pointers.

130. Write a function to count the number of prime numbers between the given limits using functions.
131. Write a function to compare two given strings using functions and pointers.

Files and Command line Arguments

132. Read data from the keyboard until '$' is encountered and transfer the same to a file.
133. Read from a file until end of file is encountered and display the same on the screen.
134. Read data character by character from the keyboard and write the same in the file.
135. Read the data one character at a time from the file and display it on the screen.
136. Copy the content of one file to another.
137. Merge the contents of two files.
138. Copy the uppercase letters in one file as lowercase letters in another.
139. Read data from the keyboard, transfer to an array and then to a file.
140. Transfer the contents of a file to an array and display the same on the screen.
141. Display the contents of a file line by line with line number.
142. Display the contents of a lengthy file page by page.
143. Encrypt the contents of a given file and display the encrypted file.
144. Decrypt the contents of a given file and display the same (use the output file of the above program as input for this program).
145. Count the number of uppercase, lowercase alphabets, special characters and digits in a given file.
146. Display the contents of more than one file at a given time (cat all the files).
147. Check for the number of occurrences of a given pattern in more than one file at a given time (Grep all the files).
148. Count the number of words, lines and characters in a given file and display the statistics.
149. Display the lines in which a particular pattern occurs with line number.
150. Display the lines in which a particular pattern does not occur with line number.

REFERENCES

Thiagarajan, B and Rajalakshmi Pa. (2009). *Computational Biology*. MJP Publishers, Chennai.

Kanetkar, Y. (1991). *Let Us C*, 4th Edition. BPB Publications, New Delhi.

Kanetkar, Y. (2001). *Understanding Pointers in C*, 3rd Edition. BPB Publications, New Delhi.

Venugopal, K.R. and Prasad, S.R. (2007). *Mastering C*. Tata McGraw-Hill Publishing Company Limited, New Delhi.

INDEX

J

K

L

M

N

O

P

R

S

T

U

V

W

Made in the USA
Monee, IL
18 October 2026

66037587R00268